Modern approach to
INORGANIC CHEMISTRY

Modern approach to
INORGANIC CHEMISTRY

C. F. Bell, M.A., D.Phil., F.R.I.C.
Lecturer in Inorganic Chemistry,
Brunel University, Uxbridge, Middlesex

and

K. A. K. Lott, B.Sc., Ph.D.
Lecturer in Inorganic Chemistry,
Brunel University, Uxbridge, Middlesex

LONDON   BUTTERWORTHS

THE BUTTERWORTH GROUP

ENGLAND
Butterworth & Co (Publishers) Ltd
London: 88 Kingsway, WC2B 6AB

AUSTRALIA
Butterworth & Co (Australia) Ltd
Sydney: 586 Pacific Highway, Chatswood, NSW 2067
Melbourne: 343 Little Collins Street, 3000
Brisbane: 240 Queen Street, 4000

CANADA
Butterworth & Co (Canada) Ltd
Toronto: 14 Curity Avenue, 374

NEW ZEALAND
Butterworth & Co (New Zealand) Ltd
Wellington: 26–28 Waring Taylor Street, 1

SOUTH AFRICA
Butterworth & Co (South Africa) (Pty) Ltd
Durban: 152–154 Gale Street

First published 1963
Second impression 1963
Second edition 1966
Second impression 1967
Third edition 1972

Suggested U.D.C. number 546

ISBN 0 408 70370 9 standard
     0 408 70371 7 limp

Filmset by Photoprint Plates Ltd., Rayleigh, Essex
Printed in England by J. W. Arrowsmith Ltd., Bristol

Preface to the third edition

This book was first published in 1963 and although we have brought the text up to date from time to time, we believe that a major revision is now necessary. We have taken this opportunity to convert previously published material to the international system of units (SI). There is increasing usage of SI units in the teaching of chemistry at University level and we trust that this new edition will help in making these units familiar to teachers and students alike.

We have preserved the original sequence of topics, but much has been re-written and some new material added in the light of developments in the teaching of inorganic chemistry since 1963. We hope that the contents are still in keeping with our original intention of writing a book suitable for the first years of an honours degree course in chemistry, the Part I examination of the Graduateship of the Royal Institute of Chemistry and other qualifications at an intermediate level such as the Higher National Certificate and general degrees. We trust it will continue to be used by those requiring chemistry as an ancillary subject in honours degree courses in other disciplines.

We wish to acknowledge our gratitude to colleagues in the School of Chemistry, Brunel University, and to teachers of chemistry elsewhere whose encouragement and critical comments have contributed so much to the success of this work. Finally we would like especially to thank our respective wives for their sustained help and patient understanding in the preparation of this and the earlier editions.

Brunel University,   C. F. Bell
May, 1972   K. A. K. Lott

Contents

Introduction: units and nomenclature

UNITS

Over the past few years agreement has been reached on an international scale to 'go metric', utilising as a basis for the changeover the system of units known as Système Internationale d'Unités or more briefly SI. In view of this, together with the fact that it is now six years since the appearance of the Second Edition of this book, we have decided to take this opportunity of updating and rewriting approaches to various sections of the book and to adopt SI units in the text. Some people might argue that to make an immediate changeover could lead to confusion, particularly where references to older texts and to the scientific literature are made. However, 'it has been the hope that, by the summer of 1972, most public examinations in the United Kingdom will be fully metric (SI)' and therefore this book should be released at a stage when the majority of students will be familiar with the SI approach. To introduce the student completely to SI requires a book in its own right, but a monograph (*Physico-Chemical Quantities and Units* by M. L. McGlashan, M.Sc., Ph.D., D.Sc., F.R.I.C.) that deals with the subject has recently appeared in its Second Edition as Monograph No. 15 in the Royal Institute of Chemistry's Series – Monographs for Teachers; we would strongly recommend that this text should be the subject of every student's attention until the language of SI becomes commonplace.

There are a number of points on SI that may be usefully summarised in this introduction to give the student using this present edition sufficient background to follow the text if familiarity with SI has not yet been achieved.

PHYSICAL QUANTITIES, UNITS AND DIMENSIONS

Physical quantities are given names and symbols to represent them; thus, for instance, length is given the symbol l. In determining the distance between two points by means of a metre rule we speak of measuring the distance between the two points. What we have, in fact, actually done is to find the ratio of the distance l to the length of the metre rule and we say that the distance is x metre where x is a pure number and the metre is the unit of length. Such physical quantities also possess dimensions and we may use the process of dimensional analysis to check the validity of some equation that interrelates two or more of these quantities. Seven physical quantities that are dimensionally independent may be used in part or in total to define particular branches of the physical sciences; their units are termed base units. These are

Physical quantity	SI Unit	Symbol for Unit
Length	metre	m
Mass	kilogram	kg
Time	second	s
Electric current	ampere	A
Thermodynamic temperature	kelvin	K
Amount of substance	mole	mol
Luminous intensity	candela	cd

Thus, the sub-set length, mass, time and electric current has been in use for some years to define electricity and magnetism (rationalised MKSA). It should be noted, in passing, that the symbol for the physical quantity is printed in *italics* (or bold italics if a *vector* quantity) whereas the symbol for the unit is in roman upright typescript. Where more than one symbol is used to define a derived quantity the symbols should be separated by a space, e.g. A m^2 for the derived quantity known as magnetic moment. Multiples and sub-multiples of various orders have a special nomenclature and symbolisation (*Table iii*) and when written in conjunction with a base unit no space is used between the symbols, e.g. cm, ns etc., and such a multiple unit may be raised to a power without the use of parentheses, e.g. cm^2 for $(0 \cdot 01 \text{ m})^2$.

The writing of numbers is recommended to be of the following form — a long number may be grouped in threes about the decimal point and where the decimal point is to be placed before the first digit of a number it should always be preceded by a zero. Thus, two examples are 3 567·945 622 and 0·467 398. In addition it may be convenient to write a number with just one digit before the decimal point, e.g. $3 \cdot 859 \times 10^4$.

In labelling graphical axes and columns in tables, the points on the axes and the figures in the columns should be pure numbers. Hence, the axes should be labelled as a physical quantity divided by the unit used, e.g. T/K for a temperature axis and similarly for a table that contains a column of temperature values.

Tables i–iv show physical quantities with their names, unit symbols and their definitions in terms of base units.

Throughout this text we have recorded bond length in pm, wavelength in nm and wavenumber in m^{-1}. For thermodynamic data we have referred to enthalpies at 298 K as ΔH° values.

NOMENCLATURE

The Commission on Nomenclature of Inorganic Chemistry of the International Union of Pure and Applied Chemistry (I.U.P.A.C.) was established in 1921 in order to produce rules which would lead to clear and acceptable names for as many inorganic compounds as possible. The latest I.U.P.A.C. recommendations (1970 Rules) have been incorporated in the following text.

According to the Commission's definition, the principal function of a name is 'to provide the chemist with a word or set of words which is unique to the particular compound, and which conveys at least its empirical formula and also if possible its principal structural features'. Some of the main recommendations are summarised below: for further details the reader should consult *International Union of Pure and Applied Chemistry — Nomenclature of Inorganic Chemistry, 1970*, Butterworths, 1971.

NAMES AND SYMBOLS OF THE ELEMENTS

These are given in the Table on the inside front cover of the book. For a few elements, alternative names are used when forming names of their compounds. These are copper (cuprum), gold (aurum), iron (ferrum), lead (plumbum), silver (argentum), tin (stannum) and tungsten (wolfram).

The mass number, atomic number, number of atoms and ionic charge of an element are indicated by four indices placed around the symbol thus:

> left upper index   = mass number
> left lower index   = atomic number
> right lower index = number of atoms
> right upper index = ionic charge

FORMULAE AND NAMES OF COMPOUNDS

Some standardisation of formulae is desirable. The electropositive atom (cation) should always be placed first, e.g. KCl, $CaSO_4$. The chief exception is with acids. These are regarded as hydrogen salts and hydrogen is placed first in the formulae, e.g. $HClO_4$ and $HFeCl_4$.

In the case of binary compounds between non-metals, the established practice is that the constituent which should be placed first is that which appears earlier in the sequence:

Rn, Xe, Kr, B, Si, C, Sb, As, P, N, H, Te, Se, S, At, I, Br, Cl, O, F.

Examples are: XeF_2, NH_3, H_2S, Cl_2O and OF_2.

In systematic names, the name of the electropositive constituent (or that treated as such according to the above sequence) is not modified. If the electronegative constituent is monoatomic or homopolyatomic, its name is modified to end in *-ide*. Examples are: sodium chloride, oxygen difluoride, lithium nitride, arsenic selenide, aluminium boride and carbon disulphide.

If the electronegative constituent is heteropolyatomic (composed of atoms of more than one element), it is designated by the ending *-ate*. In some cases, particularly where historical precedent is strong, the endings *-ide* and *-ite* are used. Some examples of these exceptions are hydroxide, hydrogendi-fluoride, nitrite, chlorite, arsenite and dithionite.

In a polyatomic group, it is generally possible to indicate a characteristic or central atom to which other atoms, radicals or molecules (collectively known as ligands) are bound. Then the polyatomic group is known as a complex and if this is negatively charged, the name is derived from that of the characteristic or central element modified to end in -ate. Anionic ligands are indicated by the termination -o. For example, the systematic name for Na_2SO_4 is disodium tetraoxosulphate, although the common name of sodium sulphate is usually used. Similarly, Na_2SO_3 is named as disodium trioxo-sulphate. This compound is better known as sodium sulphite. These rules may not appear to be of much significance in relation to such familiar com-pounds but when applied to co-ordination compounds they are invaluable in avoiding confusion.

The oxidation number of an element in one of its compounds is indicated by a Roman numeral placed in parenthesis immediately following the name of the element. For zero, the cipher 0 is used. For example, $FeCl_2$ is iron(II) chloride and $FeCl_3$ is iron(III) chloride; P_2O_5 is called phosphorus(V) oxide (alternatively diphosphorus pentaoxide); $K_4[Fe(CN)_6]$ has the syste-matic name potassium hexacyanoferrate(II). This nomenclature is known as Stock's system and is generally used throughout this book. (An alternative is the Ewens-Bassett system in which the charge of an ion is indicated by an Arabic numeral followed by the sign of the charge, both placed in parenthesis immediately following the name of the ion. For example, $FeCl_2$ is iron(2+) chloride and $FeCl_3$ is iron(3+) chloride.)

CO-ORDINATION COMPOUNDS

In formulae, the usual practice is to place the symbol for the central atom or atoms first with the ionic and neutral ligands following and to enclose the whole complex entity in square brackets [ ]. The order for each class should be the alphabetical order of the symbols for the ligand atoms.

In names, the central atom or atoms should be placed after the ligands. Two kinds of prefixes are used to indicate the number of the co-ordinating groups. The stoichiometric proportions are denoted by Greek numerical prefixes *mono, di, tri, tetra, penta, hexa, hepta* and so on. The prefix *mono* is often omitted. When it is necessary to indicate the number of complete groups of atoms, e.g., the number of organic ligands co-ordinated to a central metal ion or atom, the prefixes *bis, tris, tetrakis etc.* (2, 3, 4, etc.) are used.

As before the oxidation number of the central atom or ion is indicated, anions are given the endings -ide, -ite, -ate and cations and neutral molecules are not given any special ending. In writing down the names, the ligands are listed in alphabetical order regardless of the number of each. Thus penta-amine is listed under 'a' and triethylamine under 't'.

The names for anionic ligands, whether inorganic or organic, end in -o. If the anion name ends in -ide, -ite or -ate, this becomes -ido, -ito and -ato respectively. For example, acetate becomes acetato and acetamide becomes acetamido. Some anions do not follow this general rule. These include

F^- (fluoro), Cl^- (chloro), Br^- (bromo), I^- (iodo), O^{2-} (oxo), H^- (hydrido), OH^- (hydroxo), S^{2-} (thio), CN^- (cyano), O_2^{2-} (peroxo) and HS^- (mercapto).

Ligands derived from organic compounds by the loss of protons are given the ending *-ato*. For example, acetylacetone after loss of proton and co-ordination to a metal is termed acetylacetonato (the systematic organic name for this ligand is 2,4-pentanedionato but this is not widely used as yet in inorganic chemistry).

There are a few important exceptions to the general practice of using the unchanged name for co-ordinated neutral and cationic ligands. These are water (called aqua in co-ordination complexes), ammonia (ammine), nitric oxide (nitrosyl) and carbon monoxide (carbonyl).

Where a ligand is able to attach itself at different points to a metal it is customary to use different names. For example, thiocyanato (-SCN) and isothiocyanato (-NCS), nitro (-NO_2) and nitrito (-ONO).

Table i. SI DERIVED UNITS WITH SPECIAL NAMES

Physical quantity	Name	Symbol	Definition
frequency	hertz	Hz	s^{-1}
energy	joule	J	$kg\ m^2\ s^{-2}$
force	newton	N	$kg\ m\ s^{-2}$ $(J\ m^{-1})$
power	watt	W	$kg\ m^2\ s^{-3}$ $(J\ s^{-1})$
pressure	pascal	Pa	$kg\ m^{-1}\ s^{-2}$ $(N\ m^{-2})$
electric charge	coulomb	C	$A\ s$
electric potential difference	volt	V	$kg\ m^2\ s^{-3}\ A^{-1}$
electric resistance	ohm	Ω	$kg\ m^2\ s^{-3}\ A^{-2}$ $(V\ A^{-1})$
electric conductance	siemens	S	$kg^{-1}\ m^{-2}\ s^3\ A^2$ (Ω^{-1})
electric capacitance	farad	F	$A^2\ s^4\ kg^{-1}\ m^{-2}$ $(A\ s\ V^{-1})$
magnetic flux	weber	Wb	$kg\ m^2\ s^{-2}\ A^{-1}$ $(V\ s)$
inductance	henry	H	$kg\ m^2\ s^{-2}\ A^{-2}$ $(V\ A^{-1}\ s)$
magnetic flux density (magnetic induction)	tesla	T	$kg\ s^{-2}\ A^{-1}$ $(V\ s\ m^{-2})$

Table ii. SI DERIVED UNITS

Physical quantity	Name	Symbol	Definition
wavenumber	reciprocal metre	$\sigma \ldots \tilde{v}$	m^{-1}
speed	metre per second	u	$m\ s^{-1}$
acceleration	metre per second squared	a	$m\ s^{-2}$
density	kilogram per cubic metre	ρ	$kg\ m^{-3}$
electric field strength	volt per metre	E	$kg\ m\ s^{-3}\ A^{-1}$ $(V\ m^{-1})$
magnetic field strength	ampere per metre	H	$A\ m^{-1}$
dipole moment	coulomb metre	p_e	$A\ s\ m$ $(C\ m)$
magnetic moment	ampere square metre	m	$A\ m^2$
heat capacity	joule per kelvin	C	$kg\ m^2\ s^{-2}\ K^{-1}$ $(J\ K^{-1})$
entropy	joule per kelvin	S	$kg\ m^2\ s^{-2}\ K^{-1}$ $(J\ K^{-1})$
molar heat capacity	joule per kelvin mole	C_m	$kg\ m^2\ s^{-2}\ K^{-1}\ mol^{-1}$ $(J\ K^{-1}\ mol^{-1})$
gas constant	joule per kelvin mole	R	$kg\ m^2\ s^{-2}\ K^{-1}\ mol^{-1}$ $(J\ K^{-1}\ mol^{-1})$
concentration	mole per cubic metre		$mol\ m^{-3}$
molality	mole per kilogram		$mol\ kg^{-1}$

Table iii. SI PREFIXES

Fraction or multiple	SI Prefix	Symbol
10^{-18}	atto	a
10^{-15}	femto	f
10^{-12}	pico	p
10^{-9}	nano	n
10^{-6}	micro	μ
10^{-3}	milli	m
10^{-2}	centi	c
10^{-1}	deci	d
10	deca	da
10^2	hecto	h
10^3	kilo	k
10^6	mega	M
10^9	giga	G
10^{12}	tera	T

N.B. The use of compound prefixes is not recommended.

Table iv. RECOMMENDED VALUES OF PHYSICAL CONSTANTS

Physical constant	Symbol	Value with estimated uncertainty
speed of light in a vacuum	c, c_0	$2 \cdot 997\ 925\ 0 \times 10^8$ m s^{-1} 1 0
permeability of a vacuum	μ_0	$4\pi \times 10^{-7}$ H m^{-1} (exactly)
permittivity of a vacuum	$\varepsilon_0 = \mu_0^{-1} c_0^{-2}$	$8 \cdot 854\ 185\ 3 \times 10^{-12}$ F m^{-1} 5 9
unified atomic mass constant	$m_u = m_a(^{12}\text{C})/12$	$1 \cdot 660\ 531 \times 10^{-27}$ kg 11
mass of proton	m_p	$1 \cdot 672\ 614 \times 10^{-27}$ kg 11
mass of neutron	m_n	$1 \cdot 674\ 920 \times 10^{-27}$ kg 11
mass of electron	m_e	$9 \cdot 109\ 558 \times 10^{-31}$ kg 54
charge of proton	e	$1 \cdot 602\ 191\ 7 \times 10^{-19}$ C 7 0
Boltzmann constant	k	$1 \cdot 380\ 622 \times 10^{-23}$ J K^{-1} 59
Planck constant	h	$6 \cdot 626\ 196 \times 10^{-34}$ J s 50
Bohr radius	a_0	$5 \cdot 291\ 771\ 5 \times 10^{-11}$ m 8 1
Rydberg constant	R_∞	$1 \cdot 097\ 373\ 12 \times 10^7$ m^{-1} 11
Bohr magneton	$\boldsymbol{m}_B = eh/4\pi m_e$	$9 \cdot 274\ 096 \times 10^{-24}$ A m^2 65
Avogadro constant	L, N_A	$6 \cdot 022\ 169 \times 10^{23}$ mol^{-1} 40
gas constant	$R = Lk$	$8 \cdot 314\ 34$ J K^{-1} mol^{-1} 35
Faraday constant	$F = Le$	$9 \cdot 648\ 670 \times 10^4$ C mol^{-1} 54

1 Atomic structure—1

THE ELECTRICAL NATURE OF MATTER

Dalton's atomic theory was put.forward at a time (1805) when the existence of comparatively few elements had been demonstrated and experimental studies upon their properties had led to the concept of an indivisible atom characteristic for each element. Towards the end of the nineteenth century, however, this concept of indivisibility began to show its failings. In particular, work by Faraday on reactions in electrolytic cells and studies by Crookes, Perrin and Thomson on the conduction of electricity in gases under reduced pressure indicated that the atom was essentially electrical in character and led Stoney (1891) to propose a unit of electrical charge known as the *electron*.

A great deal of evidence for the existence and nature of electrons comes from the investigations carried out on *cathode rays* formed in electric discharge tubes at pressures of approximately $1 \cdot 2 \ N \ m^{-2}$. At such pressures the gases were found to exhibit considerable electrical conductivity and the tubes became filled with a form of radiation that was observable only when it impinged upon the walls of the tube or a fluorescent screen. This radiation was subjected to a thorough investigation which led to the recognition of certain characteristic properties:

(a) The rays travelled from the cathode to the anode in straight lines.
(b) Deflection from their path could be effected by the application of either magnetic or electric fields.
(c) The rays possessed measurable mass and a negative charge.

It soon became apparent that these rays were actually streams of electrons and that their characteristics were independent of both the electrode material and the residual gas in the discharge tube. These facts were demonstrated by Sir J. J. Thomson (1897) in experiments involving the deflections experienced by the rays in magnetic and electric fields. The basis of the method was to apply these fields at mutually perpendicular directions to the line of travel of the rays, the individual effects of the two fields being to deflect the rays in opposite directions. When applied simultaneously, therefore, the values of the field strengths could be adjusted so that no deflection from the original path occurred. The apparatus used by Thomson is represented in *Figure 1.1(a)* and the effect of the two fields is shown in *Figure 1.1(b)*.

The electrons on entering the electric field of intensity E undergo an attraction towards the positive plate. The force of attraction is given by the product of the charge upon the electron e and the field intensity; this produces an acceleration d^2y/dt^2 which is related to the force of attraction by the Newtonian equation

$$m_e \frac{d^2y}{dt^2} = Ee \tag{1}$$

where m_e is the mass of the electron.

Integrating equation (1) twice with respect to t and evaluating the integration constants leads to the expression

$$y_E = \frac{1}{2} E \frac{e}{m_e} t^2 \tag{2}$$

and substituting for the speed $u(= x/t)$ of the electrons gives

$$y_E = \frac{1}{2} E \frac{e}{m_e} \frac{x^2}{u^2} \tag{3}$$

In the magnetic field the lines of force are in the opposite direction. An electron of charge e, moving a distance dx in a time dt, is equivalent to an

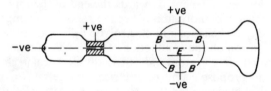

Figure 1.1(a). Thomson's apparatus for determination of e/m_e for electrons

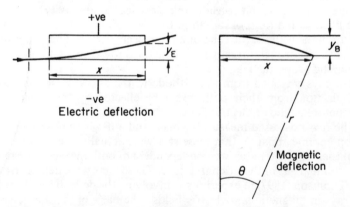

Figure 1.1(b). Electron deflection in applied electric and magnetic fields (from Adam, N. K. *Physical Chemistry*, Clarendon Press, Oxford, 1956)

electric current I such that $I\,dt = e$. The electron is thus subject to a force $BI\,dx$ tending to deflect it from its course (B being the magnetic flux density). This magnetic deflection causes the electron to be directed in a circular path of radius r such that

$$BI\,dx = m_e \frac{u^2}{r}$$

which by substituting for I and u becomes

$$Be = m_e \frac{u}{r} \qquad (4)$$

An angular deflection (θ) of the electron also occurs in a time t such that

$$\theta = \frac{u}{r} t$$

and this is accompanied by a corresponding linear deflection y_B, after travelling a distance equivalent to x along the original path. Simple geometrical considerations lead to the relation

$$x^2 = y_B(2r - y_B)$$

which for small deflections becomes

$$x^2 = 2ry_B \qquad (5)$$

Substituting from equation (5) for r in equation (4) gives the expression

$$y_B = \frac{1}{2} B \frac{e}{m_e} \frac{x^2}{u} \qquad (6)$$

By adjusting the values of the two fields it may be arranged that no deflection from the original path occurs. Hence $y_E = y_B$ or $u = E/B$. From measurements of the electron deflection made when one or other of the two fields is

Table 1.1. SOME VALUES OF e/m_e, e AND m_e

	Thomson and Millikan	Recent workers
e/m_e	5.20×10^{17} esu g^{-1}	$1.758\ 8 \times 10^{11}$ C kg^{-1}
e	4.774×10^{-10} esu	$1.602\ 1 \times 10^{-19}$ C
m_e	9.36×10^{-28} g	9.109×10^{-31} kg

switched off, the value of $u = E/B$ can be determined using either equation (3) or (6).

An evaluation of the charge on the electron was achieved by R. Millikan (1909) and the value so obtained allowed an estimate of the mass of the electron to be made as approximately 1/1820 that of the mass of the hydrogen atom.

More recent methods have enabled greater accuracy to be obtained both in the determination of e/m_e and of m_e, and these results are summarised in Table 1.1. From these figures the ratio of the mass of the electron to that of the hydrogen atom is 1/1837.

Cathode rays are not the only phenomena observable in discharge tubes that have been of fundamental importance in elucidating the electrical nature of the atom. Goldstein found, by using a perforated cathode, that another kind of ray was also produced. Such rays moved in the opposite direction to the cathode rays and it was concluded that they were composed of positively charged particles. Using different gases it was found that, unlike cathode rays, these positive rays had ratios of e/m that were dependent on the nature of the residual gas in the discharge tube. It is now known that *positive rays* are formed by ionisation of the gas molecules by the stream of electrons from the cathode. From an intensive study of positive rays has come our present knowledge of the atoms of the common elements.

By using methods similar to those for determining e/m_e of the electron it was shown that the largest value of this ratio for positive rays was obtained when hydrogen was the residual gas in the discharge tube. The positively charged particle formed in this instance had a mass 1836 times that of the electron and a charge equal in magnitude, but opposite in sign, to that of the electron. The *proton*, as this particle became known, is one of the fundamental constituents of matter.

The first systematic investigation of these rays was undertaken by Thomson (1910–1914) using his 'parabola method'. This measured the effects, recorded on a photographic plate, of the simultaneous application of electric and magnetic fields to the positive rays. Development of the plates showed parabolic traces. Each parabola corresponded to particles of the same ratio of e to m. These particles fall at different points along the parabola according to their speeds. The parabolic relationship can be derived from combination

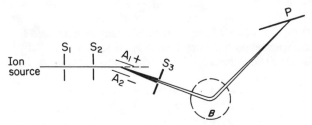

Figure 1.2. Aston's mass spectrograph

of equations (3) and (6) which gives $y_B^2 = k(e/m)y_E$. By comparison with traces from elements of known atomic weight Thomson was able to estimate the atomic weights of other elements. Thus, using neon gas in the apparatus, two lines were observed that corresponded to approximate relative atomic masses of 20 and 22; this suggested the presence of two kinds (or *isotopes*) of neon atoms.

The method of measuring masses of elements in this manner was further developed by Aston who devised the mass spectrograph, an instrument capable of focusing the positive rays as linear rather than parabolic traces. The principle of this method is shown diagrammatically in *Figure 1.2*. A fine parallel beam of positive rays from a discharge tube is selected by slits S_1 and S_2 and passes through two parallel plates A_1 and A_2 by means of which an electric field may be applied. The beam, on entering the electric field, diverges upon attraction to the negatively charged plate because of the differences in charge, mass and velocity of the individual particles. A band from this wide beam is selected by S_3 and passes into the field (B) of a powerful electro-magnet whose polarity and direction are such that the particles are bent back in the manner shown. Adjustment of the field strengths enables particles of the same e/m to be focused on a single line on the photographic plate P. The standard used in the measurement of the atomic masses was that of the oxygen atom. Lines from oxygen are obtained at positions corresponding to O_2^+, O^+ and O^{2+} respectively. Using this instrument Aston not only con-firmed the existence of the two isotopes of neon but also found isotopes for chlorine and argon. Further development was made when the method was applied to solids by producing ions from an anode coated with a suitable salt of the element. During the initial period of investigation by Aston it was apparent that the masses of individual atoms were integral, but with increased resolving power it was found that there were slight but significant deviations from the integral values*. Thus chlorine, which was found to have atoms with masses of 35 and 37 with the instruments of low resolving power, was shown to have mass values of 34·983 and 36·980 (in SI converted to the carbon-12 scale 34·968 85 and 36·965 90 m_u).

This type of instrument is particularly useful in the determination of the relative abundance of isotopes. This was first achieved by estimating the intensity of blackness of the lines on the photographic plates by means of a microphotometer. Though this method was of some use it was not particularly accurate and it has been superseded by ion-current measuring devices or mass spectrometers. In these instruments the photographic plate has been replaced by a detector for current measurement; a typical arrangement is shown in *Figure 1.3* which is a diagrammatic representation of the spectrometer

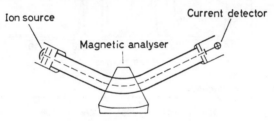

Figure 1.3. Nier's mass spectrometer

developed by Nier. A typical separation of isotopes using such an instrument is given in *Figure 1.4*.

Some selected values of isotopic masses, which can be measured to great accuracy, are quoted in *Table 1.2*.

Table 1.2. SOME ISOTOPIC MASSES

Entity	Mass/m_u
e	0·000 548
${}_0^1\text{n}$	1·008 665
${}_1^1\text{H}$	1·007 277
${}_2^4\text{He}$	4·002 604
${}_6^{12}\text{C}$	12·000 00
${}_8^{16}\text{O}$	15·994 915

* These integral values have been retained for brevity, and are referred to as mass numbers (A) of the element. The isotopic mass of a nuclide is generally expressed in terms of m_u, the unified atomic mass constant, which is defined as one-twelfth of the mass of one atom of carbon-12 = $1\cdot660\ 43 \times 10^{-27}$ kg. This quantity is more strictly a physical constant and not a unit.

The atomic weight or, as it is more correctly termed, the relative atomic mass (A_r) of an element is the number that is the weighted mean of the masses of the naturally occurring isotopes divided by m_u. Thus, for chlorine we have

$$A_r = (34.968\ 85\ m_u \times 75.53 + 36.965\ 90\ m_u \times 24.47) \div 100\ m_u$$
$$= 35.453$$

Except in the case of the hydrogen isotope of mass number one, the isotopic mass numbers of atoms are greater than the number of protons that are contained in the nucleus of the atom; this discrepancy is accounted for by the

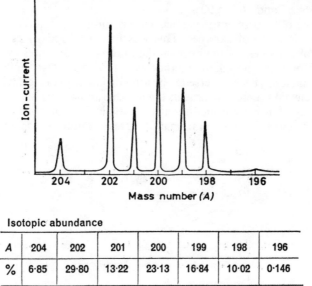

Figure 1.4. Separation of the isotopes of mercury

Isotopic abundance

A	204	202	201	200	199	198	196
%	6·85	29·80	13·22	23·13	16·84	10·02	0·146

presence of other fundamental particles in the nucleus with masses approximately that of the proton but of zero charge; these are known as *neutrons* (see p. 16). In simple terms, a nucleus is composed of *protons* and *neutrons* and an atom of mass number (A) and atomic number (Z) contains Z protons and $A - Z$ neutrons. An isotope of the element is designated as $_Z^A M$. If, however, a summation of the masses of the protons, electrons and neutrons in any one isotope (except $_1^1 H$) is made, there is a noticeable difference between this value and that obtained from physical measurement. Thus in the case of the helium atom, which may be considered to be a composite nucleus of two protons and two neutrons, together with two extranuclear electrons, the isotopic mass is $4.002\ 60\ m_u$, whereas that from a summation of the mass of two protons, two neutrons and two electrons is $4.032\ 98\ m_u$; this difference of $0.030\ 38\ m_u$ is termed the *mass excess*. Mass excesses may be accounted for by the relativity theory of Einstein which proposes that mass and energy are interconvertible. The relationship between the two is given by the equation

$$E = mc^2 \qquad (7)$$

where E is the energy, m is the mass and c is the speed of light*.

Reverting to the case of the helium atom, the mass excess of 0·030 38 m_u represents an energy release of approximately 28·2 MeV when the atom is synthesised from its components. Such energy is termed the *binding energy*, being the energy required to break down the atom into its fundamental components. Estimates of the binding energies of other atoms have been made and are shown graphically in *Figure 1.5* as binding energy per particle. This figure shows that the elements of mass number around 60, namely iron, cobalt, nickel and copper, possess the most stable nuclei; those at both ends

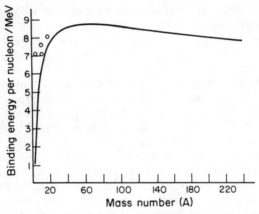

Figure 1.5. Binding energy curve. High values of B.E./A are obtained for ^{4}He, ^{8}Be, ^{12}C and ^{16}O, indicating some added stability of atoms with mass numbers that are integral multiples of four

of the Periodic Table are less stable. Energy can accordingly be released by either of two nuclear reactions:

(a) the fusion of the lightest nuclei to form heavier particles, as in the hydrogen bomb,

(b) the breaking-down or fission of the heaviest nuclei into lighter ones as in the atomic bomb and nuclear reactors.

RADIOACTIVITY

At about the same time as Thomson's researches on the electron, the phenomenon of radioactivity was discovered; this is a spontaneous breakdown of the

---

*A mass loss of one m_u means the release of energy equivalent to $1·660\,43 \times 10^{-27}$ kg $\times$ $(2·997\,9 \times 10^8$ m s$^{-1})^2 = 1·492\,3 \times 10^{-10}$ J. An energy unit (non SI) frequently employed in this field is the mega-electronvolt (MeV). The electronvolt (eV) is the energy that an electron acquires when it is accelerated through a potential difference of 1 V, which is thus $1·602\,10 \times 10^{-19}$ C V or $1·602\,10 \times 10^{-19}$ J. Hence, one MeV is equivalent to $1·602\,10 \times 10^{-13}$ J and the loss of one m_u corresponds to an energy release of 931·5 MeV.

atomic nucleus which is unaffected by external conditions such as temperature and pressure. During the course of some work concerned with x-rays and phosphorescence, Becquerel accidentally placed a uranium compound in close contact with some photographic plates; later it was found that although these plates had been wrapped in protective paper, exposure of the plates had occurred. The conclusion reached was that the uranium salt had been emitting some form of radiation that could penetrate materials normally opaque to light. Further work by the Curies led to the discovery that the same radiation ionised air.

The Curies demonstrated that the phenomenon was also characteristic of other substances; in particular they isolated two new elements — polonium and radium — in the form of certain salts by careful separation from the uranium ore pitchblende.

In contrast to x-rays, which also ionise air and affect photographic plates, the radiations from radioactive materials were found to be capable of subdivision into various categories. This subdivision was made after a study of the deflections experienced by the radiations in strong magnetic fields when three groups were characterised α, β and γ. Two sets were found to be deflected in opposite directions, one to a greater extent than the other. Those more deflected (β) were shown from the direction of deflection to be negatively charged and quantitative studies indicated that they were electrons, often with high speeds*. The second group (α) were examined by measuring e/m and were found to be positively charged particles with an e/m almost exactly half that of the proton. Verification of their identity came when experiments by Rutherford and Royds showed these particles to be ionised helium atoms. The final set of radiations were undeflected by the applied magnetic fields and are known to be electromagnetic radiation of short wavelength. These γ-rays are formed as a consequence of the readjustment of energy levels within the nucleus after the emission of alpha or beta particles has taken place.

The property common to all three types of radiation, that of ionisation of air, has been used in instruments for their detection; the property of excitation of fluorescence in phosphors, originally used by Rutherford and Crookes, has also been incorporated in scintillation counters. In the instruments based upon ionisation, detection of the radiations is effected by the amplification and recording of the change in the potential across two electrodes caused by the presence of ionising radiation. In scintillation counting, the scintillations excited in phosphors are detected by photoelectric cells and converted to electrical impulses. A further device that was of use during initial

---

* Measurement of the value of e/m_e for electrons emitted from radioactive materials gives a value that is generally less than that found for electrons from conventional sources. Because the electrons are emitted with high energies their mass is significantly different from the rest mass. The relationship between mass at high speeds and rest mass (m_0) is

$$m = \frac{m_0}{\sqrt{1-(u^2/c^2)}}$$

where u is the speed of the particle and c is the speed of light. As the particle speed increases and approaches that of light the mass will also increase; thus with a speed of $0.1c$ the mass m is approximately $1.005m_0$ and with a speed $0.9c$ the mass is approximately $2.294m_0$. Values of e/m_e will therefore decrease as the particle energy increases.

investigations — the Wilson Cloud Chamber — depended upon the condensation of water vapour on charged particles when air saturated with water vapour was adiabatically expanded within the chamber. The water droplets thus formed left distinctive trails which were further examined under a microscope or recorded photographically. It was in such a chamber that the positive analogue of the electron — the *positron* — was first identified.

Radioactive transformations

The emission of an alpha or beta particle causes a chemical change in the radioactive atom and the activity of the parent atom decays exponentially with time. The exponential decay is plotted in *Figure 1.6* and for N radioactive nuclei the variation with time of the activity may be expressed as

$$\frac{dN}{dt} = -\lambda N \tag{8}$$

where λ is a transformation constant or the proportion of the atoms in the

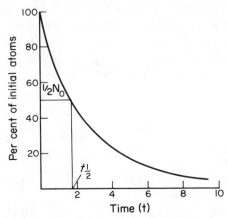

Figure 1.6. Decay curve for a radioactive material

sample that decay in unit time. Integration of equation (8) leads to

$$N = N_0 \exp(-\lambda t) \tag{9}$$

N_0 being the number of radioactive atoms at time $t = 0$. From equation (9) an extremely important constant of the radioactive atom — the half-life $t_{\frac{1}{2}}$ — may be evaluated. The radioactive half-life of the element is the time required for one half of the atoms to decay, thus

$$N = \tfrac{1}{2}N_0$$

$$\text{or} \quad \tfrac{1}{2} = \exp(-\lambda t_{\frac{1}{2}})$$

$$\text{and} \quad t_{\frac{1}{2}} = \ln2/\lambda \tag{10}$$

Values of this constant vary from millionths of a second to about 10^{10} y (strictly speaking $t_{\frac{1}{2}}$ should be quoted in seconds but normal practice is to quote values in appropriate units).

The naturally occurring radioactive isotopes or nuclides of the heavy elements are found to belong to one of three decay series, which are named after the most stable nuclide in each series. These series are given in *Figures 1.7, 1.8* and *1.9*, the parent elements being ^{232}thorium, ^{238}uranium and ^{235}uranium* with respective half-lives of $1\cdot4 \times 10^{10}$ y, $4\cdot5 \times 10^{9}$ y and $7\cdot8 \times 10^{8}$ y. An alternative nomenclature is the 4n, (4n + 2) and the (4n + 3) series, indicating that all the members of the 4n series have mass numbers that are integral multiples of four and those in the other two have mass numbers divisible by four with remainders of two and three respectively.

The emission of an alpha particle results in the formation of an atom with mass number four units less and atomic number two units less than the parent atom. The emission of a beta particle causes isobaric change or the formation of an atom with identical mass number but one greater in atomic number than the decaying atom; this type of change is also characteristic of positron emission and K-capture† which yield isobars with atomic numbers one less than the decaying atom. In the case of the heavy elements this process of radioactive decay is a series of successive ejections of particles and frequent adjustment of energy within the nucleus until an element is reached that is no longer unstable; this occurs with each of the three series at element 82.

Table 1.3. SOME NATURALLY OCCURRING RADIOACTIVE NUCLIDES NOT IN THE DISINTEGRATION SERIES

Nuclide	Type of decay	Product	$t_{\frac{1}{2}}/y$
$^{3}_{1}$H	beta	$^{3}_{2}$He	$12\cdot5$
$^{14}_{6}$C	beta	$^{14}_{7}$N	$5\cdot7 \times 10^{3}$
$^{40}_{19}$K	beta + gamma K-capture†	$^{40}_{20}$Ca $^{40}_{18}$Ar	$1\cdot4 \times 10^{9}$
$^{87}_{37}$Rb	beta + gamma	$^{87}_{38}$Sr	$6\cdot0 \times 10^{10}$
$^{152}_{62}$Sm	alpha	$^{148}_{60}$Nd	$2\cdot5 \times 10^{11}$

A fourth radioactive decay series exists for elements with mass numbers divisible by four with a remainder of one (4n + 1) and atomic numbers greater than 83. This is the Neptunium series which contains nuclides that do not occur naturally but arise from the decay of ^{241}plutonium prepared by alpha bombardment of ^{238}uranium. The end product of this series is ^{209}bismuth as shown in *Figure 1.10* (p. 14).

In addition to the nuclides of elements with atomic number greater than 82, a number of naturally occurring nuclides of elements of lower atomic number exist that are also radioactive; these are characterised by long half-lives and weak activity. Some examples are given in *Table 1.3*.

* The Actinium series takes its name from the original name *Actino*uranium (^{235}U).

† K-capture signifies the capture of an orbital electron by the nucleus. This electron generally comes from the K shell and the vacancy left by it is filled by an electron from a higher energy level, giving rise to the formation of a spectral line in the x-ray region.

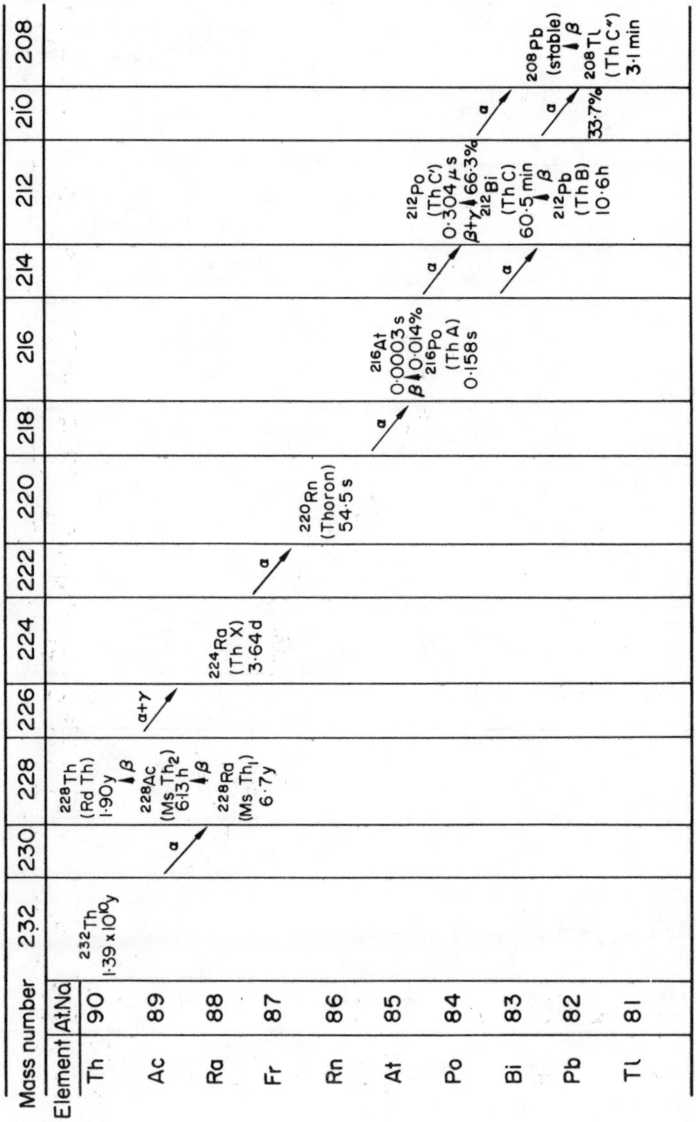

Mass number	232	230	228	226	224	222	220	218	216	214	212	210	208

Element At.No.

Th 90 — ^{232}Th 1·39 × 10^{10} y

Ac 89 — ^{228}Ac (MsTh$_2$) 6·13 h, β

Ra 88 — ^{228}Ra (MsTh$_1$) 6·7 y, ^{224}Ra (Th X) 3·64 d

Fr 87

Rn 86 — ^{220}Rn (Thoron) 54·5 s

At 85 — ^{216}At 0·0003 s, β 0·014%

Po 84 — ^{216}Po (Th A) 0·158 s, ^{212}Po (Th C') 0·304 μs

Bi 83 — ^{212}Bi (Th C) 60·5 min, β β+γ 66·3%

Pb 82 — ^{212}Pb (Th B) 10·6 h, ^{208}Pb (stable)

Tl 81 — 33·7%, ^{208}Tl (Th C'') 3·1 min, β

^{228}Th (Rd Th) 1·90 y

Figure 1.7. Thorium decay series

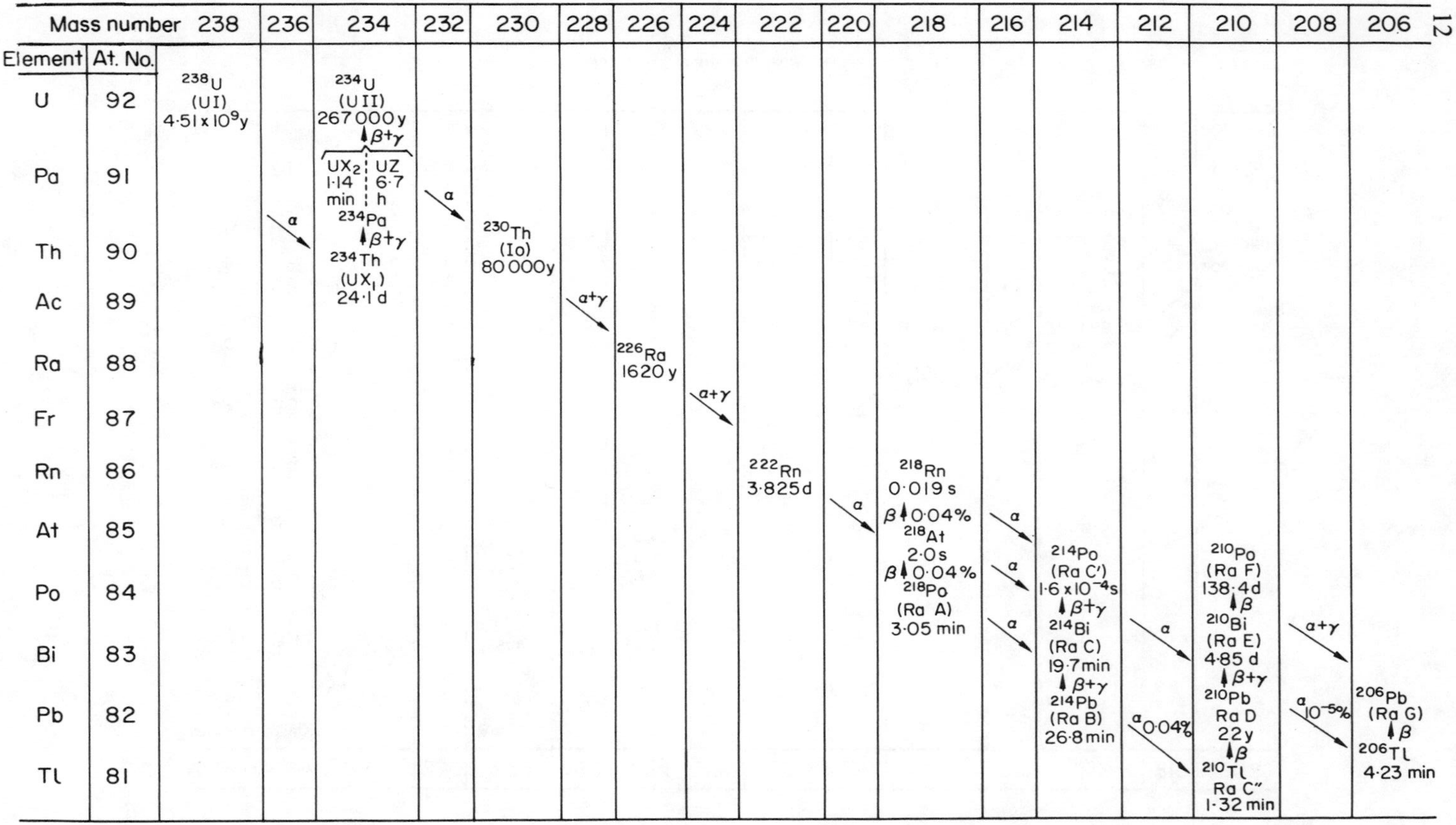

Mass number		238	236	234	232	230	228	226	224	222	220	218	216	214	212	210	208	206
Element	At. No.																	
U	92	^{238}U (UI) 4.51×10^9y		^{234}U (UII) 267 000 y $\uparrow \beta+\gamma$														
Pa	91			UX$_2$ | UZ 1.14 | 6.7 min | h ^{234}Pa $\uparrow \beta+\gamma$														
Th	90			^{234}Th (UX$_1$) 24.1 d		^{230}Th (Io) 80 000 y												
Ac	89																	
Ra	88							^{226}Ra 1620 y										
Fr	87																	
Rn	86									^{222}Rn 3.825 d		^{218}Rn 0.019 s						
At	85											$\beta \uparrow 0.04\%$ ^{218}At 2.0 s						
Po	84											$\beta \uparrow 0.04\%$ ^{218}Po (Ra A) 3.05 min		^{214}Po (Ra C') 1.6×10^{-4}s $\uparrow \beta+\gamma$		^{210}Po (Ra F) 138.4 d $\uparrow \beta$		
Bi	83													^{214}Bi (Ra C) 19.7 min $\uparrow \beta+\gamma$		^{210}Bi (Ra E) 4.85 d $\uparrow \beta+\gamma$		
Pb	82													^{214}Pb (Ra B) 26.8 min	0.04%	^{210}Pb Ra D 22 y $\downarrow \beta$	10^{-5}%	^{206}Pb (Ra G)
Tl	81															^{210}Tl Ra C'' 1.32 min		^{206}Tl 4.23 min

Figure 1.8. Uranium-238 decay series

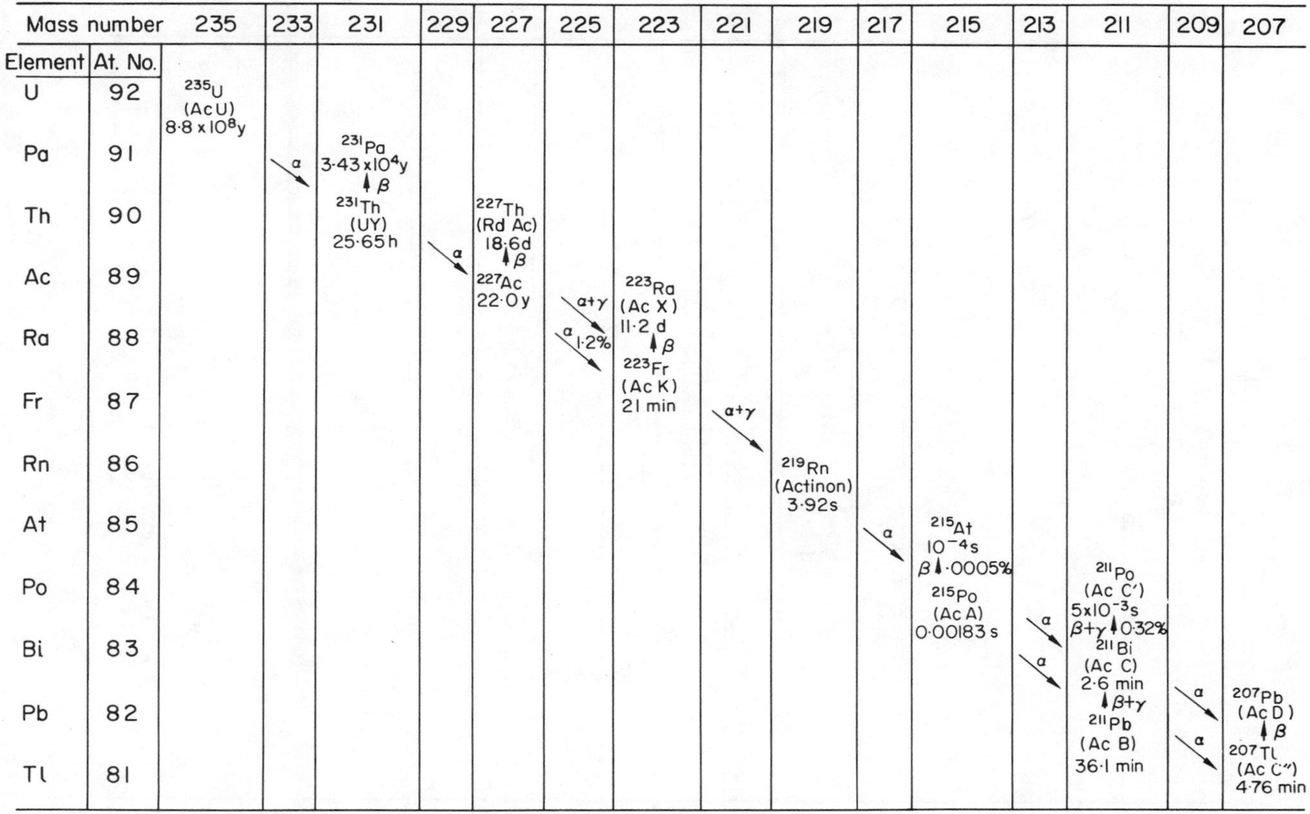

Mass number	At. No.	235	233	231	229	227	225	223	221	219	217	215	213	211	209	207
Element																
U	92	^{235}U (AcU) 8.8×10^{8} y														
Pa	91			^{231}Pa 3.43×10^{4} y $\downarrow \beta$												
Th	90			^{231}Th (UY) 25.65 h		^{227}Th (Rd Ac) 18.6 d $\downarrow \beta$										
Ac	89					^{227}Ac 22.0 y										
Ra	88							^{223}Ra (Ac X) 11.2 d $\downarrow \beta$								
Fr	87							^{223}Fr (Ac K) 21 min								
Rn	86									^{219}Rn (Actinon) 3.92 s						
At	85											^{215}At 10^{-4} s $\beta \downarrow \cdot 0005\%$				
Po	84											^{215}Po (Ac A) 0.00183 s		^{211}Po (Ac C') 5×10^{-3} s $\beta + \gamma \downarrow 0.32\%$		
Bi	83													^{211}Bi (Ac C) 2.6 min $\downarrow \beta + \gamma$		
Pb	82													^{211}Pb (Ac B) 36.1 min		^{207}Pb (Ac D)
Tl	81															^{207}Tl (Ac C'') 4.76 min $\downarrow \beta$

Figure 1.9. Actinium decay series

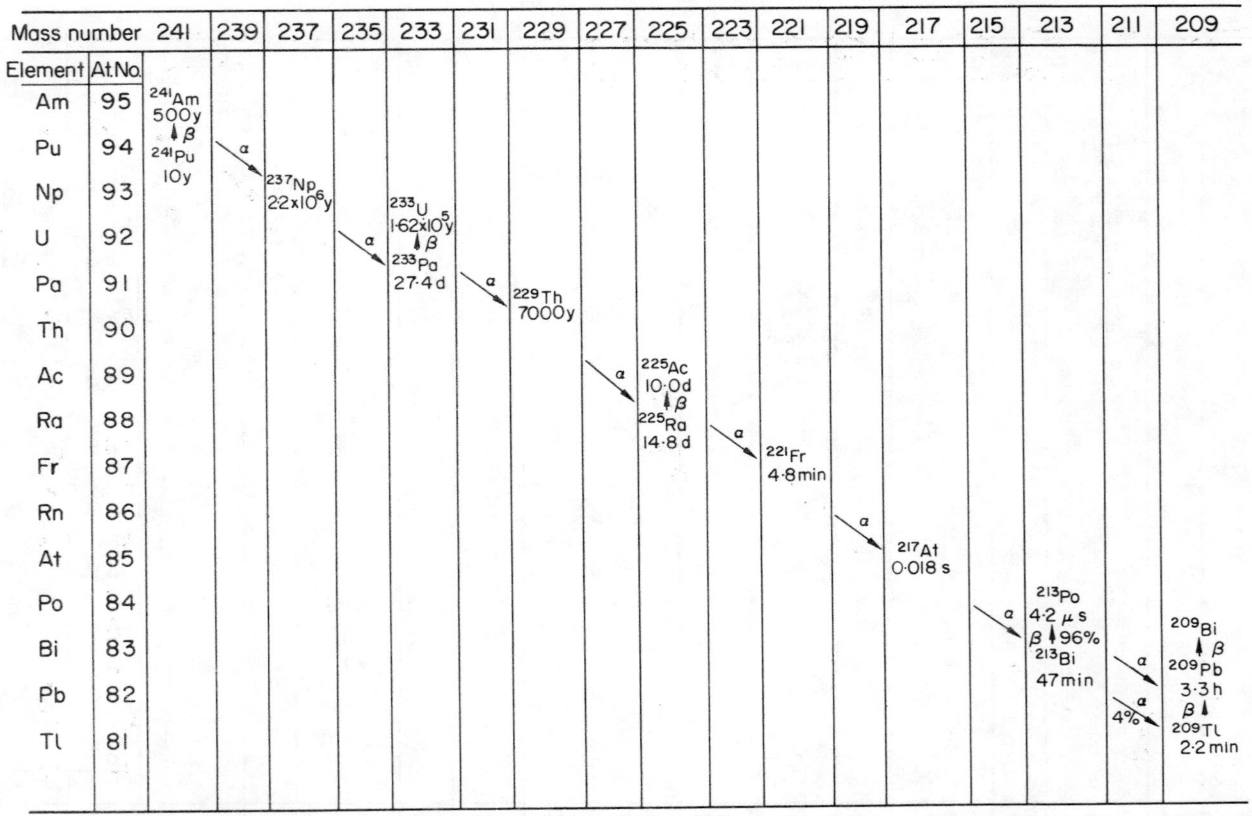

Mass number	241	239	237	235	233	231	229	227	225	223	221	219	217	215	213	211	209

Element | At.No.

Am 95 — ^{241}Am 500y β ↓ ^{241}Pu 10y

Pu 94 — ^{241}Pu 10y, α

Np 93 — ^{237}Np 22×10^{6}y, α

U 92 — ^{233}U 1·62×10^{5}y, β ↓ ^{233}Pa 27·4d

Pa 91 — ^{233}Pa 27·4d, α

Th 90 — ^{229}Th 7000y, α

Ac 89 — ^{225}Ac 10·0d, β ↓ ^{225}Ra 14·8d, α

Ra 88 — ^{225}Ra 14·8d

Fr 87 — ^{221}Fr 4·8min, α

Rn 86

At 85 — ^{217}At 0·018s, α

Po 84 — ^{213}Po 4·2 μs, β ↓ ^{213}Bi 47min, α

Bi 83 — ^{213}Bi 47min 96%; ^{209}Bi, β ↓ ^{209}Pb 3·3h, α 4%

Pb 82 — ^{209}Pb 3·3h, β ↓ ^{209}Tl 2·2min

Tl 81 — ^{209}Tl 2·2min

Figure 1.10. Neptunium decay series

Scattering of alpha particles

In 1911, Rutherford carried out experiments on the effect of placing extremely thin sheets of metal foil in the paths of streams of alpha particles. The majority of the particles, detected by scintillations produced on a zinc sulphide screen, passed through the foils suffering little or no deflection but a few underwent deflections through wide angles (*Figure 1.11*). From these observations he concluded that the atoms of which the metal foils were composed possessed a positively charged body of finite mass occupying but a small fraction of the total atomic volume. Since the majority of alpha particles were not deflected to any great extent the proposal was made that the remainder of the atomic volume was a void but that in this space there were sufficient electrons to keep the atom electrically neutral. A quantitative examination of the results indicated that the particles within the atom in fact only occupy about 10^{-12} of the total volume and that the value of the positive charge on the nucleus was numerically equal to approximately one half the relative atomic mass of the metal of which the foil was composed. In 1913 van den Broek proposed that the atomic number of the element, itself approximately one half the relative atomic mass in many instances, was in fact identifiable

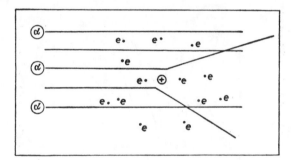

Figure 1.11. Alpha particle bombardment of atoms in metal foils

with the number of positive charges on the nucleus. This was verified shortly afterwards by Moseley from studies upon the x-ray spectra of elements of known atomic number.

Rutherford, in addition, investigated the action of alpha particles upon various gases. Using nitrogen, he observed from Cloud Chamber experiments that occasionally collision occurred between an alpha particle and a nitrogen atom resulting in the formation of a proton (*Figure 1.12*). This reaction is represented as

$$^{14}_{7}\text{N} + ^{4}_{2}\text{He} \longrightarrow ^{1}_{1}\text{H} + ^{17}_{8}\text{O}$$

and is an artificial disintegration of the atomic nucleus. It may be considered generally as a two-stage process; firstly the coalescence of the alpha particle and the nitrogen nucleus with the formation of a compound nucleus and secondly the transformation of this nucleus, which is nearly always unstable, to a stable state by the emission of either a particle or gamma radiation.

This initial discovery was quickly followed by those of other workers in the same field and reactions such as

$$^{19}_{9}\text{F} + ^{4}_{2}\text{He} \longrightarrow ^{1}_{1}\text{H} + ^{22}_{10}\text{Ne}$$
$$^{23}_{11}\text{Na} + ^{4}_{2}\text{He} \longrightarrow ^{1}_{1}\text{H} + ^{26}_{12}\text{Mg}$$

were found to occur with elements of low atomic number.

The capture of an alpha particle did not lead unequivocally to the ejection of a proton by the compound nucleus. Thus in the case of the bombardment of beryllium, Bothe and Becker found a particularly penetrating type of radiation was emitted that caused protons to be ejected from paraffin wax placed in its path. This particular type of radiation was identified by Chadwick (1932) as characteristic of a particle of unit mass and zero charge — the *neutron*.

$$^{9}_{4}\text{Be} + ^{4}_{2}\text{He} \longrightarrow ^{1}_{0}\text{n} + ^{12}_{6}\text{C}$$

The bombardment of a positively charged nucleus by a positively charged particle, such as the alpha particle, immediately poses experimental problems. A potential barrier is set up by electrostatic repulsion and if a disintegration reaction is to be achieved this potential barrier must be overcome. In order to overcome this repulsion the particle must possess a high kinetic

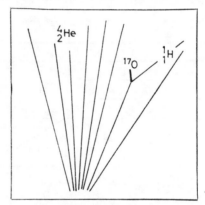

Figure 1.12. Representation of formation of the ^{17}O isotope by bombardment of ^{14}N by alpha particles

energy; that of the naturally produced alpha particles is in the range of 4–8 MeV and is sufficient only for reaction with the lightest elements.

Particles with the required kinetic energies are obtained by means of accelerating devices. The first success in this field was by Cockcroft and Walton (1932) who devised an instrument capable of producing high energy protons by accelerating them through a potential difference of the order of 5×10^5 V. These protons achieved disintegrations of the lithium isotope, mass number seven, into two alpha particles:

$$^{7}_{3}\text{Li} + ^{1}_{1}\text{H} \longrightarrow 2^{4}_{2}\text{He}$$

Further research led to the development of the linear accelerator, the cyclotron and the synchrocyclotron enabling the production of protons with energies in excess of 400 MeV and alpha particles with energies greater than

50 MeV. Using these instruments, a large variety of induced disintegrations has been accomplished by bombardment of atoms with particles such as the proton, the deuteron (the ionised atom of hydrogen with mass number 2) and the alpha particle.

Neutrons, on the other hand, have no energy barrier to overcome and consequently are extremely useful in the field of radiochemistry. In particular, they are employed to a large extent for the production of artificially radioactive nuclides in the atomic pile.

Typical reactions induced by positively charged particles and yielding stable nuclides are given below:

(a) Proton induced

$$^{19}_{9}F + ^{1}_{1}H \longrightarrow ^{16}_{8}O + ^{4}_{2}He \quad \text{otherwise} \quad ^{19}_{9}F\,(p, \alpha)^{16}_{8}O$$
$$^{27}_{13}Al + ^{1}_{1}H \longrightarrow ^{24}_{12}Mg + ^{4}_{2}He \quad\quad\quad ^{27}_{13}Al(p, \alpha)^{24}_{12}Mg$$

(b) Deuteron induced

$$^{16}_{8}O + ^{2}_{1}H \longrightarrow ^{14}_{7}N + ^{4}_{2}He \quad \text{otherwise} \quad ^{16}_{8}O\,(d, \alpha)\,^{14}_{7}N$$

(c) Alpha induced

$$^{27}_{13}Al + ^{4}_{2}He \longrightarrow ^{30}_{14}Si + ^{1}_{1}H \quad \text{otherwise} \quad ^{27}_{13}Al(\alpha, p)\,^{30}_{14}Si$$

Artificial radioactivity

The nuclear reactions dealt with above result in the formation of stable nuclides. This does not occur with every nuclear reaction. In many instances the nuclides first formed exhibit radioactivity. This production of radioactivity artificially was originally recognised by Curie–Joliot during the bombardment of boron, magnesium and aluminium with alpha particles. This resulted in the formation of neutrons and positrons, the emission of the former being terminated by the removal of the bombarding source but not the positron emission which could still be detected and was observed to decay exponentially with time.

Taking aluminium as an example, this phenomenon was explained by the initial production of an unstable nuclide that decayed by positron emission to a stable isotope of silicon. This hypothesis was confirmed (i) by dissolving the aluminium, after alpha bombardment, in hydrochloric acid. Evaporation to dryness gave a residue that had no activity, but the gas evolved during this chemical treatment, i.e. phosphine, was active; (ii) by dissolving the aluminium in aqua regia to give phosphoric acid which could be precipitated as radioactive phosphate.

The three nuclear reactions above were found to be as follows:

$$^{10}_{5}B + ^{4}_{2}He \longrightarrow ^{13}_{7}N \quad + \quad ^{1}_{0}n$$
$$\downarrow \quad t_{\frac{1}{2}} = 10 \text{ min}$$
$$^{13}_{6}C \quad + \quad ^{0}_{+1}e$$

$$^{24}_{12}Mg + ^{4}_{2}He \longrightarrow ^{27}_{14}Si \quad + \quad ^{1}_{0}n$$
$$\downarrow \quad t_{\frac{1}{2}} = 7 \text{ min}$$
$$^{27}_{13}Al \quad + \quad ^{0}_{+1}e$$

$$^{27}_{13}Al + ^{4}_{2}He \longrightarrow ^{30}_{15}P \quad + \quad ^{1}_{0}n$$
$$\downarrow \quad t_{\frac{1}{2}} = 3 \text{ min}$$
$$^{30}_{14}Si \quad + \quad ^{0}_{+1}e$$

Other transmutations effected include:

$$\,_3^7\text{Li} + \,_2^4\text{He} \longrightarrow \,_4^{10}\text{Be} \quad + \quad \,_1^1\text{H}$$
$$\downarrow \quad t_{\frac{1}{2}} = 2{\cdot}7 \times 10^6 \text{ y}$$
$$\,_5^{10}\text{B} \quad + \quad \,_{-1}^{0}\text{e}$$

$$\,_{12}^{25}\text{Mg} + \,_2^4\text{He} \longrightarrow \,_{13}^{28}\text{Al} \quad + \quad \,_1^1\text{H}$$
$$\downarrow \quad t_{\frac{1}{2}} = 2 \text{ min}$$
$$\,_{14}^{28}\text{Si} \quad + \quad \,_{-1}^{0}\text{e}$$

The positron or electron emission may produce a stable nuclide without further changes but frequently an excited nuclide is first formed and this loses energy by emission of gamma radiation.

Neutron irradiation is particularly important for production of radioactive nuclides. This irradiation is carried out in an atomic pile where the neutrons

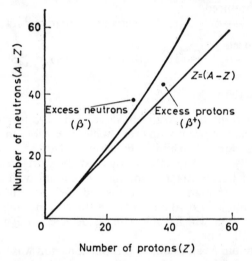

Figure 1.13. Nuclear stability curve. At low atomic numbers, the number of protons is approximately equal to the number of neutrons in stable nuclei. Radioactive nuclei are represented by points to the right and the left of the curve. (β^-) and (β^+) indicate electron and positron emitters respectively

are of two types (i) slow and (ii) fast, i.e. they have low or high kinetic energies. The reactions induced are (n, γ) and (n, α) respectively and the nuclides obtained are predominantly electron emitters. Types of reaction that may take place are (n, γ), (n, α), (n, p) and (n, 2n) and examples of these are

$$\,_{11}^{23}\text{Na (n, }\gamma) \,_{11}^{24}\text{Na} \longrightarrow \,_{12}^{24}\text{Mg} + \,_{-1}^{0}\text{e} \quad (t_{\frac{1}{2}} = 15 \text{ h})$$
$$\,_9^{19}\text{F (n, }\alpha) \,_7^{16}\text{N} \longrightarrow \,_8^{16}\text{O} \;+ \,_{-1}^{0}\text{e} \quad (t_{\frac{1}{2}} = 7{\cdot}4 \text{ s})$$
$$\,_7^{14}\text{N (n, p) }\,_6^{14}\text{C} \longrightarrow \,_7^{14}\text{N} \;+ \,_{-1}^{0}\text{e} \quad (t_{\frac{1}{2}} = 5600 \text{ y})$$
$$\,_{15}^{31}\text{P (n, 2n) }\,_{15}^{30}\text{P} \longrightarrow \,_{14}^{30}\text{Si} \;+ \,_{+1}^{0}\text{e} \quad (t_{\frac{1}{2}} = 3 \text{ min})$$

Whether the radioactive nuclide so produced is an electron or positron emitter may be deduced from *Figure 1.13* which is a plot of the number of protons against neutrons for stable nuclei of the elements. For the light elements this line has a slope of one, but with increasing atomic number the neutrons are in excess and the slope consequently is greater than one for the heavy nuclei. Nuclei with more protons than the stable nuclei lie to the right of this stability curve and decay by positron emission; those with excess neutrons lie to the left of the curve and decay by electron emission.

NUCLEAR FISSION

A study of the binding energy curve reveals that large quantities of energy are released when heavy nuclei are broken down into lighter ones. Spontaneous fission is only a rare phenomenon, but in 1939 it was found that fission could be induced in uranium by bombardment with neutrons, with the formation of a barium isotope as one of the products. Under such bombardment fission of ^{235}U occurs when the neutrons have relatively low velocities and fission of ^{238}U occurs with high velocity neutrons. In addition to the fission of the lighter isotope of uranium, slow neutrons also cause ^{233}U and ^{239}Pu to undergo fission.

The break-up of the compound nucleus into primary fission products occurs almost always into particles of unequal mass as indicated by *Figure 1.14* for

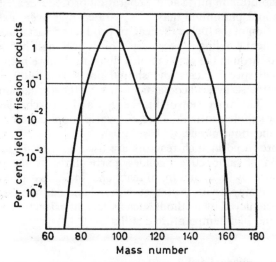

Figure 1.14. Distribution of fission products from ^{235}U. The fission products which are produced in greatest quantity are those of mass numbers in the region of 90 and 140

the fission of ^{235}U by slow neutrons; concentration of the fission products occurs in the region of mass numbers 90 and 140 as indicated by the two maxima. The fission products in general have an excess of neutrons and decay by electron emission; also formed by the fission process are a number of neutrons. A possible mode of decay may be

$$^{235}_{92}U + ^{1}_{0}n \longrightarrow ^{236}_{92}U \longrightarrow ^{94}_{38}Sr + ^{140}_{54}Xe + 2\,^{1}_{0}n + \gamma$$

and the two nuclei undergo electron emission to yield eventually $^{94}_{40}Zr$ and $^{140}_{58}Ce$ respectively.

The fact that two or more neutrons which may cause further fission are produced for every neutron consumed suggested that a *chain reaction* might be established which could be used as a source of large amounts of energy. Calculations show that a critical size of fissile material is required for a self-sustaining chain reaction to occur. When this size is exceeded the reaction can, in the absence of efficient neutron absorbers, proceed at a rate which is uncontrollable and terminate in an explosion. When pure ^{235}U is used in the atomic pile or nuclear reactor it is necessary to control the reaction by the introduction of neutron absorbers such as boron or cadmium. These make it possible to maintain a steady neutron level. Natural uranium contains only 0·7 per cent of this lighter isotope, the remainder being ^{238}U, and if used in the pile without separation from the latter, proves inefficient. Some of the heavy isotope is converted via neutron capture to ^{239}U and thence by electron emission to ^{239}Np and finally to ^{239}Pu, which is itself fissionable. In the pile, therefore, the fissile material gradually becomes converted to fission products which can absorb neutrons and cause the reaction to be no longer self-sustaining. The fuel rods are removed from time to time, stored for approximately 100 days to allow the short-lived fission products to decay, and then chemically treated by dissolution in nitric acid. Separation procedures are then applied; thus uranium and plutonium may be separated by solvent extraction.

Since the fission of the fuel in the reactor produces a vast amount of energy in the form of heat, cooling is necessary. This is accomplished by circulating a suitable fluid through the reactor; such fluids must have high thermal conductivities, high specific heats and should not significantly capture or be affected by neutrons. Typical coolants include water, air, carbon dioxide and molten sodium. The conversion of this energy via a heat exchanger to electrical power, thereby supplementing the present world supplies, is one of the major peacetime applications of nuclear fission.

Neutrons produced in such reactors are used, as already mentioned, for the production of radioactive nuclides, particularly by the (n, γ) type of reaction and also by (n, α) and (n, p) reactions. In the last two instances the product is non-isotopic with the target material and has to be separated from it; this may be achieved by methods such as solvent extraction, ion exchange, precipitation and high temperature distillation.

Applications of radionuclides

Radionuclides have many applications in the fields of analysis and biochemistry, in determining the nature of reaction mechanisms, in industrial processes and medicine to name but a few.

An example of their application in analysis is the method of radioactivation analysis. In this technique the element to be determined is converted to one of its radioisotopes by irradiating the sample, usually with slow neutrons. By comparing the activity of the sample with that from a standard sample (containing a known amount of the element), that has been irradiated similarly, an estimation of the concentration of the element in the unknown may be made.

In many instances such an estimation is non-destructive, that is to say no chemical treatment of the sample is necessary before the determination is made. Where other elements in the sample are also activated, non-destructive analysis may still be possible if these elements can be distinguished by their energy or half-lives. Otherwise, chemical treatment has to be introduced to separate the element, whose concentration is required, in suitable chemical form. An example of this type of analysis is the determination of trace impurity manganese in aluminium. Neutron activation converts ^{55}Mn to ^{56}Mn which is β-active with a half-life of 2·58 h. Although the aluminium is also activated at the same time, the isotope produced is of very much shorter half-life. The sample may be left to 'cool down' before the manganese activity is determined. This technique may be used to determine trace elements in semi-conductors and other ultra-pure industrial materials.

The mechanisms of many chemical reactions have been clarified by use of radionuclides. Thus, in organic chemistry, many compounds labelled with ^{14}C are utilised; here specific sites in molecules are labelled and the products of a chemical reaction may be investigated to determine in which of them the labelled carbon is to be found. Different mechanisms may thus be distinguished. As an example of this use in inorganic chemistry we may cite the reaction

$$S + SO_3^{2-} = S_2O_3^{2-}$$

where, by using labelled sulphur S* and sulphite ion of normal isotopic composition, the thiosulphate ion obtained is $S*SO_3^{2-}$ Decomposition of the labelled thiosulphate by acid shows the non-equivalence of the two sulphur atoms since all the labelled sulphur reappears as elemental sulphur.

$$S*SO_3^{2-} + 2H^+ = S* + H_2SO_3$$

In industry, radionuclides may be used for instance in measurement of wear and in control of thickness of sheet materials, e.g. plastic and steel. In thickness gauges the pressure of the rollers may be adjusted automatically on the basis of intensity of radiation from a γ-ray source placed on one side of the sheet. An ionisation chamber, diametrically opposite the source, records the radiation penetrating the sheet; this will depend upon the sheet thickness. Any variation in thickness will immediately affect the reading on the instrument measuring the ionisation and roller pressure may be altered accordingly. Wear of metals by friction in lubricated machinery may be investigated by coating the surface of the components with radioactively labelled metal. The amount of radioactivity in the oil will depend upon the wear of the metal surfaces.

In the physiological field the rate at which calcium is assimilated by bone structures from food eaten by humans has been determined by feeding on food containing radiocalcium. Similarly, the speed of water circulation in the body may be determined by drinking water containing some tritium oxide.

^{60}Co and ^{137}Cs, both large radiation sources, are used in the treatment of tumours, either to destroy them or to prevent further growth. They both have the advantage over conventional x-ray treatment in that they may be used to treat the tumours without exposing the body to excessive radiation which results in damaged skin tissue; their γ-radiation is more penetrating.

NUCLEAR FUSION

The second source of nuclear power is from the fusion of lighter elements into heavier nuclei. This process is the source of energy from the sun and also the basis of the hydrogen bomb.

Various reactions have been proposed for the production of stellar energy. One, involving six steps and known as the carbon cycle (a), is thought to be responsible for the energy of the brightest stars. With the cooler stars the energy is thought to arise from a smaller cycle that involves proton–proton reaction (b).

(a)
$$^{12}_{6}C + ^{1}_{1}H \longrightarrow ^{13}_{7}N$$
$$^{13}_{7}N \longrightarrow ^{13}_{6}C + ^{0}_{+1}e$$
$$^{13}_{6}C + ^{1}_{1}H \longrightarrow ^{14}_{7}N$$
$$^{14}_{7}N + ^{1}_{1}H \longrightarrow ^{15}_{8}O$$
$$^{15}_{8}O \longrightarrow ^{15}_{7}N + ^{0}_{+1}e$$
$$^{15}_{7}N + ^{1}_{1}H \longrightarrow ^{12}_{6}C + ^{4}_{2}He$$

(b)
$$^{1}_{1}H + ^{1}_{1}H \longrightarrow ^{2}_{1}H + ^{0}_{+1}e$$
$$^{2}_{1}H + ^{1}_{1}H \longrightarrow ^{3}_{2}He$$
$$^{3}_{2}He + ^{1}_{1}H \longrightarrow ^{4}_{2}He + ^{0}_{+1}e$$

These reactions suffer from the disadvantage that they are realised only at extremely high temperatures (about 10^8 K to 10^9 K). Such temperatures are difficult to obtain on Earth and as yet have been obtained only momentarily by a fission process which is used to promote the fusion process in the hydrogen bomb.

TRANSURANIC ELEMENTS

The discovery of elements beyond uranium in the Periodic Table arose from nuclear bombardment experiments. Thus the bombardment of ^{238}U by fairly fast neutrons produces an extremely unstable isotope of uranium that decays by emission of an electron to an isotope of element 93, named neptunium

$$^{238}_{92}U + ^{1}_{0}n \longrightarrow ^{239}_{92}U \xrightarrow{t_\frac{1}{2} = 23 \text{ min}} ^{239}_{93}Np + ^{0}_{-1}e$$

The isotope of neptunium so obtained is itself unstable and decays further by electron emission to element 94, named plutonium

$$^{239}_{93}Np \xrightarrow{t_\frac{1}{2} = 2.3 \text{ d}} ^{239}_{94}Pu + ^{0}_{-1}e + \gamma$$

This isotope of plutonium has a long half-life (2.4×10^4 y) decaying by alpha emission to ^{235}U.

Element 95, americium, is produced by several methods:

(a) Bombardment of ^{238}U by accelerated alpha particles (40 MeV)

$$^{238}_{92}U + ^{4}_{2}He \longrightarrow ^{241}_{94}Pu + ^{1}_{0}n$$

$$^{241}_{94}Pu \xrightarrow{t_\frac{1}{2} = 10 \text{ y}} ^{241}_{95}Am + ^{0}_{-1}e$$

and

(b) by the process ^{239}Pu $(n, \gamma) \to\ ^{240}$Pu $(n, \gamma) \to\ ^{241}$Pu which then decays as in (a).

The decay of ^{241}Am occurs by alpha particle emission of half-life 5000 y.

Isotopes of element 96 (curium) have been obtained by alpha particle bombardment of ^{239}Pu (α, n) and by the reaction ^{241}Am $(n, \gamma) \to\ ^{242}$Am which decays by electron emission to ^{242}Cm.

Accelerated alpha particles have also been used in the production of element 97, berkelium, and element 98, californium. The reactions involved are

$$^{241}_{95}\text{Am} + ^{4}_{2}\text{He} \longrightarrow\ ^{243}_{97}\text{Bk} + 2^{1}_{0}\text{n}$$

$$^{243}_{97}\text{Bk} \xrightarrow[t_{\frac{1}{2}} = 4 \cdot 5\text{h}]{K\text{-capture}}\ ^{243}_{96}\text{Cm}$$

and

$$^{242}_{96}\text{Cm} + ^{4}_{2}\text{He} \longrightarrow\ ^{244}_{98}\text{Cf} + 2^{1}_{0}\text{n}$$

Bombardment with heavy ions* has also been of use in effecting trans-mutations of this type. Thus accelerated stripped carbon atoms (C^{6+}) have enabled the 246 isotope of californium to be prepared by the reaction

$$^{238}_{92}\text{U} + ^{12}_{6}\text{C} \longrightarrow\ ^{246}_{98}\text{Cf} + 4^{1}_{0}\text{n}$$

Eleven isotopes have been reported for this element.

Element 99, einsteinium, has been prepared as follows

$$^{238}_{92}\text{U} + ^{14}_{7}\text{N} \longrightarrow\ ^{248}_{99}\text{Es} + 4^{1}_{0}\text{n}$$

and element 100, fermium, has been prepared by a similar technique using accelerated oxygen ions. These two elements have also been obtained by irradiation of ^{239}Pu in an intense neutron flux when repeated neutron capture and electron emission occurs. Eleven isotopes of the former and eight of the latter have been made.

The preparation of element 101, mendelevium, was achieved in 1955 by bombarding small amounts of ^{253}Es with 41 MeV alpha particles; this resulted in the formation of ^{256}Md by an (α, n) reaction, the product decaying by spontaneous fission.

Elements 102, 103 and 104 were all first obtained from nuclear reactions with heavy ions.

Nobelium, element 102, was made by the bombardment of a curium isotope with carbon ions. This is an α-emitter and has a very short half-life (3 s). It has been identified by the detection of its daughter isotope, $^{250}_{100}$Fm $(t_{\frac{1}{2}} = 1800\text{ s})$

$$^{246}_{96}\text{Cm} + ^{12}_{6}\text{C} \longrightarrow\ ^{254}102 + 4^{1}_{0}\text{n}$$

Another isotope of 102 has been synthesised by the bombardment of a plutonium isotope with accelerated oxygen ions

$$^{241}_{94}\text{Pu} + ^{16}_{8}\text{O} \longrightarrow\ ^{253}102 + 4^{1}_{0}\text{n}$$

* The process of bombardment with heavy ions is useful in that it allows an element of much higher atomic number to be synthesised by a single step; otherwise a process such as repeated neutron capture would have to be used.

This isotope is also short-lived ($t_{\frac{1}{2}} = 10\text{–}20$ s)

Element 103, for which the proposed name is lawrencium, has been made by the bombardment of californium isotopes with accelerated boron ions. This reaction produces $^{257}103$ ($t_{\frac{1}{2}} = 8$ s).

Element 104, kurchatovium, results from the bombardment of plutonium with neon ions

$$^{242}_{94}\text{Pu} + ^{22}_{10}\text{Ne} \longrightarrow {}^{260}104 + 4\,^{1}_{0}\text{n}$$

This reaction raises the atomic number by ten units in a single step. The isotope of 104 has a half-life of about 0·3 s and undergoes spontaneous fission into isotopes of ytterbium and selenium.

We can see that the isotopes of 102, 103 and 104 are very unstable and short-lived and it is unlikely that isotopes of these elements which are significantly more stable will be prepared.

It has been predicted theoretically that an island of nuclear stability should exist around element 114 and heavy ion accelerators are being developed in the U.S., Russia and in Europe to explore this region of the Periodic Table. Meanwhile, the search for very heavy elements in nature continues and reports of their discovery are made from time to time. For example, a group of minerals has recently been discovered in Russia which shows an abnormally high content of actinium and this is believed to be a decay product of element 108. It may be that some traces of this element will be identified in the future.

SUGGESTED REFERENCES FOR FURTHER READING

ASTON, F. W. *Mass Spectra and Isotopes*, 2nd edn, Arnold, London, 1941.
GLASSTONE, S. *Sourcebook on Atomic Energy*, 2nd edn, Van Nostrand, Princeton, 1958.
JENKINS, E. N. *An Introduction to Radiochemistry*, Butterworths, London, 1964.
TOLANSKY, S. *Introduction to Atomic Physics*, 4th edn, Longmans Green, London, 1961.

2   Atomic structure—2

QUANTUM THEORY AND ATOMIC SPECTRA

Towards the end of the nineteenth century the introduction of a wave theory of light by Huygens enabled explanations for certain properties of light such as diffraction and interference to be given; these phenomena had been previously inexplicable on the Newtonian corpuscular theory. In addition to this, scientists became aware of certain inadequacies in other branches of classical theory; thus both Wien and Rayleigh were unable to give completely satisfactory interpretations of the phenomenon of 'black-body radiation'. In 1900, however, Planck derived an empirical relationship that completely satisfied the data from experiments on 'black-body radiation' and by introducing a concept of 'quantisation of energy' he was able to justify the relationship theoretically. It had been shown that at a specific temperature the spectrum of radiation from a 'black body' was unique in its characteristics and that the energy varied throughout the spectrum and possessed a maximum value at one particular wavelength for one temperature of emission. Planck proposed that the 'black body' was composed of many vibrating bodies or oscillators that were emitting or absorbing energy in packets or 'quanta' of energy. Such quanta had energy values dependent on the frequency of radiation v and were emitted as photons of energy E, where

$$E = hv \qquad (1)$$

in which h is Planck's constant, equal to $6 \cdot 6252 \times 10^{-34}$ J s.

An analysis of the radiation emitted by bodies, using a spectrometer, results in the production of a spectrum. Spectra of this type may be obtained by examination of atoms or molecules excited to states of higher energy by the application of heat, by electrical discharge or by bombardment with high speed electrons. A return of the system to its lower energy states results in the emission of energy as radiations of definite wavelengths. With gases these emission spectra take the form of a series of sharply defined lines distributed throughout the spectral region; for instance, the analysis of radiation emitted by hydrogen gas when subjected to an electrical discharge at low pressures results in just such a spectrum (*Figure 2.1*), some of the lines being visible to the naked eye. This spectrum of atomic hydrogen is the simplest one known but at first no satisfactory explanation of the spectral lines was forthcoming. Success at interpreting the lines was achieved by Balmer (1885) who showed that the wavelengths of certain lines in the spectrum were related by the equation

$$\lambda = \left(\frac{n^2}{n^2-4}\right) \times B \qquad (2)$$

where n is any integer from three upwards and B had the value $3\cdot6456 \times 10^{-7}$ m (*Table 2.1*).

Other lines have since been shown to be related by a general formula

$$\lambda = \left(\frac{n_b^2}{n_b^2-n_a^2}\right) \times B \qquad (3)$$

n_a and n_b being integers ($n_b > n_a$) and B takes a variety of values dependent on the series. Different series of lines are recognised in the spectrum depending

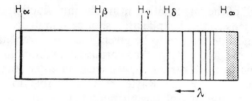

Figure 2.1. Visible spectrum of atomic hydrogen

on the values of n_b and n_a. These are:

			$10^9\ B/m$
Lyman series	$n_a = 1$;	$n_b = 2, 3, 4$, etc.	91·20
Balmer series	$n_a = 2$;	$n_b = 3, 4, 5$, etc.	364·56
Paschen series	$n_a = 3$;	$n_b = 4, 5, 6$, etc.	820·33
Brackett series	$n_a = 4$;	$n_b = 5, 6, 7$, etc.	1458·00
Pfund series	$n_a = 5$;	$n_b = 6, 7, 8$, etc.	2280·00

Expression (2) may be rearranged to give the frequency of radiation emitted. Using the relationship $v = c/\lambda$, where c is the speed of light and v is the frequency of radiation, leads to the expression

$$v = \frac{c(n^2-4)}{3.645\,6 \times 10^{-7}\,n^2}$$

$$= Rc\left\{\frac{1}{2^2}-\frac{1}{n^2}\right\} \qquad (4)$$

where R is known as the Rydberg constant.

Attempts to explain the line spectrum of hydrogen by means of the Rutherford concept of the atom proved unsuccessful. This model, it will be recalled, consisted of a central positive nucleus surrounded by electrons sufficient in number to give electroneutrality to the atom; these were assumed to rotate about the nucleus in a manner resembling the motion of planets about the sun, similar forces being envisaged to maintain the electrons in orbit as keep the planets rotating. In the case of the hydrogen atom there is one electron and the picture is a reasonably simple one. When, however, this model was used in an attempt to interpret the atomic spectrum certain flaws became apparent; thus the uniform motion of the electron about the nucleus, with its acceleration, arising from centrifugal forces, towards the centre of the orbit, would according to the theory of Maxwell require radiation to be emitted continuously. As we have already seen, the spectrum is discontinuous.

Bohr theory of atomic spectra

In 1913, Bohr proposed his theory for the explanation of the spectrum of atomic hydrogen using as his foundation the quantum theory of Planck. In this theory two assumptions were made:
(a) that the electrons in any atom could exist in a number of orbits and rotate about the nucleus in these without emitting radiation. The

Table 2.1. COMPARISON OF WAVELENGTHS OF THE FIRST FOUR LINES OF THE BALMER SERIES (I) OBSERVED AND (II) CALCULATED FROM THE BALMER FORMULA

Line	n	$\lambda_{obs.}/$nm	$\lambda_{calc.}/$nm
H_α	3	656·21	656·208
H_β	4	486·07	486·08
H_γ	5	434·01	434·00
H_δ	6	410·12	410·13

energy is quantised by placing the electrons in orbits of constant energy or 'stationary states'. Furthermore, Bohr extended Planck's theory by postulating that the electrons had angular momenta equal in value to integral multiples of $h/2\pi$.
(b) that radiation was emitted only when an electron jumped from one stationary state to another of lower energy. In making such a transition, a quantum of energy equal to $h\nu$ would be emitted where

$$h\nu = E_{n_b} - E_{n_a} \tag{5}$$

and $E_{n_b} > E_{n_a}$, these being the energies of the two quantised energy states.

Let us now consider the application of these two basic assumptions to the case in hand.

The angular momentum of an electron, mass m_e, travelling in a circular orbit of radius r about the nucleus, with speed u, is $m_e ur$. Utilising the first of the above assumptions,

$$m_e ur = nh/2\pi \tag{6}$$

where n is any integer.

The total energy (E) of the electron is the sum of its kinetic and potential energies. The former is equal to $\frac{1}{2}m_e u^2$ and the latter at a distance r from the nucleus is $-e^2/4\pi\varepsilon_0 r$, where e is the electronic charge and ε_0 is the permittivity of a vacuum. Hence

$$E = \frac{1}{2}m_e u^2 - e^2/4\pi\varepsilon_0 r \tag{7}$$

In order for the electron to remain in its orbit, the centrifugal force $(m_e u^2/r)$ must be equal to the force of electrostatic attraction between the nucleus and the electron $(e^2/4\pi\varepsilon_0 r^2)$. Hence

$$\frac{m_e u^2}{r} = \frac{e^2}{4\pi\varepsilon_0 r^2}$$

or
$$\tfrac{1}{2}m_e u^2 = \frac{e^2}{8\pi\varepsilon_0 r} \tag{8}$$

Substituting in equation (7) leads to

$$E = -\tfrac{1}{2}m_e u^2 \tag{9}$$

but from equation (6) it can be seen that

$$\tfrac{1}{2}m_e u = \frac{nh}{4\pi r}$$

and substituting this in equation (8)

$$u = \frac{e^2}{2\varepsilon_0 nh}$$

Hence
$$E = \frac{-\tfrac{1}{2}m_e e^4}{4\varepsilon_0^2 n^2 h^2}$$

$$= -\frac{m_e e^4}{8\varepsilon_0^2 n^2 h^2} \tag{10}$$

Alternatively it may be said that the energy required to remove the electron from its orbit to infinity is (10) but with a positive sign; the larger this value, the more stable the system, i.e. the most stable system is found for $n = 1$. Also the radii of various orbits are given by

$$r = \frac{\varepsilon_0^2 n^2 h^2}{\pi m_e e^2} \tag{11}$$

i.e.
$$r \propto n^2$$

For $n = 1$ and 2, the values of the radii are respectively 53 pm (frequently written as a_0) and 212 pm (by substituting the values of m_e, e and h in equation (11)).

Application of the second assumption allows us to calculate the energy of the emitted radiation. If the electron is excited to a state of principal quantum number (n) equal to n_2 and returns to one of lower energy where $n = n_1$, then a quantum of energy is emitted equal to the difference in energy between the two states, thus

$$h\nu = E_{n_2} - E_{n_1}$$

or
$$\nu = \frac{m_e e^4}{8\varepsilon_0^2 h^2}\left\{\frac{1}{n_1^2} - \frac{1}{n_2^2}\right\} \tag{12}$$

Equation (12) will be seen to be identical with the rearranged form of Balmer equation (4) where $R = m_e e^4/8\varepsilon_0^2 h^3 c$

Substituting $\bar{v} = v/c$ in (12) gives the energy of emitted radiation in wave-numbers or reciprocal metre (m^{-1})

$$\bar{v} = \frac{m_e e^4}{8\varepsilon_0^2 h^3 c} \left\{ \frac{1}{n_1^2} - \frac{1}{n_2^2} \right\}$$ (13)

The energy values of various transitions are thus dependent upon the values of n_1 and n_2; these are shown in the energy level diagram (*Figure 2.2*) which also shows the origin of the various spectral lines. Agreement between experiment and theory proves to be close; thus the experimental value of the Rydberg constant is $1.096\ 775\ 8 \times 10^7$ m^{-1} compared with the calculated value using equation (13) of $1.097\ 373\ 1 \times 10^7$ m^{-1}. Though this is a small dis-crepancy, it is possible to account for it. The approach used above has been one based on a fixed nucleus, whereas in fact rotation about a common

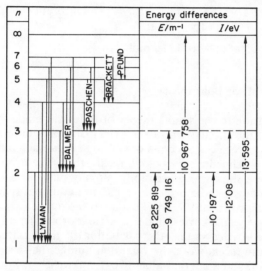

Figure 2.2. Energy level diagram for the hydrogen atom showing transitions responsible for the various spectral series

centre of gravity should have been considered. This leads to a readjustment of equation (13) which becomes

$$\bar{v} = \frac{m e^4}{8\varepsilon_0^2 h^3 c} \left\{ \frac{1}{n_1^2} - \frac{1}{n_2^2} \right\}$$ (14)

where the electron mass has been replaced by a reduced mass m given by

$$m = \frac{m_e \times m_p}{m_e + m_p}$$

m_p being the mass of the hydrogen nucleus. The values of the Rydberg constant are thus dependent upon the nuclear mass, and the value of $R = 1.097\ 373\ 1 \times 10^7$ m^{-1} is therefore that of a nucleus of infinite mass and is written as R_∞. This variation of R with mass of the nucleus has important consequences. Thus the spectral investigation of residues from the evaporation of liquid hydrogen showed a faint spectral line close to the H_α line. This line may be accounted for by the presence of an isotope of hydrogen of mass number two which has a Rydberg constant of $1\cdot097\,074\,2 \times 10^7$ m^{-1}. This would give a line approximately 0·18 nm distant from the H_α line. This isotope effect is quite common in spectral observations.

The energy for any one level of the hydrogen atom may be written as

$$E_n = \frac{-R_Hhc}{n^2} \tag{15}$$

and thus the difference between the *ground state* and the first *excited state* is $\frac{3}{4}R_Hhc = \frac{3}{4}E_{n_1}$; this energy difference is referred to as the first *excitation potential* of the hydrogen atom. The ground state has an energy E_{n_1} which may be expressed in different units

e.g. $E_{n_1} = -2\cdot179 \times 10^{-18}$ J $= -1\cdot096\,775\,8 \times 10^7$ m^{-1}
$= -13\cdot595$ eV (per atom)

(which is equivalent to $-1\,312$ kJ mol^{-1}).

Modifications of the Bohr theory

The first extension to the simple theory of Bohr was made by Sommerfield who proposed the subdivision of the principal quantum levels into sets of sub-levels and introduced a second or subsidiary quantum number (k) which together with n defined a series of elliptical orbits. The orbits of different k but same n were shown to possess only slightly different energies; electron transitions involving these orbits gave rise to additional spectral lines. The ratio of n/k defined the ratio of the major to the minor axis of the elliptical orbit; thus $n = k$ would correspond to a circular orbit. The value of $k = 0$ was excluded since it necessitated the oscillation of the electron through the nucleus. In more recent work this quantum number has been replaced by l, where $l = (k-1)$ and takes the values of 0, 1, 2, 3 to $(n-1)$. Not all transitions between energy states are possible; they are governed by a selection rule that states 'such transitions, either in emission or absorption spectra, are governed by the principle that n may change arbitrarily but l may change only by ± 1'.

Further spectral lines are produced when the atoms emitting the radiation are placed in a strong magnetic field (Zeeman effect) and in order to explain this splitting of energy levels a third quantum number was introduced which defined the possible orientations that the plane containing the electron orbit could assume with respect to the applied magnetic field. This quantum number can take the values $-l \rightarrow 0 \rightarrow +l$ and is designated m, the magnetic quantum number.

Finally a fourth quantum number, the spin quantum number s, was introduced to account for certain double lines in the spectra of alkali metals (Uhlenbeck and Goudsmidt). The electron is regarded as spinning about some axis through its centre thereby possessing a magnetic moment. An electron in a particular energy level can, in fact, have two slightly differing energy values depending on the alternative orientations of this moment, that is, on the two possible directions of spin. This spin can be described in terms of s which has values $\pm\frac{1}{2}$.

The Bohr–Sommerfield theory proved to be but temporarily successful in its interpretation of atomic spectra for, although it could be satisfactorily applied to hydrogen-like species or one-electron systems, inaccurate results were obtained when systems containing more than one electron were considered. It became obvious that if further advances were to be made an entirely new approach to the problem would be necessary.

WAVE MECHANICS

The new method originated in a hypothesis by de Broglie (1924) who suggested that an electron in motion should have wave properties associated with it that could be described by the equation

$$\lambda = h/m_e u \tag{16}$$

where λ is the wavelength of the associated wave property and u is the speed of the electron. The wave nature of the electron was shown by Davisson and Germer (1927) who succeeded in diffracting a beam of electrons by means of a crystal lattice in a manner similar to the diffraction of x-rays. The first application of this hypothesis to the problem of atomic structure was made by Schrödinger (1925) and later by Dirac and by Heisenberg. Their work laid the foundations of the methods of wave mechanics which is a more sophisticated approach than that of Bohr, though it may be shown quite readily that the Bohr theory is a special solution of this more general treatment. The basic assumption made is that the electron orbits can only exist where the waves associated with them reinforce one another, that is to say, a standing wave is set up. In an orbit of radius r therefore

$$n\lambda = 2\pi r \tag{17}$$

Substituting from equation (16) gives

$$m_e u r = nh/2\pi$$

in other words quantisation is a natural outcome of the wave theory.

The method of wave mechanics involves a reconciliation of the wave and particle properties of the electron and describes the system in terms of ψ, a function of the co-ordinates of the electron and time. Such a function is a solution of the wave equation that describes the amplitude of the electron as a wave. An integration of the square of this wave function over definite limits in space allows an estimate to be made of the probability of finding the electron within these limits. The idealised system of orbits envisaged by Bohr is now replaced by this probability concept. The dual nature of the electron means that any measurement of its position immediately places some uncertainty upon its momentum and vice versa. This point is embodied in the *Heisenberg Uncertainty Principle* which states that 'it is impossible to determine simultaneously with any degree of precision both the momentum and the position of an electron'. Heisenberg showed that if Δx were the uncertainty in determining the position and Δp that in determining the momentum, then

$$\Delta x \cdot \Delta p \geqslant h/2\pi$$

Instead of speaking of an electron moving in an orbit of fixed radius the wave-mechanical approach uses the term orbital to describe a certain volume within which the electron has a probability of being found.

The equation showing the variation of the amplitude, ψ, with space and time was first given by Schrödinger as

$$\nabla^2\psi + \frac{8\pi^2 m}{h^2}(E - V)\psi = 0 \qquad (18)$$

where $\nabla^2\psi$ is the sum $\partial^2\psi/\partial x^2 + \partial^2\psi/\partial y^2 + \partial^2\psi/\partial z^2$, E and V are the total and potential energies respectively of the electron.

Solutions to equation (18) are numerous, but for a standing wave only certain solutions are permissible, being rigorously governed by a set of

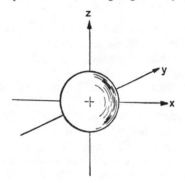

Figure 2.3. Boundary surface for 1s *electron*

boundary conditions that require solutions to be finite, continuous and single-valued. The equation may be solved by transforming to a system of spherical polar co-ordinates in place of the cartesian co-ordinate system of x, y and z. The wave equation then becomes a function of three variables, r (radial) and θ and ϕ (angular); this conversion allows the equation to be broken down into three simpler equations that depend on r, θ and ϕ separately. Solutions to these equations may only be obtained by the introduction of certain constants, the values of which are determined by the boundary conditions. Although the mathematics is rather complex, it can be shown that these constants are integral and may be identified with the three previously defined quantum numbers n, l and m.

For the hydrogen atom in the ground state the solution of the Schrödinger equation gives a wave function that is dependent upon r only; it is in fact spherically symmetrical and the values of n, l and m are 1, 0 and 0 respectively. Various methods of representation of the nature of this wave function have been used, perhaps the simplest is that of drawing (*Figure 2.3*) a boundary surface or contour of constant ψ value such that the electron has a certain chance, say 95 per cent, of being found within this surface. This boundary surface represents the 1s atomic orbital. This type of representation proves sufficiently adequate for many purposes and will be used frequently in subsequent chapters.

Another useful approach is to plot the radial density or $4\pi r^2\psi^2$. dr (the probability of the electron being found in a volume between r and $r + dr$ from the nucleus) against r. This is given in *Figure 2.4* for the ground state of the hydrogen atom; it is interesting to note that the radial density reaches a

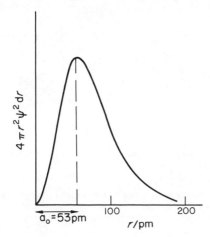

Figure 2.4. Radial distribution function of the 1s electron of hydrogen

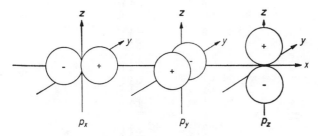

Figure 2.5. Boundary surfaces for p electrons

maximum at a distance a_0 from the nucleus which is equal to the value of the radius of the first Bohr orbit.

Other solutions of the Schrödinger equation for the hydrogen atom may be obtained that represent excited states of the atom. For instance, a second solution occurs for $n = 2$, $l = 0$ and $m = 0$; this again is dependent upon r only and represents a spherically symmetrical orbital in which the average electron density is further away from the nucleus than in the 1s, the maximum radial density occurring at $r = 216$ pm. This is the 2s orbital of the hydrogen atom. For $n = 2$ there are also solutions where $l = 1$ and $m = \pm 1$ and 0; these depend upon the radial functions in addition to r and are unsymmetrical. Boundary surfaces are drawn for these in *Figure 2.5*, one lobe has a positive ψ while the other has a negative ψ; these three orbitals, referred to as the 2p,

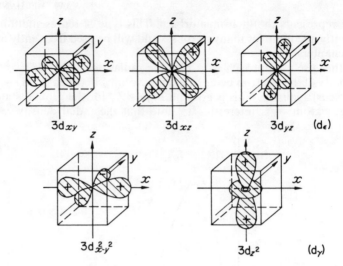

Figure 2.6. Boundary surfaces for d *electrons*

Table 2.2. QUANTUM NUMBERS FOR THE VARIOUS ELECTRONS IN THE FIRST THREE SHELLS

n	l	m	s	Nomenclature	
				Electron	Shell
1	0	0	$+\frac{1}{2}$ $\Big\}$	1s	K
1	0	0	$-\frac{1}{2}$		$\sum e = 2$
2	0	0	$+\frac{1}{2}$ $\Big\}$	2s	
2	0	0	$-\frac{1}{2}$		
2	1	0	$+\frac{1}{2}$		L
2	1	0	$-\frac{1}{2}$		$\sum e = 8$
2	1	$+1$	$+\frac{1}{2}$	2p	
2	1	$+1$	$-\frac{1}{2}$		
2	1	-1	$+\frac{1}{2}$		
2	1	-1	$-\frac{1}{2}$		
3	0	0	$+\frac{1}{2}$ $\Big\}$	3s	
3	0	0	$-\frac{1}{2}$		
3	1	0	$+\frac{1}{2}$		
3	1	0	$-\frac{1}{2}$		
3	1	$+1$	$+\frac{1}{2}$	3p	
3	1	$+1$	$-\frac{1}{2}$		
3	1	-1	$+\frac{1}{2}$		M
3	1	-1	$-\frac{1}{2}$		$\sum e = 18$
3	2	0	$+\frac{1}{2}$		
3	2	0	$-\frac{1}{2}$		
3	2	$+1$	$+\frac{1}{2}$		
3	2	$+1$	$-\frac{1}{2}$		
3	2	-1	$+\frac{1}{2}$	3d	
3	2	-1	$-\frac{1}{2}$		
3	2	$+2$	$+\frac{1}{2}$		
3	2	$+2$	$-\frac{1}{2}$		
3	2	-2	$+\frac{1}{2}$		
3	2	-2	$-\frac{1}{2}$		

are mutually perpendicular and are directed along the x, y and z axes respectively. Solutions for $n = 3$ besides giving the 3s and 3p orbitals also give five d orbitals characterised by $l = 2$, as shown in *Figure 2.6*. With $n = 4$, in addition to s, p and d orbitals, seven f orbitals are formed, characterised by $l = 3$.

Each orbital may accommodate a maximum of two electrons having spin quantum numbers of $+\frac{1}{2}$ and $-\frac{1}{2}$; such electrons are termed paired, having antiparallel spins. Electrons having parallel spins are termed unpaired and cannot be found in the same orbital. These concepts are of fundamental importance in the interpretation of the properties of chemical compounds. In the hydrogen atom itself there is only one electron in the 1s orbital. In elements of higher atomic number electrons are present in the other types of

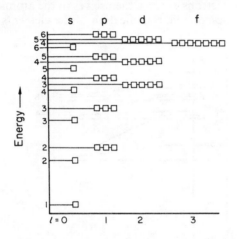

Figure 2.7. Composite energy level diagram of orbitals in atoms. Each square represents one orbital. Levels below 6s are in the order assumed by elements of low atomic number. Levels above 6s are for elements with Z ~ 50

orbital. Energy levels where there are several equivalent orbitals are spoken of as degenerate orbital levels. The s levels are non-degenerate, p levels are threefold degenerate, the d levels are fivefold degenerate and the f levels are sevenfold degenerate.

Whereas, for the hydrogen atom, atomic orbitals with the same principal quantum number are of the same energy, shielding effects of the inner electrons with the heavier atoms cause distinct separations of the levels and the individual electron energies are affected by interelectronic repulsion. Exact solutions of the Schrödinger equation cannot be obtained and various approximations have to be introduced to enable the equation to be solved. This still leads to the electrons being defined by a set of quantum numbers and the shapes of the orbitals resemble those of the hydrogen atom but are of different size.

The number of electrons that may be accommodated in the various principal quantum shells and their notation for $n = 1, 2, 3$ are given in *Table 2.2*.

Relative displacements of the energy levels have been determined from spectroscopic observations and for the elements of low atomic number ($Z < 20$) the ascending order of energy is

$$1s < 2s < 2p < 3s < 3p < 4s < 3d < 4p < 5s < 4d < 5p < 6s < 5d < 4f$$

as shown in *Figure 2.8*. With increasing atomic number the relative positions of the levels change until for the heaviest elements the order is as given in *Figure 2.8*.

PERIODIC CLASSIFICATION AND ELECTRONIC CONFIGURATION

The way in which electrons arrange themselves in the atomic orbitals available is the key to the periodic classification of the elements. There are two

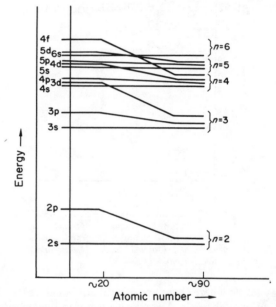

Figure 2.8. Variation of the energy of atomic orbitals with atomic number (adapted from Herzberg, G. *Atomic Spectra and Atomic Structure*, Dover Publications, New York, 1944)

fundamental rules relating to electronic configurations. The first is that electrons are accommodated in the orbitals of lowest energy first; combined with this rule is Hund's rule of *Maximum Multiplicity* which states that when electrons are present in a number of degenerate orbitals they occupy all the orbitals singly, with parallel spins, before pairing in any one orbital occurs. Thus if there are three electrons to be accommodated in the 2p level, the arrangement will be

| 1 | 1 | 1 |

and not

1↓	1	

The second principle is one of an empirical nature, deduced from a careful study of atomic spectra and correlation of the data with quantum theory. It was enunciated by Pauli (1925) — the *Pauli Exclusion Principle* — and states that no two electrons in any one atom may be described by the same set of four quantum numbers. It is obvious from this principle that electrons with parallel spins cannot be accommodated in the same orbital and therefore those in the same orbital must have antiparallel spins.

Applying these principles to the electronic configuration of the elements, the simplest case is that of the single electron in the hydrogen atom, i.e. $Z = 1$, where the electron enters the 1s orbital and the electronic configuration is written as $1s^1$.

For helium, where $Z = 2$, the electrons are paired in the orbital of lowest energy with $s = +\frac{1}{2}$ and $-\frac{1}{2}$, $n = 1$, $l = 0$ and $m = 0$. The electronic configuration is $1s^2$.

Table 2.3. ELECTRONIC CONFIGURATIONS OF THE ELEMENTS H TO Ne

Element	Electronic configuration	
H	$1s^1$	
He	$1s^2$	
Li	$1s^2 2s^1$	
Be	$1s^2 2s^2$	
B	$1s^2 2s^2 2p^1$	
C	$1s^2 2s^2 2p^1 2p^1$	} Parallel spins
N	$1s^2 2s^2 2p^1 2p^1 2p^1$	
O	$1s^2 2s^2 2p^2 2p^1 2p^1$	
F	$1s^2 2s^2 2p^2 2p^2 2p^1$	} Pairing of spins
Ne	$1s^2 2s^2 2p^2 2p^2 2p^2$	

This completes the K shell and for elements of higher atomic number the outer orbital levels start to fill up. For lithium the electronic configuration is written $1s^2 2s^1$. To obtain the electronic configuration of any element all that is required is to take that of the preceding element and to place the additional electron into the appropriate orbital with due regard for the two fundamental rules stated above. This process is termed the *Aufbau* or building-up process. Thus we may write down the electronic configurations of the first ten elements in the Periodic Table (*Table 2.3*). The electronic configurations from Na to Ca follow a similar pattern with the filling up of the 3s, 3p and 4s orbitals. After Ca, $1s^2 2s^2 2p^6 3s^2 3p^6 4s^2$, the 3d orbitals are the nearest vacant ones and the 3d level starts to fill up. The series of ten elements ending at zinc, which are thereby obtained, constitute the *First Transition Series*. Within this series there is some irregularity in filling up; chromium and copper have the respective configurations $3d^5 4s^1$ and $3d^{10} 4s^1$ and not the expected $3d^4 4s^2$ and $3d^9 4s^2$. The half-filled or completely filled inner quantum level appears to have a certain added stability; this also shows itself in the stability of certain ions of the transition metals (p. 326). After zinc the 4p level starts to fill up and is complete at Kr. The *Second Transition Series* is found after Sr and the 5p level is complete at Xe.

In the next series of elements beginning at Cs an *Inner Transition Series* is

Table 2.4. ELECTRON CONFIGURATIONS FOR ATOMS IN THEIR GROUND STATES

Element	z	1s	2s	2p	3s	3p	3d	4s	4p	4d	4f	5s	5p	5d	5f	6s	6p	6d	6f	7s
H	1	1																		
He	2	2																		
Li	3	2	1																	
Be	4	2	2																	
B	5	2	2	1																
C	6	2	2	2																
N	7	2	2	3																
O	8	2	2	4																
F	9	2	2	5																
Ne	10	2	2	6																
Na	11	2	2	6	1															
Mg	12				2															
Al	13		10		2	1														
Si	14		Ne Core		2	2														
P	15				2	3														
S	16				2	4														
Cl	17				2	5														
Ar	18				2	6														
K	19	2	2	6	2	6		1												
Ca	20							2												
Sc	21						1	2												
Ti	22						2	2												
V	23						3	2												
Cr	24						5	1												
Mn	25						5	2												
Fe	26						6	2												
Co	27		18				7	2												
Ni	28		Ar Core				8	2												
Cu	29						10	1												
Zn	30						10	2												
Ga	31						10	2	1											
Ge	32						10	2	2											
As	33						10	2	3											
Se	34						10	2	4											
Br	35						10	2	5											
Kr	36						10	2	6											
Rb	37	2	2	6	2	6	10	2	6			1								
Sr	38											2								
Y	39									1		2								
Zr	40									2		2								
Nb	41									4		1								
Mo	42									5		1								
Tc	43									(5)		(2)*								
Ru	44				36					7		1								
Rh	45				Kr Core					8		1								
Pd	46									10		0								
Ag	47									10		1								
Cd	48									10		2								
In	49									10		2	1							
Sn	50									10		2	2							
Sb	51									10		2	3							
Te	52									10		2	4							
I	53									10		2	5							
Xe	54									10		2	6							

Table 2.4 (contd.)

Element	Z	1s	2s 2p	3s 3p 3d	4s 4p 4d	4f	5s 5p	5d	5f	6s 6p	6d 6f	7s
Cs	55	2	2 6	2 6 10	2 6 10		2 6			1		
Ba	56			46			8			2		
La	57							1		2		
Ce	58					(2)				(2)		
Pr	59					(3)				(2)		
Nd	60					(4)				(2)		
Pm	61					(5)				(2)		
Sm	62					6				2		
Eu	63					7				2		
Gd	64					7		(1)		2		
Tb	65					(8)		(1)		(2)		
Dy	66					(10)				(2)		
Ho	67					(11)				(2)		
Er	68					(12)				(2)		
Tm	69					13				2		
Yb	70			54 Xe Core		14				2		
Lu	71					14		1		2		
Hf	72					14		2		2		
Ta	73					14		3		2		
W	74					14		4		2		
Re	75					14		5		2		
Os	76					14		6		2		
Ir	77					14		7		2		
Pt	78					14		9		1		
Au	79					14		10		1		
Hg	80					14		10		2		
Tl	81					14		10		2 1		
Pb	82					14		10		2 2		
Bi	83					14		10		2 3		
Po	84					14		10		2 4		
At	85					14		10		2 5		
Rn	86					14		10		2 6		
Fr	87	2	2 6	2 6 10	2 6 10	14	2 6	10		2 6		1
Ra	88			78						8		2
Ac	89										(1)	(2)
Th	90										(2)	(2)
Pa	91								(2)		(1)	(2)
U	92								(3)		(1)	(2)
Np	93								(5)			(2)
Pu	94								(6)			(2)
Am	95				86 Rn Core				(7)			(2)
Cm	96								(7)		(1)	(2)
Bk	97								(7)		(2)	(2)
Cf	98								(9)		(1)	(2)
Es	99											
Fm	100											
Mv	101											
No	102											
Lw	103											

*Configurations shown in parentheses are uncertain.

observed. In the elements Cs and Ba the electrons enter the 6s level; La has the configuration $5d^16s^2$ but element 58 (cerium) does not, however, have the configuration $5d^26s^2$ but $4f^25d^06s^2$ and thereafter the 4f level continues to fill up. The name rare earth metals is used for the elements Sc, Y, and La–Lu inclusive. The elements La–Lu are known collectively as the lanthanoids. In such an inner transition series where the outer orbital configuration is unaltered and electrons are entering an inner quantum level, chemical resemblances are particularly noticeable. Here again the filling of the inner quantum level is not entirely regular and the $4f^7$ and $4f^{14}$ arrangements appear to be of added stability. The 6p level is filled up from Tl to Rn and the 7s level fills up as expected. With elements above atomic number 90, there is a limited amount of experimental data and the electronic configurations of some elements are not conclusively settled. Beyond actinium the 5f shell is filling up and we can define a second inner transition series known as the actinoids.

The configurations of the elements are given in *Table 2.4*. From these, four distinct types of element may be recognised; they are:

(a) *The Noble Gases*: In these elements all subsidiary levels are filled and except for helium they are characterised by the configuration ns^2np^6.

(b) *The Representative Elements*: These have incomplete s and p levels with all underlying levels filled and are characterised by the configurations ns^1 to ns^2np^5.

(c) *The Transition Elements*: These are characterised by an incompletely filled d level; they are generally extended to include the elements of subgroups I B and II B.

(d) *The Inner Transition Elements*: These are distinguished from the elements of type (c) by the filling up of the inner f level.

SUGGESTED REFERENCES FOR FURTHER READING

COULSON, C. A. *Valence*, 2nd edn, Clarendon Press, Oxford, 1961.
HERZBERG, G. *Atomic Spectra and Atomic Structure*, Dover Publications, New York, 1944.
RICE, O. K. *Electronic Structure and Chemical Binding*, McGraw-Hill, New York, 1940.

3   Valency

The concept of valency as essentially electrostatic in nature was first proposed in 1812 shortly after the discovery of the phenomenon of electrolysis. In his theory, Berzelius attributed positive or negative polarities to the elements and visualised molecules as built up by partial neutralisation of electrical charges. Later work in the middle of the same century by Frankland, who surveyed numerous compounds of different elements, suggested a combining power for the elements which was looked upon as a number, the valency, such that the atomic weight of the element was the multiple of that number and the equivalent weight.

Following the discovery of the electron and the development of theories of atomic structure, it was suggested that the electrons in shells round the nucleus were involved in the formation of chemical bonds with other atoms. The Bohr theory gave the first indication of the arrangement of the electrons around the nucleus in shells and it was soon realised that for many elements the number of outermost electrons was related directly to the valency*. Thus, the alkali metals with one electron in the outermost shell have a unipositive valency, the alkaline earth metals with two outermost electrons have a dipositive valency, etc. It is therefore customary to speak of the outermost electrons as valency electrons and, as we have seen, the division of elements

---

* Confusion often arises over the term valency when the student changes from elementary to more advanced chemistry. A simple example of this arises in the case of the hexacyanoferrate(III) ion $[Fe(CN)_6]^{3-}$. The iron atom is in the ferric state, that is it has a tripositive valency, but it is bonded to six cyanide ions and so it is six covalent. To avoid the use of the term valency it is common to refer to an atom as being in a particular *oxidation state*. Thus, ferric iron is said to be in the $+3$ oxidation state and cupric copper in the $+2$ oxidation state. These are frequently written as Fe^{III} and Cu^{II} or iron (III) and copper (II).

The oxidation state of an atom is worked out from certain elementary rules.

  (i) Atoms in the elementary state are in the zero oxidation state.

  (ii) Hydrogen, except in the ionic hydrides and the hydrogen molecule, is in the $+1$ oxidation state.

  (iii) Oxygen, except in peroxides, oxygen difluoride and the oxygen molecule, is in the -2 oxidation state.

  (iv) The halogens are in the -1 oxidation state in the halides.

The oxidation state of a particular atom in a compound or ion may be worked out by breaking down the species into its component atoms and equating the algebraic sum of the component oxidation states to the charge on that species.

Thus, in $KMnO_4$ the oxidation state of the manganese atom may be computed as follows:

$$\begin{array}{ccc} K & Mn & O_4 \\ +1 & +? & -8 = 0. \quad \text{Hence} \quad Mn^{7+}. \end{array}$$

into various groups, each of which exhibits a characteristic valency or valencies, arises from their electronic configurations. The discovery of the noble gases and their remarkable inability to enter into chemical combination led to the development of an electronic theory of valency. Three main types of bond were defined: (a) electrovalent, (b) covalent and (c) dative.

The electrovalent bond was first postulated by Kossel (1916), who suggested that the atoms of elements just before or just after the noble gases in the periodic classification could increase the stability of their electron arrangement by the gain or loss of electrons respectively until a noble gas configuration was attained. An element which gains electrons is described as *electronegative* and one which loses electrons as *electropositive*. For example, in the formation of sodium chloride the sodium atom loses an electron to form a positive ion with the configuration of neon; chlorine gains one electron to form a negative ion with the configuration of argon. This may be visualised pictorially as

$$\text{Na (2. 8. 1)} \longrightarrow \text{Na}^+ \text{ (2. 8)} + e^-$$
$$\text{Cl (2. 8. 7)} + e^- \longrightarrow \text{Cl}^- \text{ (2. 8. 8)}$$

The sodium and chloride ions then form an electrovalent bond by the mutual attraction of their electrostatic charges. The electrovalent bond arises whenever atoms of one element gain electrons at the expense of the atoms of another.

In the case of the covalent bond, the essential feature of the proposal of Lewis (1916) is that electrons are shared between the atoms bonded together and that no transfer of electrons takes place. This type of bond is generally formed by the combination of two electronegative atoms with, again, the criterion of the attainment by both atoms of the electronic configuration of a noble gas. Thus the formation of a single covalent bond in the chlorine molecule may be represented as

$$\overset{\times\,\times}{\underset{\times\,\times}{\times}}\!\text{Cl}_\times + \,\cdot\,\overset{\cdot\cdot}{\underset{\cdot\cdot}{\text{Cl}}}\!: \;\rightarrow\; \overset{\times\,\times}{\underset{\times\,\times}{\times}}\!\text{Cl}\!\overset{\cdot}{\underset{\cdot}{\times}}\!\overset{\cdot\cdot}{\underset{\cdot\cdot}{\text{Cl}}}\!:$$

the shared pair of electrons constituting the covalent bond. Double and triple bonds as found for instance in ethylene and acetylene may be regarded as involving the sharing between two atoms of two and three pairs of electrons respectively:

$$\underset{\text{H}}{\overset{\text{H}}{}}\,\overset{\times}{\underset{\times}{\text{C}}}\,\begin{bmatrix}\times\\ \times\end{bmatrix}\,\overset{\cdot}{\underset{\cdot}{\text{C}}}\,\overset{\text{H}}{\underset{\text{H}}{}} \text{ or } H_2C{=}CH_2 ; \quad H_3^\times C\,\begin{bmatrix}\times\\ \times\end{bmatrix}\,C_3^\times H \text{ or } HC{\equiv}CH$$

The dative or co-ordinate bond arises when a molecule or ion donates an electron pair to an atom or ion which requires electrons to complete its shell thereby attaining a noble gas arrangement. As examples here we may quote the cyanide ion and the ammonia molecule whose electron arrangements may be written as

$$\left[\,\overset{\cdot}{\underset{\cdot}{\text{o}}}\text{C} \;\middle|\; \text{N}^\times_\times\,\right]^- \quad \text{and} \quad \overset{\text{H}}{\underset{\text{H}}{\overset{\times\circ}{\underset{\circ\times}{\text{N}}}}}\text{H}$$

In each instance, there is an electron pair or lone pair which can be donated to an acceptor molecule or atom with the formation of the dative bond. Thus hexacyanoferrate (II) may be visualised as an iron (II) ion accepting six lone pairs from cyanide ions to form a complex anion $[Fe(CN)_6]^{4-}$

i.e. $\qquad Fe\ (2.\ 8.\ 8.\ 6.\ 2) \longrightarrow Fe^{2+}\ (2.\ 8.\ 8.\ 6)$

$\qquad\qquad Fe^{2+} + 6\ CN^- \longrightarrow [Fe(CN)_6]^{4-}\ (2.\ 8.\ 8.\ 18)$

or

$$\begin{bmatrix} & \text{CN} & \\ \text{NC} & \downarrow & \text{CN} \\ & \searrow \nearrow & \\ & \text{Fe} & \\ & \nearrow \nwarrow & \\ \text{NC} & \uparrow & \text{CN} \\ & \text{CN} & \end{bmatrix}^{4-}$$

Likewise ammonia will donate its lone pair to $B(CH_3)_3$. The octet of the boron atom is thereby completed:

$$H_3C-\overset{\overset{\displaystyle CH_3}{|}}{\underset{\underset{\displaystyle CH_3}{|}}{B}} \leftarrow {}^{\times}_{\times}NH_3$$

In forming the co-ordinate bond there is effectively a partial transfer of electrons from one atom to the other; hence this bond is also referred to as a semi-polar bond. The complex between ammonia and trimethylboron can also be represented as

$$(CH_3)_3\overset{\ominus}{B}-\overset{\oplus}{N}H_3$$

THE ELECTROVALENT BOND

The discussion of the ionic bond in the résumé above has been restricted to ions possessing the configuration of noble gases. In the case of anions this is the only stable arrangement and the anions which occur are predominantly those with one or two charges. In the case of cations, however, there are configurations other than those of the noble gases that have stability.

For instance, many sub-group B elements that cannot form a noble gas configuration by electron loss can nevertheless form stable ions. Thus zinc forms a stable dipositive ion with the electron arrangement (2. 8. 18). In other cases such as the sub-group I B elements, the eighteen electron arrangement has not such great stability and variable valency is exhibited by the atoms; thus copper has the two ions Cu^+ and Cu^{2+}.

The atoms of certain other sub-group B elements, particularly those of high atomic number, possess two electrons that appear to be characterised by a certain degree of inertness. This so-called *inert-pair* effect shows up in the chemistry of these elements in the formation of ions other than the eighteen electron type. Examples of this are to be found in the chemistry of sub-groups II B to V B of the Periodic Table as indicated in *Figure 3.1*; the inert-pair effect is found with elements below the dotted line. The inert-pair effect is demonstrated by the formation of compounds in which the element shows a valency two units less than the maximum or group valency.

Group	II B	III B	IV B	V B
(Be)	B	C	N	
(Mg)	Al	Si	P	
Zn	Ga	Ge	As	
Cd	In	Sn	Sb	
Hg	Tl	Pb	Bi	

Figure 3.1. Occurrence of the inert-pair effect

With the transition elements the stability of the half-filled shell has already been mentioned; this also shows itself as a stable grouping in ion formation. This is discussed later.

Ionization energy and electron affinity

The ease with which an atom will form an ion depends upon the magnitudes of its *ionisation energy* and its *electron affinity*.

In the Bohr theory of the hydrogen atom it was pointed out that a certain amount of energy was required to remove completely the electron from its orbit to infinity. This is termed the ionisation energy of the hydrogen atom.

The *ionisation energies* of an atom can be determined from its emission spectrum. The ionisation energy is the amount of energy necessary to remove one electron completely from the gaseous atom. Experimentally, it is observed that at a definite wave number $\bar{v}_i$, the line spectrum of an atom is replaced by a continuous region of radiation. This region is the continuum and the lowest wave number for which the continuum is observed gives a direct measurement of the energy required to remove the electron.

The magnitude of the ionisation energy of an atom depends on various factors:

(i)   the distance of the electrons from the nucleus,

(ii) the effective nuclear charge, i.e. the actual nuclear charge less a co - rection for the screening effect of inner shells of electrons, and

(iii) the type of electron that is being removed, i.e. whether it is an s, p, d · or f electron.

In general, the further an electron is from the nucleus the less firmly it is held and hence the lower the value of the ionisation energy. Exact comparisons are not possible here since the atoms of two different elements do not have the same electronic configurations in the ground states. Considerations of the ionisation energies of consecutive elements in a group, e.g. I A (*Table 3.1*) where (ii) is about the same for each element and the single s electron is being removed illustrates the effect of increased atomic radius.

The ease of removal of an electron depends also on the effective nuclear charge, for in general an increase in shielding effect, other factors remaining constant, produces a decrease in ionisation energy and vice versa. This effect may be demonstrated by considering the successive ionisation energies for a particular atom where the electrons being removed are of the same type. Thus the removal of the first electron, although it has no effect on the nuclear charge, causes a decrease in screening effect as far as the next electron is concerned and consequently the second electron is harder to remove than the first; for instance, in the case of carbon the ionisation energies are 1086·2; 2352·2 and 4617·5 kJ mol^{-1} for the first, second and third electrons respectively.

The s electrons approach the nucleus more closely and on average these experience less shielding than do the p, d or f electrons of any one quantum shell. The s electrons are thus the most tightly bound in any particular atom; correspondingly the f electrons are the least strongly held. If the successive ionisation energies of an atom such as oxygen be considered, there is a steady increase in the values for the first four electrons but a considerably larger difference between the ionisation energy of the fourth and fifth electrons, the latter being an s electron (*Table 3.1*).

The lowest ionisation energies are found with the alkali metals and a gradual increase is observed in traversing the Periodic Table to the inert gases. The increase is not continuous, however, for higher values are obtained for beryllium and nitrogen than would be expected by comparison with the other elements in the same period (*Figure 3.2*). These increases are attributed

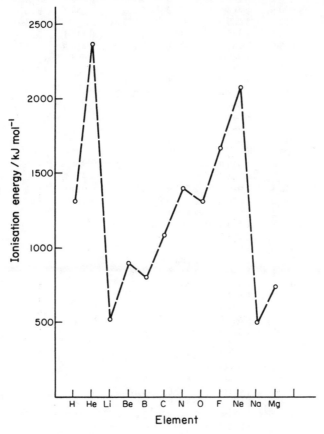

Figure 3.2. First ionisation energies for elements H *to* Mg

Table 3.1. IONISATION ENERGIES OF THE ELEMENTS

Element	Ionisation energy $\Delta U_0/\text{kJ mol}^{-1}$						
	1st	2nd	3rd	4th	5th	6th	7th
H	1 312·1						
He	2 371·5	5 248·8					
Li	520·1	7 296·1	11 809				
Be	899·1	1 756·9	14 849	21 005			
B	800·4	2 426·7	3 658·5	24 258	32 823		
C	1 086·2	2 352·2	4 617·5	6 221·2	37 819	47 254	
N	1 402·9	2 856·8	4 576·0	7 472·6	9 439·1	53 250	64 333
O	1 312·9	3 391·6	5 300·3	7 466·8	10 987	13 326	71 312
F	1 680·7	3 374·8	6 044·6	8 416·1	11 016	15 159	19 364
Ne	2 080·3	3 962·7	6 174·5	9 374·2	12 196	15 234	
Na	495·4	4 562·7	6 912·8	9 540·4	13 372	16 732	20 108
Mg	737·6	1 450·2	7 730·4	10 546	13 623	18 033	21 736
Al	577·4	1 815·9	2 743·9	10 806	14 841	18 372	23 338
Si	786·2	1 576·5	3 228·4	4 354·3	16 083	19 790	19 945
P	1 018·0	1 895·8	2 901·0	4 954·3	6 271·8	21 263	25 405
S	999·6	2 257·6	3 376·9	4 562·7	6 995·6	8 493·5	27 112
Cl	1 255·2	2 296·2	3 849·3	5 161·8	6 539·6	9 330·3	11 029
Ar	1 520·5	2 664·8	3 945·9	5 769·3	7 238·3	8 807·3	11 962
K	418·4	3 069·0	4 437·6	5 875·6		9 619·0	11 385
Ca	589·5	1 145·2	4 940·9	6 464·3	8 142 1		12 351
Sc	633·0	1 235·1	2 387·8	7 130·0	8 874·3	10 719	
Ti	659·0	1 309·2	2 715·0	4 171·9	9 627·4	11 577	13 585
V	650·2	1 413·4	2 865·6	4 631·3	6 292·7	12 435	14 569
Cr	652·7	1 590·8	(2 991·1)	(4 861·8)			
Mn	717·1	1 508·8	(3 087·8)	(5 020·8)			
Fe	762·3	1 561·1					
Co	758·6	1 645·1					
Ni	736·4	1 751·0					
Cu	745·2	1 957·7	2 846·4				
Zn	906·3	1 733·0	3 858·9				
Ga	579·1	1 979·0	2 952·2	6 155·5			
Ge	760·2	1 536·8	3 287·4	4 389·9	8 974·7		
As	946·4	1 948·9	2 701·6	4 814·5	6 029·1		
Se	940·6	2 074·4	3 270·6	4 121·7	7 020·8	7 853·4	
Br	1 142·2	2 084·1	2 479·4	(4 824·1)			
Kr	1 350·6	2 369·8	3 550·5	(6 560·5)			
Rb	402·9	2 653·5	(4 535·4)	(7 719·5)			
Sr	548·8	1 064·4					
Y	615·5	1 179·9	1 986·2				
Zr	659·4	1 246·4	2 315·4	3 261·0			
Nb	664·0	1 339·7	2 335·1				
Mo	687·8	1 473·2					
Tc	697·5	1 434·7					
Ru	710·4	1 600·8					
Rh	720·1	1 535·9					
Pd	803·7	1 873·6					
Ag	730·9	2 072·3	3 463·9				
Cd	867·3	1 630·5	3 666·4				
In	558·1	1 816·7	2 692·0	5 576·9			
Sn	705·8	1 411·7	2 942·6	3 801·6	7 778·1		
Sb	833·5	1 833·0	2 383·2	4 245·1	5 355·5		
Te	869·0	2 074 4	2 942·6	3 637·6	5 790·7	(6 970·5)	
I	1 007·1	1 833·0					
Xe	1 170·3	2 046·4	3 087·4	(4 439·2)	(7 334·6)		

Table 3.1 (contd.)

Element	Ionisation energy $\Delta U_0/\text{kJ mol}^{-1}$						
	1st	2nd	3rd	4th	5th	6th	7th
Cs	373·6	2 421·7	(3 376·5)	(4 920·4)	(5 594·0)		
Ba	502·5	964·8					
La	541·4	1 102·9	1 968·2				
Ce	(666·5)	1 428·0					
Pr	(555·6)						
Nd	607·9						
Pm							
Sm	540·2	1 080·3					
Eu	547·3	1 084·4					
Gd	594·1	1 157·7					
Tb	(650·2)						
Dy	(658·1)						
Ho							
Er							
Tm							
Yb	599·6	1 167·3					
Lu	593·3	1 418·4					
Hf	530·5	1 437·6					
Ta	743·1						
W	769·9						
Re	759·4						
Os	839·3						
Ir	887·8						
Pt	868·2	1 790·8					
Au	889·5	1 977·8	(2 893·2)				
Hg	1 006·3	1 809·2	3 309·5	(6 945·4)			
Tl	589·1	1 970·2	2 865·6	4 872·3			
Pb	715·5	1 447·7	(3 077·8)	4 063·1	6 694·4		
Bi	702·9	1 861·9	2 452·7	4 351·4	5 372·3		
Po	813·3						
At							
Rn	1 037·2						
Fr							
Ra	509·2	978·2					
Ac			2 836·3				
Th							
Pa							
U	385·9						

To obtain I/eV these figures should be divided by 96·5.

to the stability of the filled s and half-filled p levels respectively. In a transition or inner transition series the variation in first ionisation energies is not so great because the addition of an electron to an inner quantum shell gives a high shielding effect which virtually compensates for the increased nuclear charge. Values of ionisation potentials are listed in *Table 3.1*.

The *electron affinity* of an atom is the energy associated with the formation of an anion from the gaseous atom; with the halogens this process is exoergic, that is energy is given out when the anion X^- is formed, but with the elements of Groups V and VI the addition of electrons to form the anions is an endoergic process, that is energy is absorbed. Thus oxygen forms the ion O^{2-};

the formation of O^- is associated with an energy change of approximately $142 \cdot 3$ kJ mol^{-1}, the process being exoergic, but the addition of the second electron is an endoergic process and the total affinity is approximately $640 \cdot 2$ kJ mol^{-1}. The magnitude of the electron affinity is dependent upon various factors. In particular, the atomic size and effective nuclear charge are important; in general, the electron affinity decreases with increasing atomic radius and increases with decreased screening by inner electrons. Also the value will depend to a certain extent upon the type of orbital that the added electron enters; other things remaining constant, the affinity is greatest for an electron entering an s orbital and decreases for p, d and f orbitals. In fact in the common monatomic anions the electrons enter a p orbital (except for H − where the 1s orbital is occupied).

Electron affinities may be derived by indirect means involving thermodynamic data for binary compounds and the Born–Fajans–Haber cycle.

$$
\begin{array}{ccc}
MX(s) & \longleftarrow & M^+(g) + X^-(g) \\
\uparrow & & \uparrow \\
M(s) + \tfrac{1}{2}X_2(g) & \longrightarrow & M(g) + X(g)
\end{array}
$$

Figure 3.3. Born–Fajans–Haber* cycle for MX

This cycle is shown in *Figure 3.3* for the compound MX where M may be any univalent metal such as sodium and X is a halogen.

The cycle relates the standard molar enthalpy of formation ΔH_f°, that is the heat evolved when one mole of the compound MX is formed from the elements in their stable physical states at $298 \cdot 15$ K and at a pressure of $101 \cdot 325$ kPa, with the enthalpy changes of the individual steps of such a reaction, which are as follows:

(a) Atom vaporisation of the metal to give gaseous metal atoms
$$ M(s) \longrightarrow M(g) \quad \Delta H_{at}^\circ \quad (+ve) $$

(b) Dissociation of the X_2 molecule to form isolated X atoms
$$ \tfrac{1}{2}X_2(g) \longrightarrow X(g) \quad \Delta H_d^\circ \quad (+ve) $$
(this is equal to one half the molar bond dissociation energy)

(c) Ionisation of the gaseous metal atoms
$$ M(g) \longrightarrow M^+(g) \quad \Delta H_I^\circ \quad (+ve) $$

(d) Conversion of the gaseous halogen atoms to gaseous halide ions
$$ X(g) \longrightarrow X^-(g) \quad \Delta H_E^\circ \quad (-ve) $$

(e) Formation of one mole of MX as an ionic solid by bringing together, from an infinite distance apart, in the gaseous phase, one mole of M^+ and one mole of X^-
$$ M^+(g) + X^-(g) \longrightarrow MX(s) \quad \Delta H_{LE}^\circ \quad (-ve) $$
(This step at 0 K is defined as the lattice energy U_0 of the substance MX. Since most cycles refer to 298 K an adjustment to the lattice enthalpy at 298 K should be made but approximate calculations may be made by using directly the value of the lattice energy; the difference is a matter of about 5 kJ mol^{-1} in this case.)

The cycle is therefore complete and we may write, applying Hess's Law

$$ \Delta H_f^\circ = \Delta H_{LE}^\circ + \Delta H_I^\circ + \Delta H_E^\circ + \Delta H_{at}^\circ + \Delta H_d^\circ $$

* MORRIS, D. F. C. and SHORT, E. L. *Nature, Lond.*, 224 (1969) 950.

Hence values of electron affinities may be derived from calculated lattice energies and experimentally determined thermochemical quantities.

Table 3.2. ELECTRON AFFINITIES OF VARIOUS ELEMENTS

Element	Electron affinity $\Delta H_E^\circ/kJ\ mol^{-1}$
F	−335·6
Cl	−356·1
Br	−332·6
I	−303·8
O	+639·7
S	+395·4
Se	+422·6

Note 1: Negative values indicate that the process $X \rightarrow X^-$ is exoergic.

Note 2: The process for oxygen, sulphur and selenium is $X \rightarrow X^{2-}$ and is endoergic for the formation of X^{2-} since energy has to be expended in order to overcome the repulsion due to the first electron.

* Data from Cubiciotti, D. *J. chem. Phys.* 31 (1959) 1646.
† Data from Morris, D. F. C. *Proc. roy. Soc.* A-242 (1957) 116.
‡ Data from Morris, D. F. C. *Acta cryst.* 11 (1958) 163.
§ Data from Bevan, S. C., and Morris, D. F. C. *J. chem. Soc.* 106 (1960) 516.

An average value for the electron affinity is calculated from thermochemical data on a number of different compounds containing X^-. Typical values are given in *Table 3.2.*

Thermochemistry

The heat of a reaction, defined as the energy absorbed or evolved as the reaction proceeds, is a quantity of great importance in inorganic chemistry. The standard molar enthalpy of formation ΔH_f° refers to the formation of one mole of a compound from its component elements at 298·15 K and a pressure of 101·325 kPa; this is one kind of enthalpy of reaction that is particularly useful. The enthalpy change is always related to the elements and their compounds in their standard states, that is, the most stable physical state at 298 K. The magnitude of the heat of reaction is affected by the physical form of the participants. For instance, the heat evolved when one mole of gaseous SO_3 is formed from its elements is different from that given out when the same quantity of solid SO_3 is formed by an amount which equals the heat sublimation of SO_3.

The heat of formation is positive for an exothermic compound and negative for an endothermic compound. It must be remembered that ΔH°, the enthalpy change for a reaction, is numerically the same as the heat of reaction but has the opposite sign. Thus ΔH° is positive for heat absorbed (an increase in H°, the enthalpy, means that the enthalpy of the products is greater than that of the reactants) and negative for heat evolved.

Heats of reaction can be measured experimentally by calorimetry, but only rarely can the heats of formation of inorganic compounds be determined directly in this way. More commonly, the heat changes for related chemical reactions are measured and these values used in calculation of the required heats of formation. The validity of the calculation depends on the thermo-chemical law, Hess's Law, which states that the heat liberated or absorbed in a chemical reaction is independent of the number and nature of the steps by which a reaction is brought about.

The enthalpy change, $\Delta H°$, can be determined in several ways. For example, in a reversible reaction it can be calculated from the equilibrium constants, K, determined over a range of temperature. This involves use of the van't Hoff Isochore:

$$\frac{d}{dT}(\ln K) = \frac{\Delta H_f°(T)}{RT^2}$$

where $\Delta H_f°(T)$ refers to the enthalpy of formation measured at the absolute temperature, T K.

The enthalpy change, $\Delta H°$, can be used in conjunction with $\Delta S°$, the entropy change at 298 K, in order to calculate $\Delta G°$, *the standard free energy change in the reaction*. $\Delta G°$ is related to the other two quantities by the equation

$$\Delta G° = \Delta H° - T\Delta S°$$

The importance of the free energy concept is that $\Delta G°$ is always negative for a reaction which is thermodynamically possible and that the magnitude of $\Delta G°$ determines the affinity of the reactants for one another. The rate of reaction, if indeed it proceeds at all at a measurable rate, is, of course, deter-mined by kinetic factors.

Thermochemical data are necessary for the calculation of *experimental lattice energies* of ionic compounds by the Born–Fajans–Haber cycle. Not only is the enthalpy of formation of the compound required but also the enthalpy of atomisation of one or more of the reactants.

From heats of formation data, one can also arrive at values for *bond energies*, which serve to measure the strengths with which atoms are joined together. For diatomic molecules, the heat of formation of the molecule from its atoms is the same as the bond dissociation energy, which is the energy required to break the bond. These two quantities are not usually the same in the case of polyatomic molecules.

Bond dissociation energies can be found by a variety of experimental methods, some of which are:

(i) *Calorimetric method* — In some cases it is possible to measure the exo-thermic recombination of atoms or radicals.

(ii) *Electron-impact method* — Electrons of known energy are allowed to interact with molecules to produce ionisation and dissociation. Under certain conditions, the lowest electron energy required to do this is equal to the sum of the dissociation energy and the ionisation energy of the fragment. This is an experimental method for obtaining dissociation energies in polyatomic molecules.

(iii) *Spectroscopic method* — The ultra-violet spectrum of certain diatomic molecules affords a very accurate method for determining dissociation energies. For example, the ultra-violet spectrum of iodine is composed of bands gradually becoming closer together with increasing energy until the absorption is continuous. The convergence limit of the band spectrum to continuous absorption gives an accurate measure of the energy required to dissociate the excited iodine molecule.

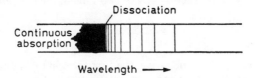

An illustration of the different values obtained for the average bond energy calculated from heat of formation data and the bond dissociation energy, measured, is given by the figures for a polyatomic molecule such as ammonia. Thus the following heats of reaction can be used to calculate the average bond energy for N—H in ammonia:

$$\tfrac{1}{2}N_2(g) + \tfrac{3}{2}H_2(g) = NH_3(g); \quad \Delta H^0 = -46.2 \text{ kJ mol}^{-1}$$
$$2N(g) = N_2(g); \quad \Delta H^0 = -944.8 \text{ kJ mol}^{-1}$$
$$2H(g) = H_2(g); \quad \Delta H^0 = -435.9 \text{ kJ mol}^{-1}$$

Hence it may be shown that

$$N(g) + 3H(g) = NH_3(g); \quad \Delta H^0 = -1172.4 \text{ kJ mol}^{-1}$$

and the average N—H bond energy is therefore 390.8 kJ mol^{-1}. The bond dissociation energy, i.e. that required to effect the reaction $NH_3 = {}^\times NH_2 + {}^\times H$, is 462.8 kJ mol^{-1}. In this text, the values given for the bond energies in polyatomic molecules are always the average values calculated from thermo-chemical data*

Lattice energy

In the Born–Fajans–Haber cycle, the quantity U_0 was introduced as the lattice energy.

The calculation of lattice energies has been made possible from the work of Born and Landé and others who developed a theory of the structure of crystalline solids from a consideration of the interaction of pairs of ions both in the gas and the solid phase. Between any two oppositely charged ions i, j in the gas phase there exists an attractive and repulsive potential. The attractive potential (U_{att}) is given by

$$U_{att} = \frac{z_i z_j e^2}{4\pi\varepsilon_0 r} \tag{1}$$

where z_i and z_j are the charge numbers on the two ions and r their distance apart. The repulsive potential (U_{rep}) may be represented as

$$U_{rep} = \frac{b_{ij} e^2}{4\pi\varepsilon_0 r^n} \tag{2}$$

* The values are taken from COTTRELL, T. L. *The Strength of Chemical Bonds*, 2nd edn. Butterworths, London, 1958.

where b_{ij} is a constant characteristic for the compound under consideration and n has the values between 5 and 14 depending upon the electron configurations of the ions (see *Table 3.3*).

Hence the energy of interaction (U) of two ions in the gas phase is given by

$$U = \frac{z_i z_j e^2}{4\pi\varepsilon_0 r} + \frac{b_{ij}e^2}{4\pi\varepsilon_0 r^n} \tag{3}$$

(*Note*: since either z_i or z_j is negative, the first term in (3) is always negative.)

The energy of an ion pair in the crystal is obtained by summation of the energies of interaction between all the isolated ion pairs. For any single ion pair in the crystalline state of a binary compound, this gives

$$U = -\frac{A_a z^2 e^2}{4\pi\varepsilon_0 r} + \frac{B e^2}{4\pi\varepsilon_0 r^n} \tag{4}$$

where A_a is the Madelung constant, which has values dependent upon the type of crystal lattice and B is the repulsion coefficient and z is the highest common

Table 3.3. VALUES OF n FOR DIFFERENT ELECTRONIC CONFIGURATIONS

Electronic configuration	n value
He	5
Ne	7
Cu^+ Ar	9
Ag^+ Kr	10
Au^+ Xe	12
Rn	14

Note: Where ions are not isoelectronic the average of the two values is used

factor of the charges on the two ions. For example, $z = 1$ for NaCl, CsCl, CaF_2 and $CdCl_2$, $z = 2$ for ZnS and TiO_2, and $z = 3$ for AlN. In these examples, z is the same as the number of charges on the negative ions of the lattice. This is not always so; for example, in the case of Cu_2O and Al_2O_3, $z = 1$. The significance of the Madelung constant may be understood by deriving the constant for a hard-sphere electrostatic model of a lattice of the sodium chloride type. Consider the composite ions M^+ and X^-, which are taken in the ideal case to be of definite incompressible size, being brought together in the lattice. The geometry of the structure assumed is that occupying the minimum volume and of least electrostatic energy. The regular geometrical array of the ions is dependent upon the relative radii of cation and anion. In the case of sodium chloride the structure is as shown in *Figure 4.24* wherein any positive ion (z_i) is surrounded by six nearest neighbour ions (z_j) of opposite charge each at a distance r_c from z_i. Second nearest neighbours — twelve positive ions — occur at distances of $\sqrt{2}r_c$ away; eight negative ions occur at a further distance of $\sqrt{3}r_c$, six further positive ions at $2r_c$ and so on.

The energy of interaction between ions is equal to the product of their respective charges divided by their distance apart. Thus, for the positive ion z_i, this electrostatic energy is given by

$$E = \frac{6e^2}{4\pi\varepsilon_0 r_c} z_i z_j + \frac{12e^2}{4\pi\varepsilon_0 \sqrt{2}r_c} z_i^2 + \frac{8e^2}{4\pi\varepsilon_0 \sqrt{3}r_c} z_i z_j + \frac{6e^2}{4\pi\varepsilon_0 2r_c} z_i^2 + \ldots$$

$$= -\frac{e^2}{4\pi\varepsilon_0 r_c}|z|^2 \left\{ 6 - \frac{12}{\sqrt{2}} + \frac{8}{\sqrt{3}} - \frac{6}{2} + \ldots \right\}$$

The term in parentheses converges to a limiting value of 1·7476 and, being purely a numerical value, is independent of both the nature and the charge of the composite ions. This limiting value of 1·7476 is the Madelung constant for compounds having the sodium chloride lattice. Values of the constant for other crystal lattices are derived by similar methods. Values of A are given in *Table 3.4*.

Table 3.4. VALUES OF THE MADELUNG CONSTANT FOR DIFFERENT CRYSTAL STRUCTURES

Madelung constant	Crystal structure
1·7476	Sodium chloride
1·7627	Caesium chloride
1·6381	Zinc blende
1·6413	Wurtzite
5·0388	Fluorite
4·816	Rutile

B may be determined from the observed equilibrium internuclear distance r_c in the crystal and equating dU/dr to zero. This gives

$$B = \frac{r_c^{n-1} A_a z^2}{n} \tag{5}$$

Hence

$$U = -\frac{A_a z^2 e^2}{4\pi\varepsilon_0 r_c} \left\{ 1 - \frac{1}{n} \right\} \tag{6}$$

and the lattice energy, which refers to 1 mole, is obtained by multiplying (6) by the Avogadro constant N_A

Hence

$$U_0 = -\frac{N_A A_a z^2 e^2}{4\pi\varepsilon_0 r_c} \left\{ \frac{n-1}{n} \right\} \tag{7}$$

Small corrections to this relationship may be made by including van der Waals forces between the ions and the zero point vibrational energy of the ions.

Theoretical values of lattice energies may thus be derived from the Born–Landé expression (equation 7) using this hard-sphere model. For sodium chloride the relevant values of the constants in the equation are

$$N_A = 6\cdot0222 \times 10^{23} \text{ mol}^{-1}$$
$$A_a = 1\cdot7476$$
$$z = 1$$
$$e = 1\cdot60210 \times 10^{-19} \text{ C}$$
$$n = \tfrac{1}{2}(7+9)$$
$$r_c = 281 \text{ pm}$$

Whence $$U_0 = 767 \text{ kJ mol}^{-1}$$

This treatment thus gives close agreement with the experimentally determined value of 769 kJ mol^{-1}.

The lattice energies, calculated for different compounds assuming complete ionic character, show in many instances distinct deviations from the values calculated from thermodynamic data.

These deviations may be seen by comparing the theoretically derived values of lattice for potassium chloride and silver chloride with those determined for the same two compounds using the Born–Fajans–Haber cycle.

Thermochemical quantities for potassium chloride and silver chloride are shown below:

KCl		AgCl	
−437·2		−127·2	$\Delta H_f^\circ/\text{kJ mol}^{-1}$
84·5		284·5	$\Delta H_{at}^\circ/\text{kJ mol}^{-1}$
418·4		730·9	$\Delta H_I^\circ/\text{kJ mol}^{-1}$
	−356·1		$\Delta H_E^\circ/\text{kJ mol}^{-1}$
	121·4		$\Delta H_d^\circ/\text{kJ mol}^{-1}$

Applying Hess's Law and moving round the cycle in a clockwise direction

$$\Delta H_f^\circ - \Delta H_{LE}^\circ - \Delta H_I^\circ - \Delta H_E^\circ - \Delta H_{at}^\circ - \Delta H_d^\circ = 0$$

or

$$\Delta H_{LE}^\circ = -\Delta H_I^\circ - \Delta H_E^\circ - \Delta H_{at}^\circ - \Delta H_d^\circ + \Delta H_f^\circ$$

Hence:

$$\text{KCl } \Delta H_{LE}^\circ/\text{kJ mol}^{-1} = -705\cdot4 \text{ (cf } -694\cdot0 \text{ calculated)}$$
$$\text{AgCl } \Delta H_{LE}^\circ/\text{kJ mol}^{-1} = -907\cdot9 \text{ (cf } -849\cdot4 \text{ calculated)}$$

The data in *Table 3.6* would suggest that deviations also occur for other compounds. These deviations indicate a change in the degree of polarity of the molecule from one composed of rigid, spherically symmetrical ions.

Fajans considered the mutual polarisation or deformation of interacting ions in a molecule A^+B^-. Polarisation of the anion occurs by the attraction of the positive field of the cation for the electron cloud system of the anion and from its repulsion of the anion nucleus. A similar polarisation of the cation may occur. Because of this mutual polarisation the polarity of the molecule decreases. The stronger the polarising fields and the greater the ion polarisability, the greater is the effect on the polarity. A number of rules have been proposed to determine where polarisation is favoured. Polarisation is likely to be large when:

(a) the anion or cation is highly charged. An anion with a charge of unity will exert less repulsion on its outer electrons than one with a higher charge; a cation with a charge greater than one will attract electrons more strongly than one with a charge of unity. Though little comparison is possible for the effect of anionic charges, since these are largely restricted to −1 and −2, the effect of increased cation charge is readily seen from the melting points of some anhydrous chlorides:

Cation	r/pm	m.p./K
Na^+	102	1073
Mg^{2+}	72	985
Al^{3+}	53	453

The decrease in melting point corresponds with increasing polarisation of chloride ion.

(b) the cation is small and the anion is large. A small cation will have a large polarising power because of the concentration of its positive charge over a

Table 3.5. LATTICE ENTHALPIES OF THE ALKALI HALIDES

| Crystal | Lattice enthalpy | | Δ |
| | Experimental | Theoretical | |
	$-\Delta H^{\circ}_{LE}/kJ\ mol^{-1}$		
NaF	907·9	907·9	0
NaCl	769·9	761·5	8·4
NaBr	736·4	723·8	12·6
NaI	690·4	677·8	12·6
KF	803·3	803·3	0
KCl	702·9	694·5	8·4
KBr	673·6	665·2	8·4
KI	636·0	627·6	8·4
RbF	769·9	765·7	4·2
RbCl	673·6	669·4	4·2
RbBr	652·7	644·3	8·4
RbI	615·0	606·6	8·4
CsF	719·6	728·0	−8·4
CsCl	644·3	627·6	16·7
CsBr	623·4	606·7	16·7
CsI	589·9	573·2	16·7

Data from Morris, D. F. C. *Acta cryst.*, 9 (1956) 197.

Table 3.6. LATTICE ENTHALPIES OF SOME METAL HALIDES

| Crystal | Lattice enthalpy | | Δ |
| | Experimental | Theoretical | |
	$-\Delta H^{\circ}_{LE}/kJ\ mol^{-1}$		
CaF_2	2 610·8	2 602·5	8·3
PbF_2	2 489·5	2 443·5	46·0
HgF_2	2 740·5	2 619·2	121·3
MgF_2	2 907·9	2 861·9	46·0
$MgBr_2$	2 405·8	2 146·4	259·4
MgI_2	2 297·0	1 991·6	305·4
FeF_2	2 912·1	2 849·3	62·8
CoF_2	2 962·3	2 878·6	83·7
NiF_2	3 046·0	2 903·7	142·3
CuF_2	3 041·8	2 623·4	418·4
$FeBr_2$	2 539·7	2 217·5	322·2
FeI_2	2 464·4	2 062·7	401·7
AgF	966·5	916·3	50·2
$AgCl$	916·3	849·4	61·9
$AgBr$	907·9	824·2	83·7
AgI	895·4	795·0	100·4

Data from Morris, D. F. C., and Ahrens, L. H. *J. inorg. nuclear Chem.*, 3 (1956) 263 and Morris, D. F. C. *J. inorg. nuclear Chem.*, 4 (1957) 8.

small area. A large anion will have a high polarisability because the outermost electrons are well shielded by inner electrons from the nuclear field of the anion. The effect of size upon polarisation is well illustrated by the group II A halides (*Figures 3.4* and *3.5*).

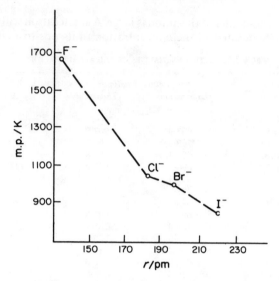

Figure 3.4. Melting points of the anhydrous calcium halides

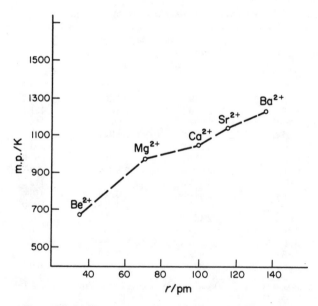

Figure 3.5. Melting points of the anhydrous chlorides of Group II A metals

(c) the cation has an electronic configuration other than that of an inert gas. This point may be demonstrated by a consideration of the corresponding compounds of the alkali metals and the coinage metals, in which the latter have the 18e configuration in the cation M^+. The univalent halides of copper, silver and gold are water-insoluble and of lower melting point than the water-soluble halides of sodium, potassium and rubidium (*Figure 3.6*). This effect may be explained by the fact that the coinage metals in their univalent cations possess d electrons which do not provide such an efficient screening of the nuclear charges as do the s and p electrons for the alkali metals. The univalent cations of the coinage metals are thus able to exert a stronger polarising effect

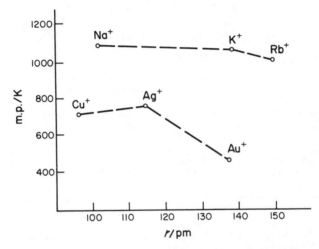

Figure 3.6. Melting points of the chlorides of the Group I A and I B metals

on an anion. A similar effect is observed for the transition metals in general as shown by the differences between calculated and experimental lattice enthalpies (*Table 3.6*).

Further applications of lattice enthalpies

Besides their usefulness in calculations of electron affinities and in the prediction of deviations from purely ionic character in crystalline materials, lattice enthalpy values may be used for a number of other purposes. These include
(a) calculation of proton affinities,
(b) prediction of stabilities of hypothetical compounds, and
(c) explanation for the occurrence of highest oxidation states of metals when in combination with oxygen and fluorine.
Some of these calculations require a number of assumptions to be made such as real and hypothetical compounds having completely ionic or similar non-ionic energy contributions and isomorphous crystal lattices possessing similar

lattice energies. Although such assumptions are not perfectly correct the errors are likely to be small, thus enabling reasonable predictions to be made.

(a) Calculation of proton affinities

The classical example here is that of the proton affinity of ammonia (ΔH_p°). A thermodynamic cycle may be written down for the ammonium halides, viz.:

$$NH_4X(s) \longleftarrow NH_4^+(g) + X^-(g)$$
$$\uparrow \qquad\qquad NH_3(g)+H^+(g) \uparrow$$
$$\uparrow \qquad\qquad\qquad \uparrow$$
$$\tfrac{1}{2}N_2(g)+2H_2(g)+\tfrac{1}{2}X_2(g) \longrightarrow NH_3(g)+H(g)+X(g)$$

For the chloride, bromide and iodide the relative figures are:

	NH_4Cl	NH_4Br	NH_4I	
Lattice enthalpy of $NH_4X(s)$	$-676\cdot13$	$-644\cdot34$	$-608\cdot77$	$\Delta H_{LE}^\circ/kJ\ mol^{-1}$
Heat of formation of $NH_4X(s)$	$-315\cdot47$	$-270\cdot29$	$-202\cdot09$	$\Delta H_f^\circ/kJ\ mol^{-1}$
Formation of $H^+(g)$	$+1\ 535\cdot95$			$\Delta H^\circ/kJ\ mol^{-1}$
Formation of $X^-(g)$	$-234\cdot72$	$-220\cdot92$	$-197\cdot07$	$\Delta H^\circ/kJ\ mol^{-1}$
Heat of formation of $NH_3(g)$	$-46\cdot02$			$\Delta H_f^\circ/kJ\ mol^{-1}$
Proton affinity of ammonia	$-894\cdot55$	$-894\cdot95$	$-886\cdot18$	$\Delta H_p^\circ/kJ\ mol^{-1}$

(Where the formation of H^+ (g) involves the steps of dissociation and ionisation, and the formation of X^- (g) involves the steps of vaporisation (Br_2, I_2) dissociation and electron affinity.)

Figures for NH_4F are not considered because they give misleading results due to hydrogen bonding in the crystal.

(b) Stabilities of hypothetical compounds

Here we may consider the possibility of the formation of salts such as NaF_2, $CaCl$ and other subhalides of elements in Group II A of the Periodic Table. These calculations may be carried out by taking the value of the lattice enthalpy for the compound concerned to be approximately that of the stable compound formed by the adjacent element in the Periodic Table.

Thus, $\Delta H_{LE}^\circ(NaF_2)$ is taken to be approximately equal to $\Delta H_{LE}^\circ(MgF_2)$ and $\Delta H_{LE}^\circ(CaCl)$ comparable to $\Delta H_{LE}^\circ(KCl)$.

For NaF_2 we may look at the possibility of the reaction

$$NaF_2 \rightleftharpoons NaF + \tfrac{1}{2}F_2$$

This reaction may be broken down into several steps:

$$NaF_2(s) \longrightarrow Na^{2+}(g)+2F^-(g)$$
$$\uparrow \qquad\qquad\qquad \downarrow$$
$$Na^+(g) +F^-(g)+F(g)$$
$$\downarrow \qquad\qquad \downarrow$$
$$NaF(s)+\tfrac{1}{2}F_2$$

Taking the lattice enthalpy of NaF_2 to be approximately that of MgF_2 (-2907.9 kJ mol^{-1}) we may now proceed to calculate the enthalpy change for the reaction as written:

$$\Delta H^\circ = 2907.9 - 4560.6 + 334.7 - 907.9 - 77.4 = -2303 \text{ kJ mol}^{-1}$$

(approximately)

An enormous amount of energy, arising principally from the high second ionisation energy of the sodium atom (4560.6 kJ mol^{-1}), would be released when NaF_2 disproportionates to the stable monofluoride and elementary fluorine and so the difluoride is too unstable to exist.

Similar calculations may be made for the subhalides of Group II A, the enthalpies of formation of which are given below (*Table 3.7*). The figures in-

Table 3.7. HEATS OF FORMATION OF SUBHALIDES OF GROUP IIA. $\Delta H_f^\circ/$kJ mol^{-1}
(ESTIMATED RE CORRESPONDING GROUP IA HALIDE VIA BORN–FAJANS–HABER CYCLE)

	F	Cl	Br	I
Be	-33.47	$+163.18$	$+217.57$	$+297.06$
Mg	-280.33	-121.34	-75.31	-4.18
Ca	-288.69	-150.62	-125.52	-62.76
Sr	-313.80	-196.65	-163.18	-100.42
Ba	-292.88	-196.65	-163.18	-104.60

Table 3.8. ENTHALPY CHANGES FOR THE DISPROPORTIONATION REACTION
$2MX(s) = MX_2(s) + M(s)$ $\Delta H^\circ/$kJ mol^{-1}

	F	Cl	Br	I
Mg	-535.6	-368.2	-322.2	-255.2
Ca	-656.9	-485.3	-456.1	-393.3
Sr	-577.4	-443.5	-397.5	-368.2
Ba	-619.2	-431.0	-439.3	-439.3

dicate that beryllium chloride, bromide and iodide are unstable with respect to decomposition into their component elements. The remainder, however, may be stable (ΔH_f° is negative) but consideration must also be given to the matter of disproportionation into the stable dihalide and metal according to

$$2MX(s) \rightleftharpoons MX_2(s) + M(s)$$

For this reaction the enthalpies of reaction are given in *Table 3.8*. From these figures it will be seen that none of the subhalides is stable.

(c) Highest oxidation states for metals

Combination with fluorine enables many metals to exhibit their highest oxidation states. For the halides MX_n the Born–Fajans–Haber cycle enables the relation

$$\Delta H_{LE}^\circ(MX_n, s) = -\Delta H^\circ(M^{n+}, g) + n\Delta H^\circ(X^-, g) + \Delta H_f^\circ(MX_n, s)$$

to be realised. This may be rewritten as

$$\Delta H_f^\circ(MX_n, s) = \Delta H_{LE}^\circ(MX_n, s) + \Delta H^\circ(M^{n+}, g) + n\Delta H^\circ(X^-, g)$$

For the compound MX_n to be formed, the reaction must be exoergic, i.e.

$$\Delta H_f^\circ(MX_n, s) < 0$$

Therefore $-\Delta H_{LE}^\circ(MX_n, s) < \Delta H^\circ(M^{n+}, g) + n\Delta H^\circ(X^-, g)$

For the halogens the term $\Delta H^\circ(X^-, g)$ is not much different from $-240\,kJ$ mol^{-1} and hence the determining factor in deciding the value of n is the lattice enthalpy. Values of lattice energies of ionic halides are found to be highest in the case of the fluorides. Hence values of n will be highest when $X = F$.

Solubility of ionic compounds

The process of solution of an ionic compound must involve the breakdown of the crystal lattice to produce solute species. In polar solvents, these species are the individual solvated ions. The energy needed for lattice breakdown is provided by that released in the solvation process.

Polar solvents are effective for the dissolution of ionic compounds because their high dielectric constants lead to diminished attractive forces between the ions*.

The process of solution of an ionic halide MX can be described in terms of an enthalpy cycle. The lattice breaks up to give the isolated, gaseous ions and then solvation takes place:

$$MX(s) \longrightarrow M^+(aq) + X^-(aq)$$
$$M^+(g) + X^-(g)$$

The enthalpy of solution ΔH_s° is related to the lattice enthalpy (ΔH_{LE}°) and the solvation energies $\Delta H_h^\circ(M^+, aq)$, and $\Delta H_h^\circ(X^-, aq)$ by the equation

$$\Delta H_{LE}^\circ - \Delta H_h^\circ(M^+, aq) - \Delta H_h^\circ(X^-, aq) = \Delta H_s^\circ$$

ΔH_s° may be positive or negative depending on the relative magnitudes of the other quantities in the equation.

The van't Hoff isochore

$$\frac{d}{dT}(\ln K) = \frac{\Delta H_f^\circ(T)}{RT^2}$$

relates the variation of an equilibrium constant, for a particular reaction, with temperature. A positive ΔH° means that the equilibrium constant will increase with temperature whereas a negative value for ΔH° indicates the constant will decrease as temperature increases. The heats of solution of a number of salts are given in *Table 3.9* and it will be seen that this quantity is frequently negative. An immediate interpretation of this fact is that the salts

---

* The force between two charges q_1 and q_2, a distance l apart *in vacuo* is $q_1q_2/4\pi\varepsilon_0 l^2$; in a medium of dielectric constant ε_r the force is reduced to $q_1q_2/4\pi\varepsilon_r l^2$.

would become less soluble on increasing the temperature whereas of course the majority of salts have increased solubility at higher temperatures. This apparent anomaly is explained by the fact that ΔH_S° itself varies with temperature and increases with temperature because the ions are less strongly solvated by water molecules; consequently slightly negative values of ΔH_S° become positive.

Electronegativity and partial ionic character

Attempts to put partial ionic character on a more quantitative basis have been made by various workers. These have all involved the electronegativity of the atoms concerned. The electronegativity of an atom has been described

Table 3.9. ENTHALPIES OF SOLUTION FOR VARIOUS SALTS $\Delta H_s^\circ/\text{kJ mol}^{-1}$

	OH	F	Cl	Br	I	NO$_3$	CO$_3$	SO$_4$	PO$_4$
Li	−20·92	+4·60	−37·24	−49·37	−63·18	−2·51	−17·57	−30·12	
Na	−42·68	+0·42	+3·77	−0·84	−7·95	−21·76	−24·69	−2·51	−78·66
K	−55·23	−17·57	+17·57	+20·08	+20·50	+35·15	−32·22	+24·27	
Ag		−20·50	+65·69	+84·52	+112·13	+22·18	+41·84	+17·57	
Cu(II)	48·12	−63·18	−51·46	−35·15	−26·36	−41·84	−16·74	−73·22	
Mg	+2·51	−17·99	−154·81	−182·00	−211·71	−85·77	−25·10	−91·21	+69·04
Ca	−16·74	+12·97	−82·84	−110·04	−120·50	−19·25	−12·13	−29·29	−64·63
Sr	−46·44	+10·46	−51·88	−71·55	−90·79	+17·15	−0·42	−8·37	−74·06
Ba	−52·30	+3·35	−12·97	−25·52	−48·12	+40·17	+4·18	+19·25	−7·95
Zn	+29·71		−71·13	−66·94	−55·23	−84·10	−15·90	−81·17	
Pb	+56·67	+6·28	+26·36	+36·82	+64·85	+37·66	+25·52	+12·55	

by Pauling as 'the power of an atom in a molecule to attract electrons to itself'. To say that fluorine is the most electronegative element implies that the electron distribution in the chemical bond X—F (X being any other element) more closely resembles that in an ion-pair X^+F^- rather than X^-F^+. Because this concept concerns atoms in molecules rather than isolated atoms a precise measurement is not easy.

One approach to the topic was made by Pauling. In it he determined the differences between the actual bond energies, $D_{AB(expt)}$, of molecules and those calculated for pure covalent bonds in compounds of the type AB on the assumption that $D_{AB(calc)} = \frac{1}{2}(D_{AA} + D_{BB})$, where the D terms are the dissociation energies of the bonds indicated by the subscripts. The difference Δ_{AB} is given by

$$\Delta_{AB} = D_{AB(expt)} - \tfrac{1}{2}(D_{AA} + D_{BB})$$

It is of necessity always positive and results from the resonance energy contributed by ionic forms (A^+B^- or A^-B^+). Pauling's electronegativity scale is based on the equation

$$\Delta_{AB} = 96·5(x_A - x_B)^2 \tag{8}$$

where x_A and x_B are the electronegativities of the elements A and B, and 96·5 is the conversion factor from electron-volt to kJ. This, however, only gave a value for the difference in electronegativities for the elements and in order to obtain comparative electronegativities of the elements the arbitrary value of 2·1 was assigned to hydrogen, giving the elements from carbon to fluorine values of 2·5 to 4·0. This method enabled values for the more electronegative elements, e.g. F, O, Cl, N, Br, S, C, I, Se, P, As and Si, to be determined; those for other elements were obtained from thermochemical data for compounds formed between the element, whose electronegativity value was required, and one of the elements whose electronegativity value was known.

The values derived by Pauling are quoted in *Table 3.10* and it will be seen that the values are highest for the element fluorine and lowest for caesium.

A second approach was that of Mulliken who considered the process of

Table 3.10. ELECTRONEGATIVITY VALUES FOR SELECTED ELEMENTS (AFTER PAULING)

			H 2·1			
Li 1·0	Be 1·5	B 2·0	C 2·5	N 3·0	O 3·5	F 4·0
Na 0·9	Mg 1·2	Al 1·5	Si 1·8	P 2·1	S 2·5	Cl 3·0
K 0·8	Ca 1·0	Sc 1·3	Ge 1·8	As 2·0	Se 2·4	Br 2·8
Rb 0·8	Sr 1·0	Y 1·2	Sn 1·8	Sb 1·9	Te 2·1	I 2·5
Cs 0·7	Ba 0·9					

electron transfer between atoms in the formation of an ion-pair X^+Y^- or X^-Y^+ The energy changes in the two processes would be $I_X + E_Y$ and $I_Y + E_X$ respectively. If the formation of X^+Y^- is the easier of the two then,

$$I_X + E_Y < I_Y + E_X$$
or
$$I_X - E_X < I_Y - E_Y$$

hence values of $I - E$ might be used as a measure of the electronegativity of a particular atom. One of the practical disadvantages of this approach is the lack of electron affinity values for all but a few elements. Also, the appropriate ionisation potentials and electron affinities to be used are not necessarily those of the isolated atoms since the atoms may be in an excited state and values will thus be different from those of the atom in the ground state.

A further approach by Alfred and Rochow, which has wider applicability, measures electronegativity in terms of the force of attraction between the atom under consideration and an electron at a distance r from it equal to the covalent radius of the atom

i.e. $$F = \frac{Z_{\text{eff}} e^2}{4\pi\varepsilon_0 r^2} \qquad (9)$$

where Z_{eff} is the effective nuclear charge (i.e. the nuclear charge, effective at a distance r, after screening effects of the inner electrons have been taken into account).

A number of rules for determining effective nuclear charges have been proposed by Slater. These are:

(a) $Z_{eff} = Z - S$, where S is the screening factor of the electrons.

(b) S may be determined from a division of the electrons into the following sub-groups:

$$1s; 2s, 2p; 3s, 3p; 3d; 4s, 4p; 4d; 4f; 5s, 5p; \text{etc.}$$

each of which has a different screening constant.

(c) S is obtained by a summation of the separate contributions:

(i) no contribution from any electrons in shells outside the one being considered,

(ii) 0·35 from every other electron in the sub-group considered (except 1s, where 0·30 is used),

(iii) if the electron being considered is in an s,p sub-group an amount of 0·85 from each electron in the next inner sub-group and 1·00 from all other electrons is counted. If, however, the electron is in a d or f sub-group then an amount of 1·00 from every electron in the inner sub-groups is counted.

As an example we may consider the case of bismuth ($Z = 83$).

$$Bi = [1s^2\ 2s^2\ 2p^6\ 3s^2\ 3p^6\ 3d^{10}\ 4s^2\ 4p^6\ 4d^{10}\ 4f^{14}][5s^2\ 5p^6\ 5d^{10}][6s^2\ 6p^3]$$

for which
$$Z_{eff} = 83 - [0·35\ (5) + 0·85\ (18) + 1·00\ (60)]$$
$$= 5·95$$

By plotting Z_{eff}/r^2 against Pauling's electronegativity values Allred and Rochow obtained a straight line plot of the form

$$x = \frac{3·59 \times 10^3\ Z_{eff}\ \text{pm}^2}{r^2} + 0·744$$

where r is in picometre.

Hence, for bismuth the corresponding electronegativity value is

$$x = \frac{3·59 \times 10^3 \times 5·95\ \text{pm}^2}{(152\ \text{pm})^2} + 0·744 =$$

taking the covalent radius of bismuth as 152 pm.

Table 3.11 shows electronegativity values determined by this method.

Generally, the electronegativity values decrease in going down a particular group, though there are exceptions (see for instance Group IV B) where alternation occurs, and increase from left to right across the Periodic Table.

Attempts to relate difference in electronegativity to percentage ionic character in the bonding have also been made. Empirical relationships have been derived that agree with the data for the hydrogen halides in which the

Table 3.11. ELECTRONEGATIVITY VALUES OF THE ELEMENTS

IA	IIA	IIIA	IVA	VA	VIA	VIIA	VIII	VIII	VIII	IB	IIB	IIIB	IVB	VB	VIB	VIIB	O
H 2·1																	He
Li 0·97	Be 1·47											B 2·01	C 2·50	N 3·07	O 3·50	F 4·10	Ne
Na 1·01	Mg 1·23											Al 1·47	Si 1·74	P 2·06	S 2·44	Cl 2·83	Ar
K 0·91	Ca 1·04	Sc 1·20	Ti 1·32	V 1·45	Cr 1·56	Mn 1·60	Fe 1·64	Co 1·70	Ni 1·75	Cu 1·75	Zn 1·66	Ga 1·82	Ge 2·02	As 2·20	Se 2·48	Br 2·74	Kr
Rb 0·89	Sr 0·99	Y 1·11	Zr 1·22	Nb 1·23	Mo 1·30	Tc 1·36	Ru 1·42	Rh 1·45	Pd 1·35	Ag 1·42	Cd 1·46	In 1·49	Sn 1·72	Sb 1·82	Te 2·01	I 2·21	Xe
Cs 0·86	Ba 0·97	RE	Hf 1·23	Ta 1·33	W 1·40	Re 1·46	Os 1·52	Ir 1·55	Pt 1·44	Au 1·42	Hg 1·44	Tl 1·44	Pb 1·55	Bi 1·67	Po 1·76	At 1·90	Rn
Fr 0·86	Ra 0·97	AS															

Rare Earths (R.E.)

IA	IIA	IIIA	IVA	VA	VIA	VIIA	VIII	VIII	VIII	IB	IIB	IIIB	IVB	VB
La 1·08	Ce 1·08	Pr 1·07	Nd 1·07	Pm 1·07	Sm 1·07	Eu 1·01	Gd 1·11	Tb 1·10	Dy 1·10	Ho 1·10	Er 1·11	Tm 1·11	Yb 1·06	Lu 1·14

Actinide Series (A.S.)

IA	IIA	IIIA	IVA	VA	VIA	VIIA	VIII	VIII	VIII	IB	IIB	IIIB	IVB	VB
Ac 1·10	Th 1·11	Pa 1·14	U 1·22	Np 1·22	Pu 1·22	Am (1·2)	Cm (1·2)	Bk (1·2)	Cf (1·2)	Es (1·2)	Fm (1·2)	Md (1·2)	No (1·2)	Lr (1·2)

ionic character is determined from dipole moments*. Work by Hannay and Smyth showed that

$$\text{Percentage ionic character} = 16(x_A - x_B) + 3 \cdot 5(x_A - x_B)^2 \qquad (11)$$

and its relation with electronegativity difference is shown in *Figure 3.7*. In *Figure 3.7* it is apparent that when the difference in electronegativity is above about 2·1 then the bonding is more than 50 per cent ionic.

As mentioned previously, the electronegativity will depend upon the nature of the atom present and it must also vary with the oxidation state of the atom

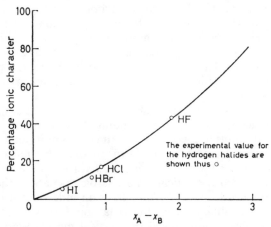

Figure 3.7. Percentage ionic character as a function of electronegativity difference $x_A - x_B$ plotted from Hannay and Smyth's formula (11)

Table 3.12. DIPOLE MOMENTS AND ELECTRONEGATIVITY DIFFERENCES FOR THE HYDROGEN HALIDES

Hydrogen halide	r/pm	$10^{30}p_e$/C m (ionic)	$10^{30}p_e$/C m (observed)	p_e(obs)/ p_e(ionic)	$x_A - x_B$
HF	92	14·68	6·36	0·43	1·9
HCl	127·5	20·21	3·43	0·17	0·9
HBr	143	22·71	2·60	0·11	0·8
HI	162	25·77	1·26	0·05	0·4

* *Dipole Moments* — In a diatomic molecule a dipole arises through unequal sharing of electric charge between the two atoms and as a consequence the molecule has a positive and negative end. Polyatomic molecules also possess dipoles except in certain instances such as CCl_4 and CO_2 where the bonds are symmetrical about the central atom. This charge displacement is measured by the dipole moment of the molecule (p_e). If two charges $+\delta$ and $-\delta$ are separated by a distance r, the dipole moment is given by $p_e = \delta \cdot r$. Dipole moments have been determined experimentally for the hydrogen halides and compared with the theoretical values derived from the observed internuclear distance (in m) in the hydrogen halide and the value of the electronic charge $1 \cdot 602 \, 10 \times 10^{-19}$ C. This then gives the dipole moment in C m; a unit of dipole moment still in common use is the debye (D) where $1 \, D = 3 \cdot 33 \times 10^{-30}$ C m. The ratio of these two dipole moments can be regarded as the fraction of ionic character in the bond assuming that the covalent bonding contributes nothing to the dipole moment. The values for the experimental and calculated dipole moments of the hydrogen halides are given in *Table 3.12*.

concerned. Consequently, the electronegativity values should not be regarded too rigidly but rather should be used for more qualitative discussion than quantitative. It may be said, however, that bonds between elements with widely variant electronegativities will be essentially ionic whereas covalent bonds will be favoured by combination between atoms of similar electronegativity values.

THE COVALENT BOND

The simple Lewis theory of the covalent bond envisaged the sharing of electrons between atoms forming the bond but gave no indication as to how or why this sharing occurred. The more modern picture of covalent bond formation considers the energy changes taking place when two atoms approach one another. The condition for the formation of the stable bond is that, at a certain inter-atomic distance, the potential energy of the system reaches a minimum.

A clear idea of the concept of resonance is essential to the understanding of the energy changes which occur on bonding. In describing a molecule in terms of covalent bonds it is possible to write down more than one formula that satisfies the requirements of the Lewis theory. The molecule is said to have various resonance forms. These are all involved in the structure of the molecule, which is itself energetically more stable than any of the component forms. The resonance forms at no time have an independent existence and so the concept is quite different from that of tautomerism where the individual forms may be isolated by chemical or physical means. Resonance simply involves the rearrangement of electrons between the various forms whilst maintaining the same number of unpaired electrons in each of the structures and the same relative positions of the nuclei.

Thus in the case of the hydrogen molecule, we may write down the following four resonance forms:

$$^1H_a - {}^2H_b \qquad {}^2H_a - {}^1H_b \qquad H_a^+ \, {}^{1,\,2}H_b^- \qquad {}^{1,\,2}H_a^- \, H_b^+$$
$$\text{(I)} \qquad\qquad \text{(II)} \qquad\qquad \text{(III)} \qquad\qquad \text{(IV)}$$

where the two hydrogen nuclei are designated H_a and H_b and the two electrons are 1 and 2. Resonance form (I) represents H_a and H_b joined by a covalent bond, electron 1 being more closely associated with H_a and electron 2 with H_b. Form (II) is a second covalent structure with the positions of the electrons interchanged. Forms (III) and (IV) are ionic contributions with both electrons associated with either H_b or H_a respectively.

Similarly, for a more complex molecule such as CO_2, we may write different resonance formulations such as:

$$\overset{\times\times}{\underset{\times\times}{\times}} O \overset{\times\times}{\times} C \overset{\times\times}{\underset{\times\times}{\times}} O \overset{\times\times}{\times} \qquad \overset{-}{\overset{\times\times}{\underset{\times\times}{\times}}} O \overset{\times}{\times} C \overset{\times\times\times}{} O \overset{+}{\times} \qquad \overset{+}{\times} O \overset{\times\times\times}{} C \overset{-}{\overset{\times\times}{\underset{\times\times}{\times}}} O \overset{\times}{}$$
$$\text{(V)} \qquad\qquad\qquad \text{(VI)} \qquad\qquad\qquad \text{(VII)}$$

Each of the resonance forms on the wave-mechanical approach has a definite energy and may be described by an approximate wave function ψ_1, ψ_{II}, etc. We may now apply one of the general principles of wave mechanics, namely that if a certain system can be described by a set of approximate wave functions then a linear combination of these is also a satisfactory description of the system. The linear combination that gives the lowest energy is taken to be the best description of the system in its normal state.

If we consider a system such as nitromethane (CH_3NO_2) where two resonance forms can be written

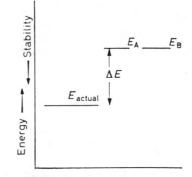

then the actual structure is more stable than either of these by an amount of energy ΔE, termed the *resonance energy* of the molecule, this being the difference in energy between that of the actual molecule and that of the most stable of the resonance forms (*Figure 3.8*).

Where multiple bonds are present, the existence of resonance forms may be deduced from the measurements of bond distances within the molecule and

Figure 3.8. Representation of the energies of resonance forms of a molecule such as nitromethane where both forms are of equal energy. The nitromethane molecule is more stable than either of these forms [(A) and (B) above] by the resonance energy ΔE

of heats of formation. Thus in the case of CO_2 the carbon–oxygen bond length is 115 pm for both bonds and this is intermediate between the values predicted for C=O and C≡O; none of the three resonance forms in itself accounts for this observation. The measured heat of formation of carbon dioxide is 1602.5 kJ mol^{-1} compared with that of 1447.7 kJ mol^{-1} predicted theoretically for the simple formulation O=C=O, indicating a resonance energy of 154.8 kJ mol^{-1}

The nature of the covalent bond

Two theories have been put forward to explain the formation of a covalent bond. The first of these is the valence-bond approach (Heitler, London, Pauling and Slater) which treats the bond formation from the standpoint of the pairing of electron spins and the maximum overlapping of atomic orbitals containing these electrons to give a region of common electron density to the combining atoms. The second approach, known as the molecular-orbital approach, considers the molecule as a whole and allocates electrons to a set of molecular orbitals in a manner similar to the allocation of electrons in an atom to a set of atomic orbitals (Hund and Mulliken). Unfortunately neither of the methods is applicable to all systems. They are both extreme cases and both therefore have their limitations.

The valence-bond approach may be illustrated qualitatively by considering the hydrogen molecule. In this case, we consider the possibility of interchange of the electrons between the two atoms. There are two possible equivalent structures which are indistinguishable once the bond has been

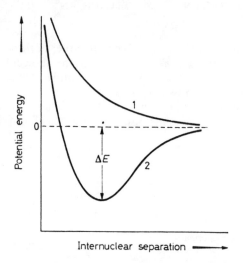

Figure 3.9. Energy variation as a function of internuclear separation for the hydrogen molecule

formed. These are given by (I) and (II) of the resonance forms. As the two hydrogen nuclei approach each other from an infinite distance the weak attractive forces are gradually opposed by strong repulsive forces at short interatomic distances. Interaction between the wave functions of the electrons occurs, for their spins may be parallel or opposed. If the electrons have parallel spins then the energy continues to rise as the atoms get closer together (line 1 in *Figure 3.9*) and no bond is formed. If, however, the spins are opposed the energy curve possesses a definite minimum which corresponds to the formation of a stable molecule (line 2, *Figure 3.9*).

This approach stresses the importance of electron spin, since in order for bonding to occur each atom must have an unpaired electron available for pairing up with the unpaired electron on the other atom.

The molecular-orbital method of Hund and Mulliken differs basically from the valence-bond approach in that it takes into consideration all the electrons of the combined atoms and considers these to be jointly held by the molecule in a set of polynuclear orbitals.

This method may be illustrated by reference to diatomic molecules. The molecular orbitals are obtained by a process known as the linear combination of atomic orbitals (LCAO). Thus the electron, when it is in close association with any one of the nuclei, may be described by a wave function that is approximately that of an atomic orbital. The wave function of a molecular orbital is then formed by taking a linear combination of the separate atomic orbital wave functions. From each pair of atomic orbitals so combined a new pair of molecular orbitals is formed. One of these is of higher energy than the other and the combination may be represented as

$$\Psi = c_1\psi_1 \pm c_2\psi_2$$
$$= \psi_1 \pm \lambda\psi_2$$

For homonuclear diatomic molecules each makes an equal contribution to the wave function of the molecular orbital and so $\lambda = 1$.

In the case of the combination of the two $1s$ wave functions, on the atoms A and B respectively, the two molecular orbitals are given by

$$\Psi_{bonding} = \psi_{A, 1s} + \psi_{B, 1s}$$
and $$\Psi_{antibonding} = \psi_{A, 1s} - \psi_{B, 1s}$$

The first of these is of lower energy than the component atomic orbitals and is termed a bonding molecular orbital; the second is of higher energy and is termed an antibonding molecular orbital. Pictorially they may be represented as in *Figure 3.10*. Both of these molecular orbitals have symmetry about the bond axis and are designated as σ orbitals. The bonding orbital is termed the $\sigma 1s$ and the antibonding orbital the $\sigma^* 1s$.

With p atomic orbitals, two different types of molecular orbital may be formed. If the bond is formed along the x axis, the two p_x atomic orbitals will give rise to a σp and a $\sigma^* p$ molecular orbital which are, again, symmetrical about the bond axis. The p_y and p_z orbitals, however, overlap to form orbitals that do not have symmetry about the bond axis. The molecular orbitals formed in this instance are termed π molecular orbitals. The combination

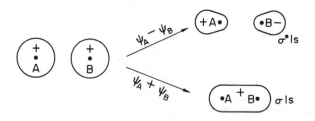

Figure 3.10. Formation of the σ1s and σ*1s molecular orbitals

is shown (*Figure 3.11*) for the $2p_y$ atomic orbitals where both bonding and antibonding orbitals are formed (π and π^* respectively).

We may go one step further and consider what happens if the two nuclei coalesce into a united atom. The correlation between the atomic orbitals on the separated atoms and those on the united atom is given in *Figure 3.12*. The formation of the united atom orbitals is easily seen from the diagrams of the formation of the molecular orbitals; thus the $\pi^* 2p$ become of similar shape to the interaxial atomic d orbitals as the internuclear distance decreases, hence these molecular orbitals become correlated with the 3d level on the united atom.

The energies of the various molecular orbitals of the homonuclear diatomic molecules formed by the elements in the first row of the Periodic Table are arranged in the following order:

$$\sigma 1s < \sigma^* 1s < \sigma 2s < \sigma^* 2s < \sigma 2p < \pi_y 2p = \pi_z 2p < \pi_y^* 2p = \pi_z^* 2p < \sigma^* 2p$$

The molecular orbitals are filled up in exactly the same manner as atomic

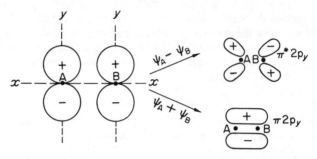

Figure 3.11. Formation of the $\pi 2p$ and $\pi^* 2p$ molecular orbitals

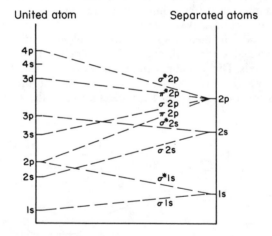

Figure 3.12. Relation between the energy levels in the separated atoms and in the united atoms

orbitals, i.e. two electrons to an orbital and in order of increasing energy with due regard for the possible degeneracy of the orbitals.

We may therefore write down the molecular orbital description of some simple diatomic molecules:

(i) H_2 $(\sigma 1s)^2$

(ii) He_2 $(\sigma 1s)^2 (\sigma^* 1s)^2$

In the case (ii) the energy of the antibonding electrons exceeds that of the bonding electrons and the molecule is therefore unstable. Effectively, whenever the number of bonding and antibonding electrons is the same there is no stable bond formed between the atoms. Excited states of the molecule may, however, be obtained by exciting one of the antibonding electrons to a higher bonding orbital. Thus the excited states of the helium molecule have been observed in discharge tubes.

(iii) Li_2 $(\sigma 1s)^2(\sigma^* 1s)^2(\sigma 2s)^2$

The lithium molecule exists in lithium vapour. The $(\sigma 2s)$ pair is the bonding pair, for the inner 1s electrons are very little affected by bonding and remain virtually as they are in the isolated atoms. These electrons are non-bonding and the configuration of the lithium molecule is commonly written as

Li_2 $K\,K\,(\sigma 2s)^2$

(iv) N_2 $K\,K\,(\sigma 2s)^2(\sigma^* 2s)^2(\sigma 2p)^2(\pi_y 2p)^2(\pi_z 2p)^2$

The bond in nitrogen is a triple one arising from one σ- and two π-type bonds.

(v) O_2 $K\,K\,(\sigma 2s)^2(\sigma^* 2s)^2(\sigma 2p)^2(\pi_y 2p)^2(\pi_z 2p)^2(\pi_y^* 2p)^1(\pi_z^* 2p)^1$

This description of the oxygen molecule tells us that there are two unpaired electrons in the doubly degenerate $\pi^* 2p$ molecular orbital level. On the valence-bond approach all electrons would be paired. The molecular-orbital approach is superior to the valence-bond description in this case as it can account for the observed paramagnetism of the oxygen molecule.

(vi) F_2 $K\,K\,(\sigma 2s)^2(\sigma^* 2s)^2(\sigma 2p)^2(\pi_y 2p)^2(\pi_z 2p)^2(\pi_y^* 2p)^2(\pi_z^* 2p)^2$

In this molecule, although several pairs of bonding and antibonding electrons are present, the fluorine atoms are joined effectively by a single bond, the $(\sigma 2p)$ pair.

For heteronuclear diatomic molecules such as CO and NO the situation becomes a little more complicated. In the linear combination of atomic orbitals, we must write

$$\Psi = \psi_1 \pm \lambda \psi_2$$

where λ is not equal to one, for in such cases the separate atomic wave functions no longer contribute equally to the molecular orbital. The energies of the orbitals of the separate atoms are not the same and a different nomenclature for the molecular orbitals is preferable. The nomenclature used is that evolved by Mulliken, who suggested that the molecular orbitals from the $(\sigma 2s)$ to the $(\sigma^* 2p)$ should be designated

$$z\sigma < y\sigma < x\sigma < w\pi < v\pi < u\sigma$$

This avoids the association, which arises in the case of the homonuclear diatomic molecules, of the molecular orbitals with the particular atomic orbitals.

The molecular orbital configurations of CO and NO can be written as

CO $KK(z\sigma)^2(y\sigma)^2(x\sigma)^2(w\pi)^4$

NO $KK(z\sigma)^2(y\sigma)^2(x\sigma)^2(w\pi)^4(v\pi)^1$

The bond in CO is effectively a triple bond (one σ- and two π-bonds). In NO there is an additional electron present in the antibonding orbital; this accounts for the observed paramagnetism of the nitric oxide molecule.

POLYATOMIC MOLECULES AND HYBRIDISATION

The formation of the covalent bond may be considered as the pairing of electron spins by approach and overlap of atomic orbitals of suitable energy and symmetry. Because the combining atoms have orbitals of definite geometric shape the bonds will be directed in space, the strength of the bond depending upon the extent to which overlapping occurs. The strongest bonds are formed when a maximum admixture of electron density is effected.

In the case of the water molecule, which is one of the simplest polyatomic molecules, we may consider bond formation as taking place by σ overlap, in the plane of the two oxygen orbitals containing the unpaired p electrons, between the hydrogen 1s orbital and the oxygen 2p orbitals. Strong overlap occurs between the 1s and 2p as illustrated in *Figure 3.13*, but effectively no

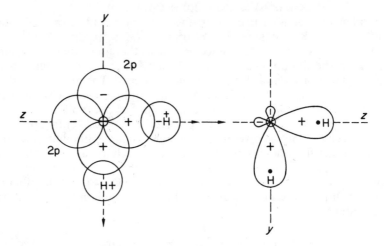

Figure 3.13. Predicted shape of H_2O *by overlap of* 2p *orbitals on oxygen with* 1s *orbitals on hydrogen*

other overlap is obtained. In such a case we speak of a localised molecular orbital, i.e. one formed over only two atoms in the molecule. According to this simple pictorial approach the expected bond angle in water is 90°; an angle of 104° 30′ is found. From a similar treatment of the formation of ammonia we would predict three N—H bonds at 90° to each other in a pyramidal arrangement formed by overlapping of the three 2p orbitals on the nitrogen atom with three 1s orbitals on hydrogen. The observed bond angle is 106° 45′.

The simple theory thus appears to be inadequate to explain the observed shapes of polyatomic molecules.

A consideration of the electronic configurations of beryllium, boron and carbon suggests that these elements should be inert, univalent and divalent respectively. The elements are, however, typically divalent, trivalent and tetravalent respectively. It is impossible to explain the valencies of these elements on the basis of the number of unpaired electrons in the atoms and the overlap of atomic orbitals. In such cases it becomes necessary to suggest that electrons from the 2s level are unpaired and one promoted to the 2p level with the formation of excited states of the atom.

The promotions for the three elements above are

Be $(1s^2 2s^2)$ to   Be $(1s^2 2s^1 2p^1)$
B $(1s^2 2s^2 2p^1)$ to   B $(1s^2 2s^1 2p^1 2p^1)$
C $(1s^2 2s^2 2p^1 2p^1)$ to   C $(1s^2 2s^1 2p^1 2p^1 2p^1)$

and in these excited states the number of unpaired electrons corresponds with the observed valencies of the atoms. However, this still does not account for the fact that all bonds are equivalent in a molecule such as BCl_3 or CCl_4. For example, if the carbon atom used three pure p orbitals and one pure s orbital in binding four atoms, one would expect in CCl_4 three fairly strong bonds directed at right angles to one another and a fourth weaker bond with no special orientation in space relative to the other three. It is necessary, therefore, to introduce a further concept, that of *hybridisation* (or mixing) of atomic orbitals to form exactly equivalent hybrid orbitals which have characteristic directions in space.

The simplest case is that of *digonal* hybridisation where two orbitals from an s and a p atomic orbital form two sp hybrid orbitals. This occurs in the

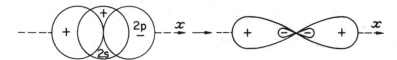

Figure 3.14. Process of digonal hybridisation

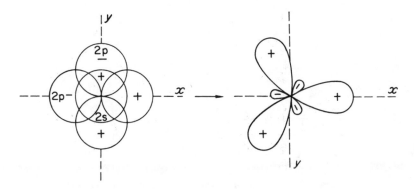

Figure 3.15. Process of trigonal hybridisation

beryllium atom wherein each hybrid orbital contains one electron and is capable of overlapping with a half-filled orbital on another atom to form a covalent bond. This type of hybridisation is shown in *Figure 3.14*. The bonds that digonal hybrids make are at 180° to each other so that molecules such as $BeCl_2$ in the gas phase are linear.

In a similar manner the 2s and two 2p orbitals may undergo *trigonal* hybridisation to give three equivalent sp² hybrid orbitals which are at an angle of 120° to each other in a plane (*Figure 3.15*).

The hybridisation of the 2s and three 2p orbitals produces four sp³ hybrid orbitals that are directed towards the four corners of a regular tetrahedron as shown in *Figure 3.16*.

The sp² hybridisation accounts for the shape of molecules such as BCl_3 and the sp³ hybridisation for the *tetrahedral* arrangement of bonds in CCl_4.

Hybridization of carbon orbitals—the formation of double and triple bonds

The organic molecules methane, ethylene and acetylene serve a useful purpose in demonstrating the three types of hybridisation above and help to illustrate the formation of double and triple bonds by orbital overlap.

(a) *Methane* (CH_4) — In this molecule the carbon orbitals are sp³ hybridised and bonding occurs by overlap of the four hybrid orbitals with four 1s atomic

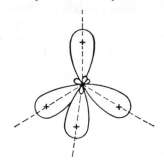

Figure 3.16. sp³ Hybrid orbitals

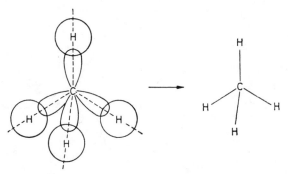

Figure 3.17. Formation of methane by overlap of carbon sp³ hybrid orbitals with 1s atomic orbitals of hydrogen

orbitals on hydrogen. This results in a regular tetrahedral distribution of σ-bonds (*Figure 3.17*).

(b) *Ethylene* (C_2H_4) — Here the orbitals of the two carbon atoms are sp² hybridised. Overlap of one of these hybrid orbitals from each carbon atom gives rise to a σ-bond between the carbon atoms. The remaining hybrid orbitals on each carbon overlap with hydrogen 1s orbitals to give four σ-type carbon–hydrogen bonds as shown. Each carbon atom also has one electron located in a 2p orbital, the axis of which is perpendicular to the plane of the hybrid orbitals. These 2p orbitals add to the stability of the molecule by overlapping laterally to form a π-type carbon–carbon bond. This overlap is greatest when the carbon–hydrogen bonds are all coplanar and gives rise to a π molecular orbital as shown in *Figure 3.18*. This molecular orbital contains two electrons and takes the form of extended lobes, one above and one below the plane of the molecule.

The overlap of the two atomic p orbitals is less than the overlap of the two sp^2 hybrid orbitals and hence the π-bond is weaker than the σ-bond. This lower stability of the π-bond accounts for the reactivity of ethylene; the two carbon atoms tend to form a more stable σ-bond with other atoms. The presence of a π-bond means that the two carbon atoms are bound more strongly than by a single σ-bond; as a result of this the interatomic distance is shorter than would be obtained if only a single bond were present.

(c) *Acetylene* (C_2H_2) – In this molecule the orbitals of the carbon atoms are sp hybridised and the remaining 2p orbitals on each are unaffected. Bonding between the two carbon atoms occurs by overlap of an sp hybrid orbital

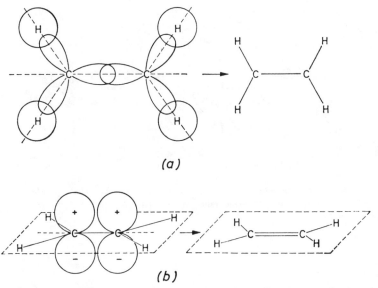

(a)

(b)

Figure 3.18. Bonding in the ethylene molecule, showing (a) formation of σ-bonds and (b) formation of π-bonds

from each carbon atom with the formation of a σ-bond. The remaining sp hybrid orbital on each carbon overlaps with the 1s orbital of hydrogen to form the carbon–hydrogen bonds. The p orbitals, each of which contains one electron, overlap, as in the case of ethylene, but in this instance two π orbitals are formed (*Figure 3.19*). The superposition of these π orbitals leads to cylindrical symmetry about the carbon–carbon axis.

In systems where more than one double bond occurs, as in conjugated molecules such as butadiene and benzene, the π electrons may be associated with more than two nuclei and are said to be delocalised. Thus in CO_2 the resonance forms may be written as

$$O=C=O \qquad {}^+O{\equiv}C{-}O^- \qquad {}^-O{-}C{\equiv}O^+$$

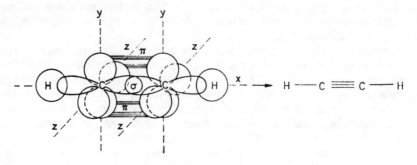

Figure 3.19. Bonding in the acetylene molecule

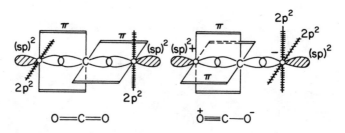

Figure 3.20. Bonding in resonance forms of carbon dioxide. (2p orbitals are drawn as lines to avoid confusion. Shading indicates unshared electron pairs)

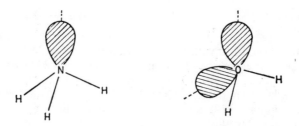

Figure 3.21. Bonding in ammonia and water showing lone-pair orbitals

Since the molecule is linear we may postulate sp hybridisation for carbon and oxygen, giving the configurations

$$C = 1s^2(sp)^1(sp)^1 2p^1 2p^1 \qquad O = 1s^2(sp)^2(sp)^1 2p^2 2p^1$$
$$O^- = 1s^2(sp)^2(sp)^1 2p^2 2p^2 \qquad O^+ = 1s^2(sp)^2(sp)^1 2p^1 2p^1$$

The resonance hybrids may therefore be pictured as in *Figure 3.20*. Delocalisation of the π-electrons occurs here.

The concept of hybridisation gives a better picture of the structure of ammonia and water. If sp^3 hybridisation is invoked for the nitrogen and oxygen atoms, the tetrahedral distribution of electron pairs would be as for methane, but for nitrogen one of these is a lone pair and for oxygen two are lone pairs. In such cases the lone pairs take up a larger volume than the bond pairs around the nitrogen and oxygen nucleus and this causes a decrease of the inter-bond angle in ammonia and water to below the tetrahedral value of 109° 28' (*Figure 3.21*). The presence of lone pairs therefore affects the bond angles of molecules; this is discussed further below.

When d orbitals are available, other types of hybridisation may occur. Some examples of hybridisation involving d orbitals are tabulated below.

Table 3.13. DIRECTIONAL PROPERTIES OF HYBRID ORBITALS

Co-ordination number	Hybridisation	Orbitals used	Stereochemistry
2	sp or dp	$s, p_z; d_z^2, p_z$	Linear
3	sp^2	s, p_x, p_y	Trigonal planar
4	dsp^2	$d_{x^2-y^2}, s, p_x, p_y$	Square planar
	sp^3 or sd^3	s, p_x, p_y, p_z	Tetrahedral
		$s, d_{xy}, d_{xz}, d_{yz}$	
5	dsp^3	$d_{z^2}, s, p_x, p_y, p_z$	Trigonal bipyramid
6	d^2sp^3	$d_{x^2-y^2}, d_{z^2}, s$	Octahedral
		p_x, p_y, p_z	

SHAPES OF INORGANIC MOLECULES AND IONS

A useful qualitative approach to the shapes of inorganic molecules and ions was made by Sidgwick and Powell in 1940, who considered the shape to be dictated by the number of lone pairs and bonding pairs of electrons associated with the central atom. The basis of this premise is that these electron pairs arrange themselves so that the least repulsion occurs between them. Thus the shape of a molecule containing a multicovalent atom is related to the size of the valency shell of electrons on that atom. Accordingly, two pairs of electrons are arranged linearly, three pairs in a trigonal plane, four pairs tetrahedrally, five pairs as a trigonal bipyramid and six pairs as an octahedron. Regular molecular shapes are obtained only when all the electron pairs are used in bonding to identical atoms. If some of the atoms are different, departures from the ideal shape occur. When the central atom carries one or more lone pairs the shape of the molecule or ion can be explained by postulating that the repulsion between electron pairs decreases in the order:

lone pair–lone pair > lone pair–bond pair > bond pair–bond pair.

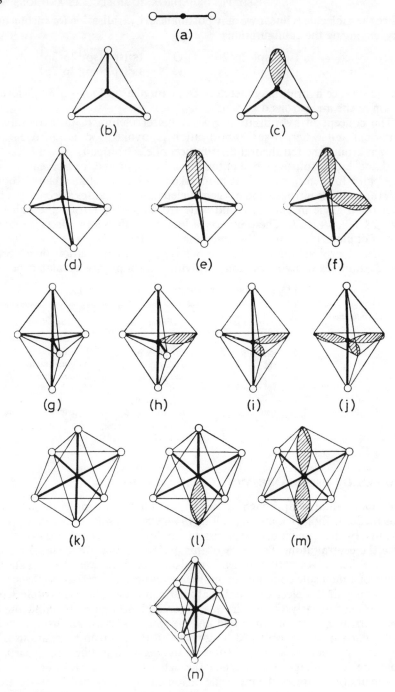

Figure 3.22. Shapes of simple molecules (adapted from Gillespie, R. J., and Nyholm, R. S. Quart. Rev. chem. Soc., Lond., 11 (1957) 339)

This makes it possible to account for the observed bond angles in molecules where bond pairs are successively replaced by lone pairs, e.g. the series CH_4, NH_3 and H_2O. In the first molecule we have bonding to four hydrogen atoms and the bond angle is regular tetrahedral, i.e. 109° 28′. The replacement of one of the bonding pairs by a lone pair in ammonia causes the bond angle to decrease to 106° 45′ since the lone pair has a greater repulsive effect upon the remaining three bond pairs. In water the replacement of a further bond pair by a lone pair causes even greater repulsion and the bond angle decreases further to 104° 30′. The decrease in bond angle observed for the hydrides of Group VI B and V B as the atomic number increases can also be interpreted by the simple Sidgwick–Powell approach.

The shapes for simple inorganic molecules and ions of the non-transitional elements are given in *Figure 3.22* and an example is given for each shape.

The shapes of molecules and ions containing multiple bonds may also be explained if it is assumed that the multiple bond occupies only one position in space. The shapes for various molecules and ions are given in *Table 3.14*.

ONE- AND THREE-ELECTRON BONDS

Although the covalent bond that is most common is of the electron-pair type, two other related bonds are encountered in certain compounds. These are the one- and the three-electron bond.

The classic example of the first of these is the hydrogen molecule ion H_2^+ which has been detected spectroscopically. Two resonance forms of equal stability may be written for this species:

$$H.H^+ \text{ and } {}^+H.H$$

It has a bond energy of 255·3 kJ mol^{-1}. On the molecular orbital approach the electron will be accommodated in the $\sigma 1s$ molecular orbital.

The three-electron bond is found in certain odd molecules which it is impossible to describe in terms of completed octets. Such molecules are generally formed between atoms that are quite close in electronegativity. Typical examples are NO, NO_2 and ClO_2, for which we may write down the following conventional formulae:

:N≡O: N with O: and O: :Cl with O: and O:

Key to Figure 3.22

Shape	Electron pairs	Lone pairs	Molecular shape	Example
(a)	2	0	Linear	$HgCl_2$
(b)	3	0	Trigonal planar	BCl_3
(c)	3	1	V-shape	$SnCl_2$ (gas)
(d)	4	0	Tetrahedral	CH_4
(e)	4	1	Trigonal pyramid	NH_3
(f)	4	2	V-shape	H_2O
(g)	5	0	Trigonal bipyramid	PCl_5 (gas)
(h)	5	1	Irregular tetrahedral	$TeCl_4$
(i)	5	2	T-shape	ClF_3
(j)	5	3	Linear	ICl_2^-
(k)	6	0	Octahedral	SF_6
(l)	6	1	Square prism	IF_5
(m)	6	2	Square planar	ICl_4^-
(n)	7	0	Pentagonal bipyramid	IF_7

Table 3.14. THE SHAPES OF MOLECULES CONTAINING MULTIPLE BONDS

Total number of bonds and lone pairs	Arrangement	Number of bonds	Number of lone pairs	Shape	Examples
2	Linear	2	0	Linear	$O=C=O$; $H-C\equiv N$
3	Triangular	3	0	Trigonal planar	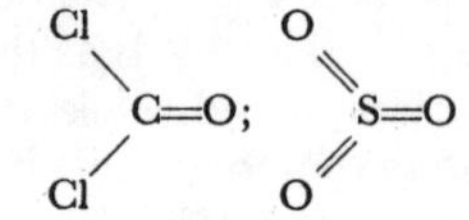 $\begin{array}{c} Cl \\ \backslash \\ C=O; \\ / \\ Cl \end{array}$ $\begin{array}{c} O \\ \backslash \\ S=O \\ / \\ O \end{array}$
	Planar	2	1	V-shaped	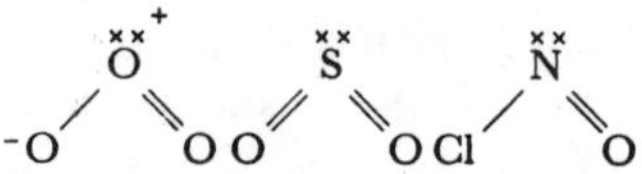 $\overset{\times\times}{\underset{-O}{O}}\overset{+}{\diagdown}O$; $\overset{\times\times}{\underset{O}{S}}\diagdown O$; $\overset{\times\times}{\underset{Cl}{N}}\diagdown O$
4	Tetrahedral	4	0	Tetrahedral	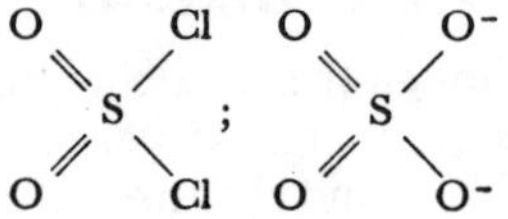 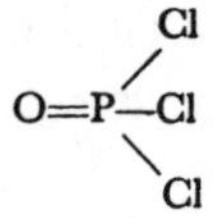 $\begin{array}{c} O \quad\ Cl \\ \diagdown\ / \\ S \\ \diagup\ \diagdown \\ O \quad\ Cl \end{array}$; $\begin{array}{c} O \quad\ O^- \\ \diagdown\ / \\ S \\ \diagup\ \diagdown \\ O \quad\ O^- \end{array}$ $\begin{array}{c} Cl \\ / \\ O=P-Cl \\ \diagdown \\ Cl \end{array}$
		3	1	Trigonal pyramidal	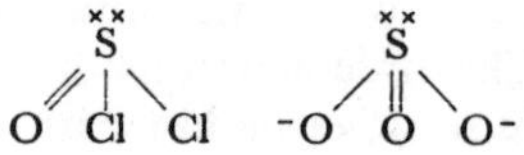 $\overset{\times\times}{S}$ over $O=$, Cl, Cl ; $\overset{\times\times}{S}$ with ^-O, O, O^-
		2	2	V-shaped	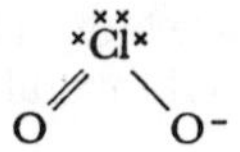 $\overset{\times\times}{\times Cl \times}$ with $O=$ and O^-
5	Trigonal bipyramidal	5	0	Trigonal bipyramidal	
		4	1	Irregular tetrahedral	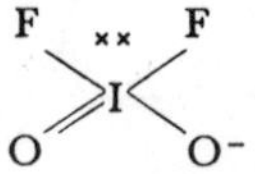 $\begin{array}{c} F \quad \overset{\times\times}{} \quad F \\ \diagdown I \diagup \\ O \quad\ O^- \end{array}$
6	Octahedral	6	0	Octahedral	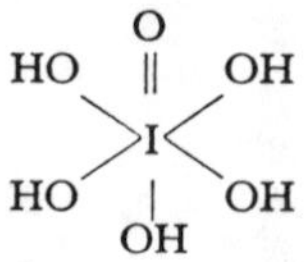 $\begin{array}{c} O \\ \| \\ HO \quad\ OH \\ \diagdown I \diagup \\ HO \quad\ OH \\ \| \\ OH \end{array}$

From Gillespie, R. J., and Nyholm, R. S. 'Inorganic stereochemistry' *Quart. Rev. chem. Soc., Lond.* 11 (1957) 339.

Nitric oxide, as we have already seen, is alternatively described in terms of the molecular orbital theory as having a single electron in the $v\pi$ molecular orbital giving rise to paramagnetism characteristic of one unpaired electron. The other odd molecules are paramagnetic and the odd electron may be considered as spread over the whole molecule in a molecular orbital.

SYMMETRY

We have seen in the preceding sections how molecules and ions possess geometric frameworks defined by the positions of the constituent atomic nuclei. Many of these nuclei may be identical and occupy physically equivalent positions within the molecular framework of a particular species. By considering the rearrangements of these nuclei by such symmetry operations as rotation about an axis, reflection in a mirror plane or the combination of some rotation and reflection, we may decide to which symmetry group a molecule belongs. Any symmetry operation that is applied to a molecule must result in motion of the molecular framework to a position where the nuclei are coincident with identical nuclei in the original orientation of the molecule.

Altogether there are five kinds of operation that may be performed:

(i) *The identity operation*: this changes nothing and leaves all nuclei in their original positions. It is denoted by the symbol E.

(ii) *Rotation about an axis of symmetry*: a rotation of a molecule about some axis by an angle α that results in coincidence of the nuclei with their counterparts in the original orientation is denoted $C(\alpha)$. If the application of $C(\alpha)$ n times returns us to the original state after rotation through 2π (i.e. $\alpha = 2\pi/n$), then the symmetry operation is written as C_n (where α is the smallest angular rotation required to give coincidence). The molecule is then said to possess an *n-fold axis of symmetry*.

For a molecule possessing various axes of differing order, that of highest order is termed the *principal* axis. Twofold axes that are perpendicular to this axis are denoted C_2' and C_2''.

Linear molecules form a special subdivision of this section for they possess an infinite number of positions of rotation about the axis that are identical with the original molecule.

(iii) *Reflection in a plane of symmetry*: this operation is denoted as σ. If the reflection takes place in a plane perpendicular to the principal axis it is written as σ_h; if in a plane that contains the principal axis, σ_v and if in a plane that both contains the principal axis and which bisects two C_2' axes it is σ_d.

(iv) *Rotary reflection or improper rotation*: this comprises rotation about an axis followed by reflection in the plane perpendicular to that axis — neither of which, in their own right, constitute a symmetry operation. The operation is denoted S_n, where n is the order of the rotation.

(v) *Inversion in a centre of symmetry*: it may be in the molecule under consideration that straight lines may be drawn from every atom, through the centre of the molecule, to meet similar atoms equidistant from the centre on the other side of it. The molecule is then said to possess a *centre of symmetry* and the operation is denoted by i.

When all the symmetry operations for a molecule are written down they constitute the *symmetry* or *point group* of the molecule. Various rules may be applied to determine the symmetry group to which a molecule belongs and the theory developed, known as *group theory*, is the basis for understanding electronic and vibrational spectra, dipole moments, optical activity and crystal structures.

These rules are as follows:

(a) Determine whether the molecule belongs to one of three special point groups without unique axes of high symmetry. These are regular tetrahedral T_d, regular octahedral O_h and icosahedral I_h.

(b) If the molecule does not conform to any one of the special groups in (a) we must search for a symmetry axis. If there is none, but there is a plane of symmetry, the point group is C_s.

(c) If the molecule possesses more than one axis of symmetry the point group is labelled by the axis of highest order.

(d) If there is an S_n operation where n is twice the order of that of the principal axis, and no other, then the point group is S_n.

(e) If there are twofold axes of symmetry perpendicular to the principal axis the point group is D; otherwise it is C.

(f) The point group C may be subdivided:
  (i) C_n —no symmetry planes;
  (ii) C_{nv}—with vertical planes of symmetry;
  (iii) C_{nh}—with horizontal planes of symmetry.

(g) The point group D may be subdivided:
  (i) D_n —no symmetry plane;
  (ii) D_{nh}—with horizontal planes of symmetry;
  (iii) D_{nd}—with dihedral planes of symmetry.

(h) For linear molecules we may find *either* $C_{\infty v}$ where there are no horizontal planes of symmetry *or* $D_{\infty h}$ where horizontal symmetry planes exist.

Some simple examples will serve to illustrate the above principles:

$$H_2O, \ NH_3, \ CH_4, \ SF_6 \text{ and } trans \ MX_4Y_2.$$

The water molecule may be written as (a)

with a principal (z) axis bisecting the HOH angle. Rotation about this axis by π interchanges H_1 and H_2 but leaves us with an indistinguishable atomic arrangement (b). There is, in addition, a vertical plane of symmetry containing the z axis and the molecular point group is C_{2v}.

The ammonia molecule may be treated similarly by writing the three protons at the vertices of an equilateral triangle and the nitrogen atom vertically above the centre of the triangle (though here it is omitted for clarity). The various symmetry operations may then be represented as in

(c) to (h) below

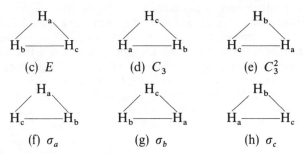

(c) E (d) C_3 (e) C_3^2

(f) σ_a (g) σ_b (h) σ_c

and the molecular point group is C_{3v}.

With methane and sulphur hexafluoride we have two of the special cases T_d and O_h respectively.

The final molecule is a special case of an octahedral arrangement with the X atoms forming a square plane about the metal atom M and the Y atoms in positions vertically above and below this plane. In this case there is a plane of symmetry perpendicular to the fourfold principal axis through YMY and the molecular point group is D_{4h}.

ELECTRON DEFICIENT MOLECULES

Elements of Group III B, in particular boron and aluminium, show a great tendency to form 4 co-ordinate compounds, although the number of valency electrons is only three; when electron donor molecules are not present to form a 4 co-ordinate compound, the molecules of the Group III B elements may themselves dimerise. Thus aluminium chloride dimerises to Al_2Cl_6 forming two electron bonds by bridging through chlorine. With other molecules of similar formulae, e.g. B_2H_6 and $Al_2(CH_3)_6$, there are insufficient electrons to write satisfactory structures involving electron-pair bonds. Such molecules are termed electron deficient and their structures are explained in terms of *multicentre* orbitals.

HYDROGEN BONDING

There is a great deal of evidence to show that when hydrogen is bonded to the very electronegative elements — fluorine, oxygen and nitrogen — there is a strong tendency for molecular association to occur. Thus hydrogen fluoride in the gas phase has a molecular weight far greater than that corresponding to the monomer HF. Association takes place in this and other compounds by hydrogen bonding. Compared with the strength of a normal covalent bond, the hydrogen bond is quite weak.

This type of bond can be explained in electrostatic terms. The bonding between hydrogen and the electronegative element has a large dipole and the electrostatic attraction between the positive end of one dipole on one molecule and the negative end of another dipole on a second molecule gives rise to molecular association. It has been suggested that this occurs by interaction between the positive end of the dipole and a lone pair of electrons on the electronegative atom. This situation in ice is shown in *Figure 3.23*. In ice (I), a form stable at low temperatures, the structure is similar to wurtzite for each oxygen is tetrahedrally surrounded by four other oxygens at distances of 276 pm and the hydrogens are positioned between the oxygen atoms, two at a distance of 99 pm (the normal O—H bond distance) and two at a distance of 177 pm. This wurtzite type structure (*Figure 3.24*) is an 'open' one and accounts for the low density of ice. In melting, this structure starts to break down causing a decrease in the volume which continues until 277·2 K. At this

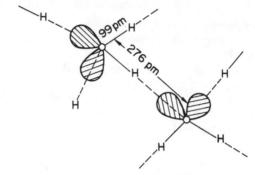

Figure 3.23. *Hydrogen bonding in ice (after Fowles, G. W. A. J. chem. Educ., 34 (1957) 187)*

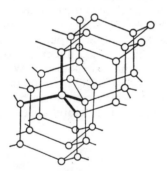

Figure 3.24. *Section of the crystal structure of ice (I) showing the tetra-hedral arrangement of the oxygen atoms*

temperature, the expansion arising from the increase in thermal vibrations starts to outweigh this decrease in volume and water shows a maximum density at 277·2 K.

For hydrogen fluoride in the vapour phase puckered rings of formula $(HF)_6$ exist between 254 and 295 K as well as simple HF molecules. The rings are hydrogen-bonded structures. In the solid state, infinite chains of hydrogen-bonded HF molecules are found.

Recent work on the absorption of the corresponding hydrogen halide by tetra-alkyl ammonium salts has shown that stoichiometric compounds are formed according to

$$R_4NX(s) + HX(g) = R_4NHX_2(s) \quad (X = Cl, Br \text{ and } I)$$

and spectroscopic evidence shows the existence of hydrogen bonding in their structures.

As a consequence of hydrogen bonding, certain other anomalous physical properties arise. The presence of hydrogen bonding is responsible for the deviations in melting points, boiling points, heats of vaporisation and heats of fusion of the simple covalent hydrides of the most electronegative elements. The values of these physical properties for hydrogen fluoride, water and ammonia are all unexpectedly high compared with those of the other elements

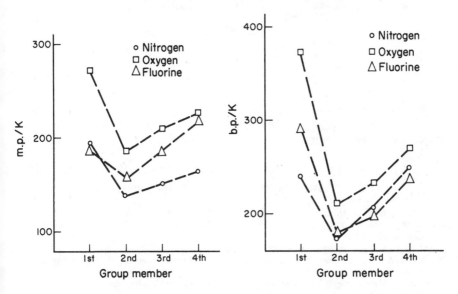

Figure 3.25. Melting points and boiling points of the hydrides of the nitrogen, oxygen and fluorine sub-groups

in the respective groups, due to molecular association. The variations in melting points and boiling points are shown in *Figure 3.25*.

The structures of certain organic molecules, e.g. *ortho*-hydroxybenzaldehyde and *ortho*-nitrophenol, are such that intramolecular rather than inter-molecular hydrogen bonding may take place. The *meta* and *para* isomers of these compounds do not possess structures where hydrogen bonds can be formed within the molecules but they do form intermolecular hydrogen bonds. The *ortho* isomers are therefore of higher volatility than either the *meta* or *para* isomers.

o-hydroxybenzaldehyde   *o*-nitrophenol

Hydrogen bonds are to be found in many crystalline compounds, for instance H_3BO_3 and $NaHCO_3$. In copper sulphate pentahydrate the structure is as shown below.

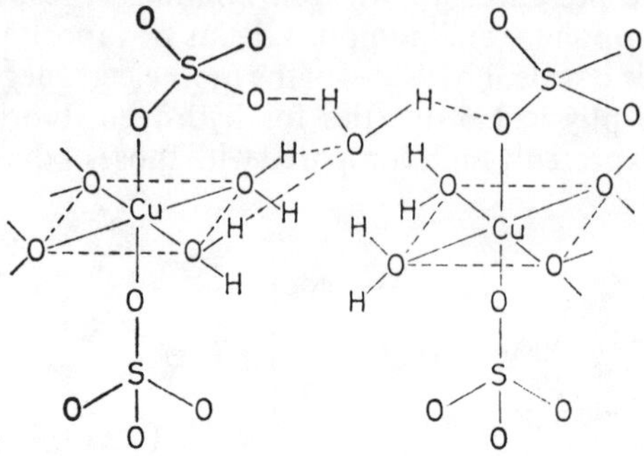

The copper atom is octahedrally surrounded by four water molecules and two sulphate ions. The fifth water molecule is hydrogen bonded to two sulphates and two waters.

SUGGESTED REFERENCES FOR FURTHER READING

BARNARD, A. K. *Theoretical Basis of Inorganic Chemistry*, McGraw-Hill, New York, 1965.

CARTMELL, E. and FOWLES, G. W. A. *Valency and Molecular Structure*, 2nd edn, Butterworths, London, 1961.

COULSON, C. A. *Valence*, 2nd edn, Clarendon Press, Oxford, 1961.

DASENT, W. E. *Inorganic Energetics*, Penguin Books Ltd., Harmondsworth, Middlesex, England, 1970.

PAULING, L. *The Nature of the Chemical Bond*, 3rd edn, Cornell University Press, New York, 1960.

SANDERSON, R. T. *Chemical Periodicity*, Rheinhold, New York, 1960.

SANDERSON, R. T. *Chemical Bonds and Bond Energy*, Academic Press, New York, 1971.

4 The structures of the elements and their compounds

THE CLASSIFICATION OF CRYSTALS

Crystals were first studied systematically by examining their external shape and optical properties. It was recognised by Stensen in 1669 that quartz crystals, however they originated, always had the same interfacial angles. The study of crystals by the contact and reflecting goniometers, instruments capable of measuring interfacial angles, amply confirmed this observation. One of the first laws of crystallography to be formulated stated the constancy of the angle between certain related faces of a crystal.

Crystals are classified into seven systems on the basis of their external shapes. The geometrical properties of a crystal are conveniently described in terms of its *crystallographic axes*. These are three, or sometimes four, lines meeting at a point. They are chosen so as to bear a definite relationship with characteristic features of the crystal, for example, the axes may coincide with or be parallel to the edges between principal faces. Where possible, the axes are chosen to be at right angles to each other. This is illustrated in *Figure 4.1*. Ox, Oy and Oz are the crystallographic axes and the interaxial angles, α, β and γ, are all equal to 90°. Next, a particular plane, say ABC, of the crystal is chosen as a standard or unit plane, in terms of which the crystal faces may be described. This plane must cut all three crystallographic axes, and is often a fourth face of the crystal. The intercepts where this plane cuts the axes, OA = a, OB = b, OC = c, are known as the crystal *parameters*. Their lengths are purely relative but their ratios are important and are used to describe the crystal. The choice of axes and unit plane is such that the measured intercepts of any other face of the crystal such as LMN on these axes may be written as la, mb, nc, where l, m and n are simple whole numbers (2, 2 and 3 respectively in *Figure 4.1*). This is a statement of another crystallographic relationship, the law of rational indices. This empirical law was discovered by Haüy about 1780 and the numbers are usually called Miller indices.

Every crystal possesses certain elements of symmetry. Externally, this appears as a repetition of the crystal faces and their angles. A crystal is said to have an *n*-fold axis of symmetry when a rotation of $360°/n$ about this axis produces an orientation which cannot be distinguished from the first. 1-, 2-, 3-, 4- and 6-fold axes are found for crystals and some of these are illustrated in *Figure 4.2(a)*.

Another possible symmetry operation is inversion. This may be visualised in terms of the normals to the crystal faces. If these are imagined to be a set of vectors with a common origin, then on inversion the vectors are reversed. Therefore, rotation inversion axes, denoted $\bar{1}, \bar{2}, \bar{3}, \bar{4}$ and $\bar{6}$, may also be used to describe the symmetry of a crystal. A 1-fold inversion axis, $\bar{1}$, means that in effect there is a centre of symmetry present and for each face on the crystal

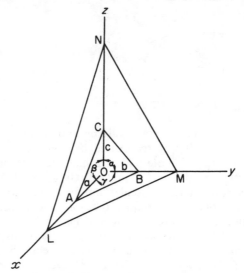

Figure 4.1. The crystallographic axes

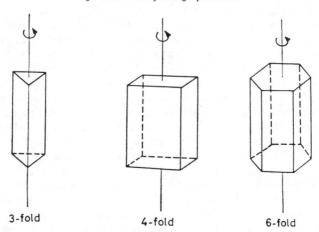

3-fold   4-fold   6-fold

Figure 4.2(a). Rotation axes of symmetry

there will be one parallel on the opposite side. The seven classes of crystal are summarised in *Table 4.1*. The rhombohedral class can be included as a special example of the hexagonal system so there are really only six distinctive crystal systems.

As an example of symmetry operations, those which may be performed on a cube may be considered:

(i) rotation about the principal axes (x, y and z in *Figure 4.2b*) by 360°; this is the identity operation giving an orientation identical with the original:
(ii) rotations about x, y or z by 180°; totalling three symmetry operations:
(iii) rotations about x, y or z by ±90°; six operations:
(iv) rotations about the six face diagonals by 180°; six operations:

Table 4.1. THE CLASSIFICATION OF CRYSTALS

Class	Lattice constants	Examples	Essential symmetry
Cubic	$\alpha=\beta=\gamma=90°$ $a=b=c$	NaCl, CsCl, CaF_2, ZnS, diamond, etc.	Four 3-fold axes
Tetragonal	$\alpha=\beta=\gamma=90°$ $a=b\neq c$	TiO_2, SnO_2, $NiSO_4$	One 4-fold or 4-fold inversion axis
Hexagonal	Three axes at 120°; a fourth at right angles $a_1=a_2=a_3\neq b$	HgS, graphite, AgI	One 6-fold or 6-fold inversion axis
Rhombo-hedral (Trigonal)	$\alpha=\beta=\gamma\neq90°$ $a=b=c$	$NaNO_3$, Al_2O_3, $CaCO_3$ (calcite)	One 3-fold or 3-fold inversion axis
Orthorhombic	$\alpha=\beta=\gamma=90°$ $a\neq b\neq c$	$PbCO_3$, $BaSO_4$, α-sulphur	Three mutually perpendicular 2-fold axis (either rotation or rotation inversion)
Monoclinic	$\alpha=\gamma=90°$ $\beta\neq90°$ $a\neq b\neq c$	$CaSO_4.2H_2O$, β-sulphur	One 2-fold or 2-fold inversion axis
Triclinic	$\alpha,\ \beta,\ \gamma\neq90°$ $a\neq b\neq c$	$CuSO_4.5H_2O$, $K_2Cr_2O_7$	One 1-fold or 1-fold inversion axis

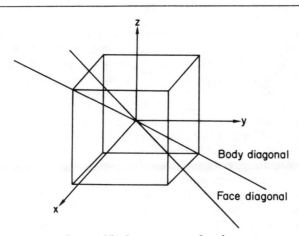

Figure 4.2(b). Symmetry axes of a cube

(v) rotations about the four body diagonals by ±120°; these diagonals are the four 3-fold axes giving the cube its essential symmetry; eight operations.

The number of operations is thus 24 and as each of these may be followed by inversion, the total number of possible symmetry operations on a cube is 48. To demonstrate the above rotational axes convincingly, a three-dimensional model should be used.

The external crystal form, closely limited by the rigid rules of symmetry, can give only a restricted expression to the structure of a crystal and many crystalline compounds are classified together although their chemical and physical properties may vary enormously.

Insight into the internal structure of crystals quickly followed the classical experiment of Friedrich and Knipping (1912) in which it was shown for the first time that x-rays were diffracted by a crystalline solid. Crystals act like a diffraction grating towards x-rays. The crystal is composed of series of parallel planes of atoms and the incident x-rays are reflected by these planes. The x-rays reflected by one plane penetrate the crystal to a different extent compared with those reflected by an adjacent plane. Thus there is a certain path difference between reflections from the two planes. When the path difference equals an integral number of wavelengths, the x-rays reinforce one another. This is the condition represented by the Bragg relationship:

$$n\lambda = 2d \sin \theta$$

where n is an integer, λ is the wavelength of the x-rays, θ is the angle of incidence and d is the perpendicular distance between the planes:

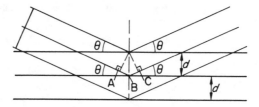

When a crystal is rotated in an x-ray beam, reflection or diffraction occurs whenever the Bragg condition is obeyed. The diffraction beam may be recorded on a photographic plate as a series of spots, each spot corresponding with a particular pattern gives the distribution of electron density in the crystal for the diffraction is caused primarily by the extra-nuclear electrons. Thus a picture of the structure can be built up consisting of sections of projections of the crystal lattice which show contours of equal electron density. The positions of the atoms are usually assumed to coincide with regions of maximum electron density.

In 1913, W. L. Bragg published the first crystal analysis, of sodium chloride, and many other structures were worked out within a short time. The crystal structures of very many solids have now been elucidated and it is quite clear that it is the internal pattern of atoms or ions which determines the external form and optical properties of a crystal. A crystal is made up of an infinite number of repeating groups. Each group constitutes a *unit cell*, this being the smallest portion of the crystal which possesses all the various kinds of symmetry which characterise the crystal as a whole.

A study of the elements and their compounds from the structural aspect is one important avenue of approach to the understanding of modern inorganic chemistry. A number of simple concepts have been applied to the correlation of structural data so that even the structures of complex, naturally occurring minerals such as the silicates are now well established.

THE STRUCTURE OF THE ELEMENTS

Atomic radii

From results of x-ray diffraction studies on solids and from electron diffraction or spectroscopic examination of gaseous molecules, a large amount of information is now available on the interatomic distances in the elements and their compounds.

Many elements, including the noble gases and most metals, are monoatomic. Discrete atoms, not chemically bound to one another, are present in the solid state. On the simple picture of atoms as spheres of a definite radius which pack together so that adjacent atoms touch, the measured interatomic

Table 4.2. THE ATOMIC RADII OF THE ELEMENTS (r/pm)
(EXCLUDING THE RARE EARTHS AND ACTINOIDS)

Group	IA	IIA	IIIB	IVB	VB	VIB	VIIB	0
								He 93
	Li 123	Be 89	B 80	C 77	N 74	O 74	F 72	Ne 112
	Na 157	Mg 136	Al 125	Si 117	P 110	S 104	Cl 99	Ar 154
	K 203	Ca 174	Ga 125	Ge 122	As 121	Se 117	Br 114	Kr 169
	Rb 216	Sr 191	In 150	Sn 141	Sb 141	Te 137	I 133	Xe 190
	Cs 235	Ba 198	Tl 155	Pb 154	Bi 152	Po 152		

Group	IIIA	IVA	VA	VIA	VIIA		VIII		IB	IIB
	Sc 144	Ti 132	V 122	Cr 117	Mn 117	Fe 116	Co 116	Ni 115	Cu 117	Zn 125
	Y 161	Zr 145	Nb 134	Mo 129	Tc —	Ru 124	Rh 125	Pd 128	Ag 134	Cd 141
	La 169	Hf 144	Ta 134	W 130	Re 128	Os 126	Ir 126	Pt 129	Au 134	Hg 144

Note: The atomic radius quoted for gold (134 pm) is less than the ionic radius given (Au^+ = 137 pm) in *Table 4.5.* For other elements, as we would expect, the positive ion radius is less than the radius of the atom. The example of gold serves to emphasise that although atomic radii are of significance relative to one another and similarly ionic radii are of significance relative to one another, the two sets of radii are not perfectly consistent with each other and the values assigned for one particular element cannot be regarded as absolute.

distance corresponds with twice the atomic radius. This concept of a fixed size for an atom conflicts with the wave-mechanical viewpoint which regards the electronic density in an atom as extending indefinitely from the nucleus and approaching zero asymptotically. Strictly speaking, it is impossible to define the size of an atom. Nevertheless, it is helpful in discussing the structure of the elements (and their covalent compounds) to assign a consistent set of radii which represent the relative sizes of the atoms. Atomic radii were assigned by Pauling from the interatomic distances observed in a number of elements. For many non-metals, where covalent bonds are present in the crystalline state, the atomic radii are determined from the measured interatomic distances. For a few non-metals such as nitrogen or oxygen, the atomic radius is calculated from interatomic distances in compounds containing a single covalent bond between non-metallic atoms, for example, the radius of the nitrogen atom is derived from the N—N distance in hydrazine, H_2N—NH_2. Values for the radii of metal atoms are half the interatomic distances in the solid state.

Values of atomic radii are given in *Table 4.2*. A study of these reveals a number of regularities:

(a) In a group, the atomic radius increases with the atomic number.
(b) This increase becomes more gradual as the atomic number becomes larger.
(c) In a period, there is a general decrease in radius from left to right as the atomic number increases.
(d) This decrease is less marked the higher the atomic number.

Considering the atom as a sphere of fixed size, the structure of the elements can be discussed from the viewpoint of the possible packing arrangements for spheres of equal size.

The close-packing of spheres

In a plane, spheres pack together in such a way that their centres are at the corners of equilateral triangles with each sphere touching six others (*Figure 4.3*). A second layer can be superimposed on the first so that each sphere is in contact with three spheres of the first layer. A and B represent the first and second layers respectively in the figure. A third similar layer of spheres may

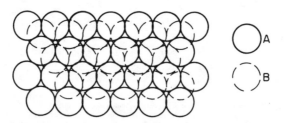

Figure 4.3. The close-packing of spheres, showing the superimposition of one layer upon another

now be added in one of two ways, either so that the spheres in it are centred immediately above those in the first or so that they are over holes in the first layer which are not occupied by second-layer spheres. In the first arrangement the structure is repeated after two layers and represents *hexagonal close-packing* (ABABABABABAB......). In the second, the structure is repeated after three layers and represents *cubic close-packing* (ABCABCABCABC......).

This is, in fact, an assemblage of spheres at the corners and face centres of a cubic unit cell (*Figure 4.4*).

In both these close-packed structures each sphere is in contact with 12 neighbours; in other words its co-ordination number is 12. The two systems are very similar energetically and some metals are dimorphic, crystallising in either the cubic (A 1) or the hexagonal (A 3) close-packed structure under different conditions. For perfect hexagonal packing the axial ratio $c/a = 1·633$.

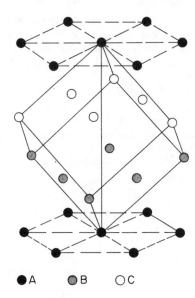

Figure 4.4. Face-centred cubic close-packing

● A   ◉ B   ○ C

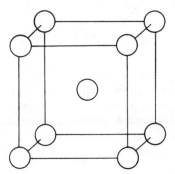

Figure 4.5. Body-centred cubic packing. This is the A2 arrangement for certain metals

However, in almost all cases of A 3 packing c/a is found to be slightly less than 1·633 and hence the six neighbours of one atom all in the same plane are slightly further from it than the three in the plane above and the three in the plane below. The axial ratio is observed to increase with temperature, thus tending to equalise the interatomic distances.

A third structure found with some metals is the *body-centred cubic arrangement* (A 2) in which the atoms are located at the corners and centre of a cube. The co-ordination number of the metal atom is therefore 8 (*Figure 4.5*).

BONDING IN THE CRYSTALLINE STATE

In the case of the elements, we can distinguish three types of bond in the solid state. These are the *metallic, covalent* and *residual* (or *van der Waals*) bonds.

The metallic bond

Most metals resemble one another chemically, for example in their tendency to form positive ions and to displace hydrogen from certain acids. They also have many physical and mechanical properties in common including good conduction of heat and electricity, paramagnetism, a high tensile strength, malleability and ductility. Two or more metals can be melted together to form alloys which differ profoundly in their properties from other chemical compounds. Particularly associated with alloys is the wide range possible in their chemical composition. These characteristic properties of metals and alloys are explicable in terms of the modern 'zone' theory of the metallic state.

A few properties of a metal were explained, at least qualitatively, by Lorentz in terms of a metal lattice consisting of positive ions with the valency

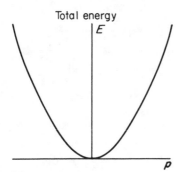

Figure 4.6. The variation of energy (E) with momentum (p) for a free electron

electrons free to move within the body of the metal. Sommerfeld extended the simple theory by the application of quantum mechanics and concluded that the mobile electrons were restricted to sets of permitted energy levels. Even at the absolute zero of temperature the energies of electrons are spread over a range of values, all the low-lying quantum states being fully occupied.

The energy of an electron, E, is related to its momentum, p, by the equations

$$E = \tfrac{1}{2}m_e u^2 = \frac{p^2}{2m_e}$$

where m_e is the mass, and u the velocity of the electron. *Figure 4.6* illustrates the parabolic relationship between E and p and *Figure 4.7* shows how the electron density (the number of electron pairs, $N(E)$, for any value of E) increases continuously with E.

Bloch pointed out that the electron should more properly be regarded as a wave moving in the periodic field of the lattice of positive ions. In these circumstances, he showed that the energy no longer varies continuously with its momentum and that there are energy discontinuities at certain critical momenta values. This is shown in *Figure 4.8*, where ΔE_1, ΔE_2 and ΔE_3 represent ranges of energy that are completely forbidden to the electron. Electrons are accordingly restricted to certain permitted bands or zones of energy. In some respects this situation is analogous to that in the isolated atom where each electron has a certain energy, corresponding with a particular quantum level, and can only move to other permitted energy levels by the gain or loss of definite quantities of energy.

This simple approach is concerned with the relation between the energy and momentum of an electron in one dimension. The extended theory deals

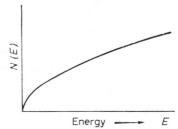

Figure 4.7. The variation of electron density, N(E), with E for a free electron

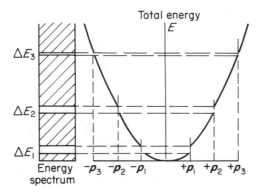

Figure 4.8. The variation of E with **p** for an electron moving in the periodic field of a lattice of positive ions

with the behaviour of electrons in a three-dimensional periodic field. It cannot be discussed here but it is interesting to note one of the conclusions of the complete theory—namely that electron densities in crystals can vary with E in one of two ways: in these, overlap may or may not be possible between permitted energy zones. *Figures 4.9(a)* and *4.9(b)* illustrate these alternatives.

The zone theory can account for many metallic properties. For example, electrical conductivity is associated either with an incompletely filled zone or two overlapping zones with vacant energy levels in the upper zone. In both cases electrons can take up energy and move to higher levels. If, however, one zone is completely filled and there is an appreciable gap between this and the next zone, which is completely vacant, the electrons cannot take up energy because there are no higher permitted energy levels within reach. As a result, the material is an insulator.

The characteristic behaviour of a semi-conductor can also be explained. This is a substance whose electrical conductivity is greatly increased by a rise in temperature or by the addition of certain impurities. Single crystals of extremely pure germanium or silicon are the most commonly used semi-conductors. Compounds such as gallium arsenide, GaAs, and indium anti-monide, InSb, are being increasingly employed because of their more favour-able electrical properties. One way in which semi-conductivity can arise is when the lower filled zone is separated from a completely empty higher zone by a small energy gap. Then it may be possible to excite thermally an electron

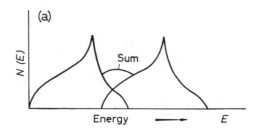

Figure 4.9(a). Overlapping between adjacent energy zones. Electrical conductivity is shown whenever the electrons available are insufficient to fill both zones completely

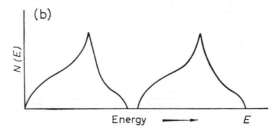

Figure 4.9(b). No overlap between adjacent zones. When the first zone is filled and the second completely empty, the substance is an insulator

from the first to the second zone. The material is an insulator at low tempera-tures but shows conductivity when heated. Alternatively, the presence of an impurity may introduce a new zone which lies between and possibly overlaps the two zones which are separated by a large gap in the pure material. The introduction of this extra zone may thus lead to conductivity in the impure material.

The covalent bond

In contrast to the close-packed structures of many elements, with their high co-ordination number of 12, some elements, especially the non-metals, have very low co-ordination numbers because of the formation of a small number of covalent bonds by each atom. The number of bonds formed depends on the number of valency electrons in the atom, that is, on the periodic group of the element, and is equal to $(8 - N)$ where N is the ordinal number of the group. The $(8 - N)$ rule is valid for many of the non-transitional elements in Groups VII to IV inclusive and the structures found are determined by the formation by each atom of one, two, three or four covalent bonds respectively.

The residual bond

This is the name given to the attraction which must exist between all atoms and molecules irrespective of whether or not they are joined by other bonds; it is often called the van der Waals bond. It is responsible for the cohesion between noble gas atoms in the liquid and solid states and for the attraction between molecules such as oxygen, nitrogen and hydrogen. Since these elements are solid only at very low temperatures and their heats of sublimation are very small (of the order of $2-8$ kJ mol^{-1}), this type of bond is very weak compared with the metallic or covalent bond. The strength of the metallic bond, for example, is indicated by the high temperature usually necessary to break down the solid structure and cause the metal to melt or vaporise.

Before the advent of quantum mechanics, no description of the residual bond could be given. However, quantum mechanics shows that it is necessary to consider the dispersion forces, as they are called, between atoms or molecules. In very simple terms, molecules are regarded as having quickly fluctuating dipoles so that when two molecules approach one another, the dipoles at that instant come into phase and there is always a resultant attractive force. There is no permanent dipole moment, however, because the moments continue to vary rapidly in phase. A similar process is visualised as leading to interatomic attraction.

THE CLASSIFICATION OF THE ELEMENTS ACCORDING TO THEIR STRUCTURE

Six main types of structure exist, and each will be considered in turn.

1. The noble gases

The elements are monatomic and the bonding is exclusively residual. Ne, Ar, Kr and Xe have the A 1 arrangement, whereas He has the A 3.

2. Hydrogen, nitrogen, oxygen and the halogens

These form stable diatomic molecules. In the case of oxygen, the paramagnetic O_2 molecules dimerise to O_4 molecules which rotate to give cubic symmetry and the A 1 structure for solid oxygen above 43 K. The structure below this temperature is uncertain. In H_2 (A 3) and N_2 (A 1 below and A 3 above 35 K), the molecules again achieve spherical symmetry by rotation. The halogens Cl_2, Br_2 and I_2 are orthorhombic with the diatomic molecules oriented as shown in *Figure 4.10*.

3. *The representative elements of groups VI, V and IV*

The structures of these exemplify the $(8 - N)$ rule. The units may be small molecules, chains or layers of atoms, or three-dimensional networks. The bonding within the units is covalent with fairly weak residual forces between them. For some elements, notably those in Groups VI and V, there is more than one way in which the atoms can combine covalently and satisfy the

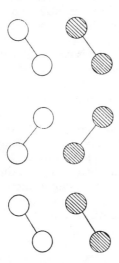

Figure 4.10. The arrangement of halogen molecules in the solid state. (Shaded circles represent atoms in a different plane from those indicated by open circles)

$(8 - N)$ rule. As a result, the element shows allotropy, that is, several different solid modifications are known.

Sulphur is well known for its many solid forms. The interrelation between the various allotropes is summarised in *Figure 4.11*. In several of these, the puckered S_8 ring (*Figure 4.12*) is the structural unit. Each sulphur atom forms two covalent bonds; the S—S distance is 204 pm and the bond angle is 107° 30′, close to the tetrahedral angle (109° 28′). The existence of a puckered S_6 ring in Engel's sulphur has been established. Zig-zag chains of high molecular weight are also formed. In these, the sulphur atoms, except the two terminal ones, again form two covalent bonds. These polymers are found, for instance, in fibrous sulphur and their formation is believed to account for the remarkable viscosity increase and colour change observed when liquid sulphur is heated over the range 438–473 K. At 393 K, sulphur is a yellow, mobile liquid. Above 433 K, it becomes brown and increasingly viscous up to 473 K. Beyond this temperature, the viscosity decreases normally to the boiling point at 717·7 K. The viscosity increase is attributed to the formation of polymeric chains of sulphur atoms up to 10^6 atoms in length. Nacreous (γ) sulphur, formed by the slow oxidation of an alcoholic solution of ammonium polysulphide, consists of S_8 rings and is very similar in its properties to rhombic sulphur. Finally, it is worth noting that the paramagnetic S_2 molecule, apparently analogous to O_2, is present in sulphur vapour and in the purple

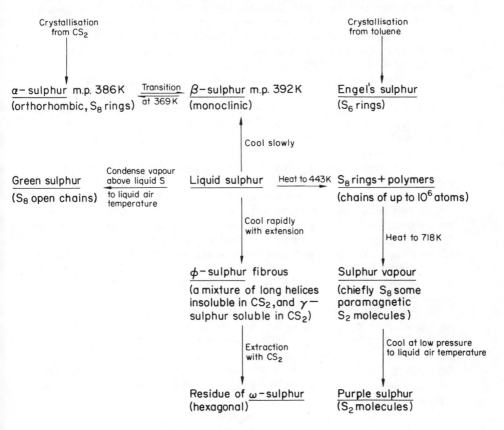

Figure 4.11. The allotropes of sulphur

Figure 4.12. S_8 puckered ring

solid allotrope. All four forms, Engel's, nacreous, fibrous and purple, are metastable and change to rhombic sulphur. In the case of Engel's and fibrous sulphur, the changes are slow because movement of atoms and the breakage and formation of bonds are involved.

Two non-metallic crystalline forms of selenium are formed. These are the red monoclinic allotropes which are usually designated α and β. They are obtained from solutions of the element in carbon disulphide. Slow evaporation produces Se_α: rapid removal of solvent gives Se_β. Puckered Se_8 rings are present in both allotropes and the difference between them is probably associated with the packing of the rings in the solid state. Metallic selenium is an opaque grey solid prepared by the crystallisation of molten selenium. The small electrical conductivity of this allotrope is enhanced greatly by exposure to light, hence the use of this element in photocells. It is of the hexagonal class and contains zig-zag chains of great length packed together with their axes parallel. Metallic tellurium is similar to metallic selenium and amorphous varieties of both elements are also known.

The different allotropes of phosphorus are given in *Figure 4.13*. Molecular-weight determinations on solutions of white phosphorus in organic solvents have shown the presence of the P_4 molecule. Density measurements on phosphorus vapour show that this molecule is also present there. In the P_4 molecule (*Figure 4.14*) each phosphorus atom is located at the corner of a tetrahedron and is bound covalently to the three other atoms. Red phosphorus is believed to have a polymeric structure in which chains of phosphorus atoms are joined by random cross-linking (*Figure 4.15*). Each atom has three nearest neighbours at 229 pm, the next being at a distance of 348 pm. The scarlet and violet modifications are also believed to be polymeric. Black phosphorus has a metallic lustre and shows some electrical conductivity. It has a characteristic orthorhombic 'double-layer' structure (*Figure 4.16*) wherein each atom has two neighbours in the same layer at a distance of 217 pm and a third in the next layer at 220 pm. Although difficult to prepare, black phosphorus appears to be thermodynamically the most stable allotrope of this element.

The most stable allotrope of arsenic, antimony and bismuth is the metallic modification in each case and is in fact the only form known for bismuth. The metallic allotrope has a layer structure (*Figure 4.17*) with each atom having three nearest neighbours and three more somewhat further away. The difference between the distances of these two sets of neighbours becomes relatively smaller as the metallic character of the element increases from arsenic to bismuth:

	Nearest atoms	*Next nearest atoms*
	(r/pm)	(r/pm)
Arsenic	251	315
Antimony	287	337
Bismuth	310	347

These three elements in their metallic allotropes are evidently intermediate between the non-metals of low co-ordination number, and the metallic structures of high co-ordination number. The yellow non-metallic allotropes of arsenic and antimony are metastable solids which have cubic symmetry. Yellow arsenic consists of tetrahedral As_4 units and probably yellow antimony is similar in structure. These readily change over to the more stable metallic modifications. Black amorphous forms of both arsenic and antimony are also known.

Two allotropes are known for carbon. Here the allotropy arises because the

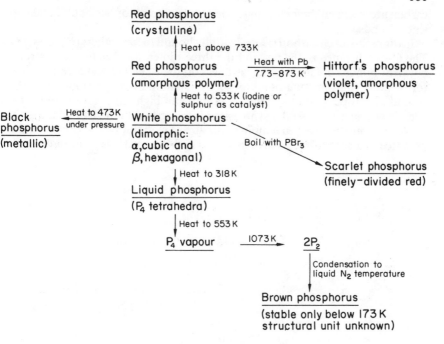

Red phosphorus
(crystalline)

↑ Heat above 733K

Red phosphorus   Heat with Pb 773–873 K →   Hittorf's phosphorus
(amorphous polymer)   (violet, amorphous polymer)

↑ Heat to 533 K (iodine or sulphur as catalyst)

Black phosphorus ← Heat to 473K under pressure   White phosphorus
(metallic)   (dimorphic: α,cubic and β, hexagonal)

Boil with PBr_3 → Scarlet phosphorus (finely-divided red)

↓ Heat to 318 K

Liquid phosphorus
(P_4 tetrahedra)

↓ Heat to 553 K

P_4 vapour   1073 K → $2P_2$

↓ Condensation to liquid N_2 temperature

Brown phosphorus
(stable only below 173 K structural unit unknown)

Figure 4.13. The allotropes of phosphorus

Figure 4.14 P_4 tetrahedron

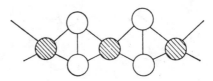

Figure 4.15. Probable chain structure (with cross-linking) of red phosphorus

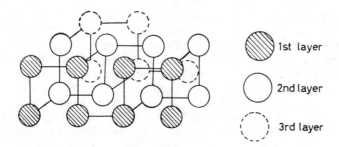

1st layer

2nd layer

3rd layer

Figure 4.16. Layer structure of black phosphorus

carbon atom can either form four single covalent bonds or one double and two single bonds.

In diamond, each carbon is tetrahedrally bound to four others (*Figure 4.18*). The C—C distance is 154 pm and the three-dimensional network of strong covalent bonds confers extreme hardness on the structure and is responsible for the very high melting point (3840 K). Diamond is more stable than the second allotrope, graphite, under conditions of high temperature and pressure. It was first successfully synthesised from carbon in the United States in 1954. Carbon, dissolved in a molten mixture of iron sulphide and iron, in the presence of a second metal such as Cr, Ni or Mn as catalyst, was induced to

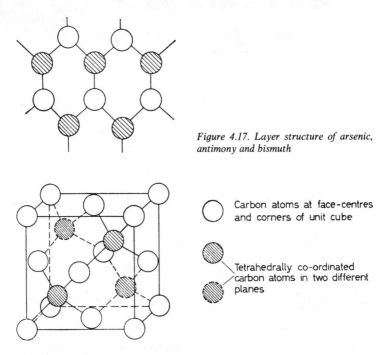

Figure 4.17. Layer structure of arsenic, antimony and bismuth

○ Carbon atoms at face-centres and corners of unit cube

◐ Tetrahedrally co-ordinated carbon atoms in two different planes

Figure 4.18. The structure of diamond

crystallise as diamonds by the application of high pressure at a high temperature. Diamonds made in this way are individually small, usually coloured and so find their main uses in industrial cutting and abrading machinery.

Graphite possesses a layer lattice (*Figure 4.19*) in which each atom has three nearest neighbours in the same plane at a distance of 145 pm. Each layer is composed of hexagons of carbon atoms. Each atom forms three coplanar σ-bonds and the remaining valency electron enters a non-localised π-orbital. The presence of 'mobile' electrons makes the structure electrically conducting. The inter-layer distance is large (335 pm) and so the binding forces between the layers are weak. The attribution of the good lubricating properties of graphite to the weakness of these forces appears, however, to be an over-simplification, for it has recently been shown that the presence of substances such as water, ammonia, acetone, benzene, etc., is essential for good lubrication. The cause of the lubricating action of graphite is not fully understood at present.

A number of derivatives of graphite, named *intercalation compounds*, are known in which foreign atoms are inserted between the carbon layers. For example, a number of homogeneous compounds of composition between $CF_{0.68}$ and $CF_{0.995}$ have been prepared by direct reaction between fluorine and graphite between 693 and 713 K. The fluorine atoms in these, situated between the graphite layers, are bound covalently to the carbon atoms. In pure graphite, the planar arrangement of atoms may be attributed to sp^2 hybridisation. In the carbon–fluorine intercalation compounds, the majority of carbon atoms form four single bonds. The carbon atoms are therefore to be regarded as sp^3 hybridised; this accords well with the experimental observation that in these compounds the carbon layers are puckered and

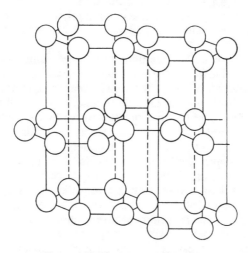

Figure 4.19. The structure of graphite

are no longer planar. A number of other intercalation compounds are known, for example, the compounds of formulae C_8M and $C_{24}M$ which are prepared by direct reaction between alkali metal vapours and graphite.

The common form of graphite is hexagonal. Metastable rhombohedral graphite has been produced by mechanical deformation of the hexagonal allotrope. Here the packing is of the type ABCABCABCABC......, where every fourth layer is vertically above the first. In hexagonal graphite, the packing of the layers is ABABABABAB...... (*Figure 4.19*). Other forms of carbon such as charcoal and soot appear to be microcrystalline with the hexagonal graphite structure.

Carbon has a number of characteristic properties, including low vapour pressure up to very high temperatures and resistance to chemical attack, which make it very useful in many industrial applications. The specialised uses to which carbon is put are closely associated with the particular physical form of the element. Thus graphite, because of its electrical conductance, is widely used as electrodes in arc furnaces and as an anode material in aqueous and fused salt electrolysis. Pyrolytic carbon and graphite, produced by the decomposition on hot graphite supports of vapourised carbon compounds such as methane, ethane, acetylene and benzene, is used for coating fuel particles in nuclear reactors.

In the last decade, much interest has centred around the preparation and development of carbon fibres. These are made by the thermal degradation of polymeric materials like polyacrylonitrile, polyvinylacetate and polyvinyl chloride at temperatures below their melting points. In this way the carbon skeleton of the polymeric material is changed into an essentially two-dimensional graphite-type structure. The carbonisation process is carried out in several stages and it is necessary to subject the material to tension in the early stages to produce an ordered structure which has the required physical characteristics. Carbon fibres produced in this way have a high tensile strength and a high modulus, properties which are combined with flexibility and low density. There are a number of applications, the most well known being as a reinforcing agent for plastics used as components in jet engines.

In the case of silicon and germanium, both these elements crystallise in the diamond structure. However, the very much reduced melting points, 1687 K and 1231 K respectively, reflect the much weaker covalent bonds present compared with those in carbon (diamond allotrope). Both silicon and germanium can be converted to other structures by the application of high pressures. For example, the diamond structure of silicon collapses to a metallic form, similar to that of white tin, under a pressure of 15 GN m^{-2} or above. When decompressed, this metallic form changes to a new, semi-conducting modification which is some 10 per cent more dense than the diamond form of silicon. On heating at atmospheric pressure, the new form changes into a hexagonal form of silicon which has about the same density as the diamond phase. Germanium behaves similarly under high pressures. In this case, decompression gives a tetragonal, semi-conducting form.

The use of very high pressures in laboratories is becoming more common and many new structures are being made. As a consequence of the closer approach of atoms at high pressure, it is not unusual for insulators to become semi-conductors or semi-conductors to change to metals. For example, silicon under a pressure of 20 GN m^{-2} behaves like a metal and has an electrical conductivity similar to that of aluminium.

Tin has two well-known allotropes. Grey tin is stable below 286·3 K and has the diamond structure. The four nearest neighbours to an atom are at a distance of 280 pm. White tin, the metallic form stable above 286·4 K, has a tetragonal lattice with each tin atom surrounded by six neighbours centred at the corners of a distorted octahedron, four at 301·6 pm and two at 317·5 pm. Lead has a typically metallic structure, A 1.

4. The transition metals

Their structures are summarised in *Table 4.3*. This includes a number of polymorphic forms which are observed at elevated temperatures. The transition metals which are noted for their malleability and ductility have the A 1 structure. This class includes Ni, Cu, Ag, Au and Pt. Those, like Ti, V, Mo and W, which crystallise in the A 2 or A 3 structure, are less malleable and more brittle. There is an important difference between the two forms of close-packed structures which is responsible for the marked contrast in the physical properties shown. In hexagonal close-packing, A 3, there is only one direction in which the atoms are arranged in close-packed sheets whereas in the cubic close-packed structure, A 1, the atoms are close-packed in no less than four directions, these being normal to the four body-diagonals of a cube (one of these diagonals is shown in *Figure 4.4*). The ductility and malleability of metals depend largely on the ease with which adjacent planes of atoms can slip or glide over one another. Gliding occurs most readily between planes of close-packed atoms; there are more of these planes in the cubic

Table 4.3. THE CRYSTAL STRUCTURES OF THE TRANSITION METALS

Group	III A		IV A			V A	
Sc	A 3		Ti	A 3		V	A 2
	A 1			A 2 > 1170K			
Y	A 3		Zr	A 3		Nb	A 2
				A 2 > 1120K			
La	A 3		Hf	A 3		Ta	A 2
	A 1						

Group	VI A		VII A		VIII	
Cr	A 2		Mn	complex	Fe	A 2
	A 1 > 2120K					A 1 (1190–1670K)
Mo	A 2		Tc	A 3	Ru	A 3
W	A 2		Re	A 3	Os	A 3

Group	VIII		VIII		I B	
Co	A 1/A 3		Ni	A 1	Cu	A 1
	A 3 > 773K					
Rh	A 1		Pd	A 1	Ag	A 1
Ir	A 1		Pt	A 1	Au	A 1

close-packed than in the hexagonal close-packed structure and so the former is the more malleable and ductile.

Manganese has three crystalline forms; two of these are complex with 58 and 20 atoms respectively in the unit cell; the third modification has a distorted A 3 structure. Cobalt is of interest because below 773 K it has a structure consisting partly of A 1 packing and partly of A 3 packing, with a random distribution of these two types of packing throughout the structure.

5. The metals of groups II B and III B, boron and aluminium

The structures of zinc and cadmium are sometimes quoted as extreme examples of the $(8 - N)$ rule. They both have distorted hexagonal structures in which each atom has six coplanar nearest neighbours and a further six, three above and three below the plane. In mercury, which has a rhombohedral lattice, the converse is the case for the six coplanar neighbours are further away than three atoms above and three below the plane.

At least three crystalline forms of boron are known. Two are rhombohedral, containing 12 and 108 atoms respectively in the unit cell. The third, containing 50 atoms in the unit cell, is tetragonal. Their structures are very complex and will not be considered here. Aluminium has the A 1 structure. Gallium, its neighbour in the same group, has a unique arrangement for a metal in that each atom has one nearest neighbour, at 244 pm, and six more at distances between 270 and 280 pm. Indium has a tetragonal structure which is a distortion of cubic close-packing. Four nearest neighbours to each atom are located at a distance of 325 pm and a further eight are at 337 pm. Thallium is dimorphic, having the A 3 structure at ordinary temperatures and the A 2 structure above 535 K.

6.   The alkali and alkaline-earth metals

Lithium, sodium, potassium, rubidium, caesium and barium all have the A 2 structure. The softness and low density of these metals compared with those having the A 1 or A 3 structures is closely related to the lower co-ordination number of 8 compared with 12. Beryllium and magnesium have the A 3 arrangement. Calcium and strontium are trimorphic and can show any of the three structures, depending on the temperature.

THE STRUCTURE OF INORGANIC COMPOUNDS

Bonding

In many inorganic compounds, the bonding is either predominantly ionic or chiefly covalent and characteristic properties are associated with each type of crystal. As in the case of the elements, residual forces will always be present.

An ionic compound may be visualised as an assemblage of positive and negative ions which extends indefinitely in three dimensions. Its chemical formula indicates the relative numbers of the different ions present in the crystal as a whole. For example, in the sodium chloride crystal, six ions of one charge are octahedrally situated around one ion of the opposite charge. The chemical formula, NaCl, indicates that equal numbers of Na^+ and Cl^- ions are present but there exists no unit of Na^+Cl^- in which one chloride ion is more closely associated with one sodium ion than with any others. The strong coulombic attractions between the oppositely charged ions result in lattice energies of appreciable magnitudes. Accordingly, ionic structures usually have a characteristic hardness and are of high melting point and boiling point.

Other typical properties include very poor electrical conductivity in the solid state which generally improves near the melting point and becomes good in the fused state. The increase in electrical conductivity with temperature, particularly marked at the melting point, is due to the breaking up of the lattice. The ions become mobile and can transport an electric current. The cause of electrical conductivity is therefore different here from that shown by metals which is due to the presence of mobile electrons. The conductivity shown in the fused state is exploited in the extraction of metals from their molten ionic compounds by electrolysis. It should be noted that electrical conductivity in the fused state does not necessarily mean that ions are present in the solid state. For instance, some of the interhalogens can be used as ionising solvents because of the formation of ions in the liquid state from the discrete molecules which exist in the solid state. Ionic compounds are generally insoluble in non-polar solvents such as benzene, carbon tetra-chloride, etc., but are frequently very soluble in polar solvents like water, ammonia, etc.

Covalent compounds are of two main types; firstly the molecular compounds in which the molecules are held together in the solid state by comparatively weak forces: and, secondly, the substances wherein covalent bonding extends throughout the crystal lattice.

The first type, exemplified by the halides of non-metals and of some metals in their high valency states, have soft crystals and low melting and boiling points. Strong covalent bonds are present within the molecule only. The solubility of this type of compound is usually much greater in non-polar than in polar solvents unless, of course, chemical reaction with the solvent (solvolysis) occurs. Another property typical of many molecular compounds is the absence of electrical conductivity in the fused state, provided that ionisation does not occur on melting.

Compounds of the second class, typified by silicon carbide (SiC), silica (SiO$_2$), etc., are hard, infusible and insoluble substances. The three-dimensional network of strong covalent bonds present in such substances requires a great deal of energy to destroy it.

A purely ionic and a completely covalent bond represent the two extremes of structure which are rarely observed in practice and the vast majority of compounds contain bonds that are intermediate in character. For example, layer lattices are observed when polarisation of the anion by the cation is considerable and they can be regarded as structurally intermediate between a three-dimensional array of ions and a molecular lattice.

Ionic crystals

In many compounds the bonding between the atoms is largely electrostatic and their structures in the solid state can be described in terms of an ionic model in which the ions are treated as charged compressible spheres. This approach provides a valuable starting point for the structural study of inorganic compounds.

Univalent and crystal radii

In ionic crystals the ions may be regarded as in contact with one another and so the measured internuclear distance corresponds with the sum of the radii of the cation and anion. As in the case of atomic radii, it is advantageous to have some idea of the relative sizes of ions and several attempts have been made to arrive at a suitable set of ionic radii.

The set most widely used is that due to Pauling. He took the experimental values of the internuclear distance in a number of crystals – namely NaF, KCl, RbBr and CsI – and deduced from them a set of ionic radii which closely reproduce the observed internuclear distances in many other compounds.

The four alkali halides have the following equilibrium internuclear distances: NaF, 231 pm; KCl, 314 pm; RbBr, 343 pm; CsI, 385 pm. The first three have the same crystal structure, namely that of sodium chloride, whereas CsI has that of caesium chloride. In each case, the compound is composed of a pair of isoelectronic ions.

The size of an ion is determined by the force of attraction between the nucleus and the outermost electrons. Pauling assumed this to be inversely proportional to the effective nuclear charge, defined as the actual nuclear charge, Ze, from which has been subtracted a screening correction, Se, to allow for the effect of electrons in the intervening shells. The screening constants, S, are variously obtained by theoretical calculation and from molecular refraction and x-ray experimental data.

The ionic radius, r, is given by

$$r = \frac{C_n}{Z-S}$$

where C_n is a constant for a particular isoelectronic sequence. For ions having the neon structure, Pauling assigned the value of 4·52 to S: the effective nuclear charges for Na^+ ($Z = 11$) and F^- ($Z = 9$) are then 6·48 e and 4·48 e respectively.

The observed Na—F distance of 231 pm is split up in the inverse ratio of these charges:

$$\frac{r_{Na^+}}{r_{F^-}} = \frac{4\cdot48}{6\cdot48}$$

and so the values $r_{Na^+} = 95$ pm and $r_{F^-} = 136$ pm are obtained.

Similarly, the radii of the other ions, $r_{K^+} = 133$ pm, $r_{Cl^-} = 181$ pm, $r_{Rb^+} = 148$ pm, $r_{Br^-} = 195$ pm, $r_{Cs^+} = 169$ pm and $r_{I^-} = 216$ pm, are calculated.

The calculations may be extended to ions carrying multiple charges which also have the noble-gas configurations (such as Ca^{2+}, S^{2-}, etc.) by using the above equation for r and the values for the constants, C_n, given by the radii of the alkali metal and halide ions. In this way the *univalent radii* of these multi-charged ions are estimated and these represent correctly their relative sizes compared with the alkali and halide ions. They cannot, however, be added to give the observed internuclear distances in crystals containing multivalent ions, since univalent radii, because of the way in which they are derived, are 'the radii the multivalent ions would possess if they were to retain their electronic distribution but to enter into coulombic interaction as if they were univalent' (Pauling).

The *crystal radii* of multivalent ions, so-called because the sum of two crystal radii is equal to the actual internuclear distance in a crystal containing the ions, are calculated from the univalent radii as follows:

The total potential energy, U, of an ion pair in a crystal is given by

$$U = \frac{-A_a z^2 e^2}{4\pi\varepsilon_0 r} + \frac{Be^2}{4\pi\varepsilon_0 r^n}$$

where the ions have charges of $+ze$ and $-ze$ respectively, A_a is the Madelung constant and B is the Born coefficient*.

Hence
$$\frac{dU}{dr} = \frac{A_a z^2 e^2}{4\pi\varepsilon_0 r^2} - \frac{nBe^2}{4\pi\varepsilon_0 r^{n-1}}$$

At the equilibrium distance r_c, the forces of attraction and repulsion are equal, $dU/dr = 0$ and $r_c = (nB/A_a z^2)^{1/(n-1)}$.

If the coulombic forces corresponded with those of univalent ions, that is $z = 1$, with an unchanged value for B, the equilibrium internuclear distance r_1 would be

$$r_1 = \left(\frac{nB}{A_a}\right)^{1/(n-1)}$$

The univalent and crystal radii are therefore related to one another by

$$r_c = r_1 z^{-2/(n-1)}$$

Approximate values for the Born exponent, n, can be calculated from the results of experiments on the compressibility of crystals and so r_c values can be determined.

Using these expressions, Pauling calculated sets of univalent and crystal radii. His values are given in *Table 4.4*. The crystal radii refer to ions with a

Table 4.4. CRYSTAL RADII ACCORDING TO PAULING (r_{ion}/pm)

I		II		III		IV		V		VI		VII	
Li^+	60	Be^{2+}	31	B^{3+}	20			N^{3-}	171	O^{2-}	140	F^-	136
Na^+	95	Mg^{2+}	65	Al^{3+}	50	Si^{4+}	41	P^{3-}	212	S^{2-}	184	Cl^-	181
K^+	133	Ca^{2+}	99	Sc^{3+}	81	Ti^{4+}	68						
Rb^+	148	Sr^{2+}	113	Y^{3+}	93	Zr^{4+}	80						
Cs^+	169	Ba^{2+}	135	La^{3+}	115	Hf^{4+}	81						
Cu^+	96	Zn^{2+}	74	Ga^{3+}	62	Ge^{4+}	53			Se^{2-}	198	Br^-	195
Ag^+	126	Cd^{2+}	97	In^{3+}	81	Sn^{4+}	71			Te^{2-}	221	I^-	216
Au^+	137	Hg^{2+}	110	Tl^{3+}	95	Pb^{4+}	84						

co-ordination number of 6 in an octahedral environment. Pauling recognised that the radius of an ion varies with its co-ordination number and that corrections must be applied to allow calculation of internuclear distances in crystals where the co-ordination number of the cation differs from 6. Another correction is necessary to allow for the effect of cation–cation and

---

* B is a term expressing the repulsion between the ions and it varies with co-ordination number. Consequently, the crystal radius also changes. The values of crystal radii given in *Table 4.4* are calculated for a co-ordination number of 6 and must be decreased by about 5 per cent if tetrahedral and not octahedral radii are required.

anion–anion repulsions on the magnitude of ionic radii. Other authors, notably Goldschmidt, Zachariasen and Ahrens, have produced sets of ionic radii, all showing some differences from one another.

The question therefore arises as to which is the best set of ionic radii of those which have been calculated. In this context, 'best' may reasonably be interpreted to mean that set which most closely reproduces the observed internuclear distances in crystals. Besides the co-ordination number of the ion and the effect of strong repulsion forces between ions of the same sign, internuclear distances and therefore ionic radii depend on the degree of covalency in the bonding and, in the case of transition metal ions, on the distribution of electron spin in the cation. All these factors have been taken into account by Shannon and Prewitt in their derivation of an empirical set of 'effective' ionic radii, based on approximately 100 measured internuclear distances in oxides and fluorides. Their values for 6 co-ordination are given in *Table 4.5*. The radius of the 6 co-ordinated oxide ion is assumed to be 140 pm and that of the fluoride ion is taken as 133 pm.

The ionic radii show a number of interesting trends:

(a) In moving from left to right across a period, a sequence of isoelectronic cations or anions shows a marked decrease in radius for example from Na^+ to Si^{4+} and from N^{3-} to F^-. As the atomic number increases the increasing nuclear charge acts on the same number of electrons and so the radius of the ion decreases.

(b) In moving down a group of non-transitional elements, the ionic radius increases both for cations and anions. When passing from one element to the next of higher atomic number an extra shell of electrons is interposed between the nucleus and the outermost valency electrons. The nuclear charge also increases in the same direction but evidently the effect of this is more than offset by the additional screening effect of the extra electrons.

The combination of (a) and (b) results in a close similarity between the ionic radii of elements having a diagonal relationship to one another in the Periodic Table. This is most apparent for the electropositive metals of low atomic number as shown by the chemical resemblances between lithium and magnesium and between beryllium and aluminium.

(c) For the sequence of transition metals from titanium to nickel, an increase in atomic number of one unit is generally accompanied by a small decrease in ionic radius. The decrease in radius is rather variable in magnitude on passing from one metal to the next and for those ions which can show alternative spin configurations, low- or high-spin, there is a difference of some 10–15 per cent between the two radii. The low-spin ion has the smaller size and the reason for the difference may be appreciated by reference to the section on crystal field theory in Chapter 6.

(d) With the elements from cerium to lutetium, the 4f sub-shell is filling up and there is a concomitant decrease in ionic radius known as the *lanthanide contraction*. The decrease from one element to the next is very small and as a result these metals are related very closely to one another in their properties. The lanthanide contraction has important consequences outside Group III. For example, the ionic radius of Hf^{4+} is very near to that of Zr^{4+}, the effect of the lanthanide contraction being to counterbalance almost exactly the increase in ionic radius with atomic number normally found in a group.

Table 4.5. 'EFFECTIVE' RADII (r/pm) OF IONS IN 6 COORDINATION (DATA FROM
SHANNON, R. D., AND PREWITT, C. T., *Acta Cryst.*, B 25 (1969) 925)
LS = LOW-SPIN; HS = HIGH-SPIN

Li^+	74	Be^{2+}	35	B^{3+}	23			N^{3-}	171	O^{2-}	140	F^-	133
Na^+	102	Mg^{2+}	72	Al^{3+}	53			P^{3-}	212	S^{2-}	184	Cl^-	181
K^+	138	Ca^{2+}	100	Sc^{3+}	73	Si^{4+}	26	(Pauling		Se^{2-}	198	Br^-	196
								radii for					
								X^{3-})					
Rb^+	149	Sr^{2+}	116	Y^{3+}	89	Ti^{4+}	60			Te^{2-}	221	I^-	220
Cs^+	188	Ba^{2+}	136	La^{3+}	106	Zr^{4+}	72						
Cu^+	96	Zn^{2+}	74	Ga^{3+}	62	Hf^{4+}	71						
Ag^+	115	Cd^{2+}	95	In^{3+}	79	Ge^{4+}	54						
Au^+	137	Hg^{2+}	102	Tl^{3+}	88	Sn^{4+}	69						
						Pb^{4+}	77						

	Ti^{2+}	86		Ti^{3+}	67		
	V^{2+}	79		V^{3+}	64		
	Cr^{2+}	73	LS	Cr^{3+}	62		
		82	HS	Mn^{3+}	58	LS	
	Mn^{2+}	67	LS		65	HS	
		82	HS	Fe^{3+}	55	LS	
	Fe^{2+}	61	LS		64	HS	
		77	HS	Co^{3+}	52	LS	
	Co^{2+}	65	LS		61	HS	
		73	HS	Ni^{3+}	56	LS	
	Ni^{2+}	70			60	HS	

		La^{3+}	106·1			Ac^{3+}	118	
		Ce^{3+}	103·4	Ce^{4+}	80	Pa^{3+}	113	Pa^{4+} 98
		Pr^{3+}	101·3			U^{3+}	106	U^{4+} 97
		Nd^{3+}	99·5			Np^{3+}	104	Np^{4+} 95
		Pm^{3+}	97·9			Pu^{3+}	100	Pu^{4+} 93
		Sm^{3+}	96·4			Am^{3+}	101	Am^{4+} 92
Eu^{2+}	117	Eu^{3+}	95·0			Cm^{3+}	98	
		Gd^{3+}	93·8					
		Tb^{3+}	92·3	Tb^{4+}	76			
		Dy^{3+}	90·8					
		Ho^{3+}	89·4					
		Er^{3+}	88·1					
		Tm^{3+}	86·9					
		Yb^{3+}	85·8					
		Lu^{3+}	84·8					

This accounts for the great chemical similarity between hafnium and zirconium. The effect of the lanthanide contraction extends to some of the later groups of transition metals so that niobium and tantalum resemble one another very closely and molybdenum and tungsten have many properties in common.

An analogous *actinide contraction* is shown by the elements from actinium onward and again results in marked similarities between them.

(e) The effective ionic radius increases with the co-ordination number of the ion. For example, data for the Ca^{2+} ion in different environments are as follows:

Radius (in pm)	100	107	112	118	128	135
Co-ordination number	6	7	8	9	10	12

and for the oxide ion:

Radius (in pm)		135	136	138	140	142
Co-ordination number		2	3	4	6	8

Modern x-ray diffraction techniques have made it possible to determine the actual distribution of valency electrons in an ionic crystal such as sodium chloride. Experimentally, the variation of electron density with distance along an internuclear axis between sodium and chloride ions is measured and it is found that this density drops to a low minimum value at some point on this axis. This observation confirms that the bonding within the crystal is essentially ionic and it provides a direct way of assigning radii to the constituent ions. Thus, from the electron distribution in sodium chloride, the radii derived are 117 pm for Na^+ and 164 pm for Cl^-. Similarly the following values have been found for the other alkali metal and halide ions:

Li^+	93	F^-	116
K^+	149	Br^-	180
Rb^+	164	I^-	204
Cs^+	183		

With the exception of LiF, these figures reproduce to within 1 per cent the observed internuclear distances in alkali metal halides having the sodium chloride structure. The internuclear distance observed in LiF is 201 pm. The 'map' of the electron density distribution in this crystal indicates that the ions deviate from spherical symmetry to a greater extent than the ions in sodium chloride and that overlap of the electron clouds is greater. Nevertheless the bonding can be regarded as almost completely ionic and it is possible to calculate the radii – $Li^+ = 91$ pm and $F^- = 109$ pm – from the position of minimum electron density. The radii found in this way are significantly different from those given by Pauling. They relate directly to the situation in a real crystal and do not, of course, depend on the kind of assumptions which underlie the derivation of Pauling's values.

Measurements taken from the electron density map of another crystal, magnesium oxide, give an oxide ionic radius of 109 pm. This is considerably smaller than the value of 140 pm on which Pauling based his values for empirical crystal radii and it leads to larger values for the radii of the alkaline–earth metal ions, namely: $Mg^{2+} = 102$; $Ca^{2+} = 131$; $Sr^{2+} = 148$ and $Ba^{2+} = 168$ pm.

It is important to remember that if the ions show appreciable departure from spherical symmetry or if there is significant overlap between their electron clouds, then the values assigned to ionic radii become much more speculative.

FACTORS WHICH DETERMINE THE OCCURRENCE OF SIMPLE IONIC STRUCTURES

In a simple approach to ionic structures we make use of the following concepts:

(a) Ions are spherical and have a definite size.

(b) Each ion tends to surround itself with as many ions of opposite charge as possible. This co-ordination number will depend on the relative sizes of the anion and cation and on the requirement that for the arrangement to be stable, the central ion must be in contact with each of its neighbours. Anions are generally larger than cations and so there will be a limit to the number which can surround and touch one cation. As a result, the co-ordination number of the cation will determine the type of structure found. In a simple binary compound AX, the co-ordination numbers of the cation and anion must be the same to ensure electroneutrality.

(c) The arrangement of anions about a central cation will usually be the most symmetrical in space. The electrostatic repulsion between the anions is then at a minimum. This concept leads us to expect that when the co-ordination number of the cation A is 2 with respect to the anion X, then the grouping AX_2 will be linear with an X ion on either side of A. For a co-ordination number of 3, least repulsion occurs when A is at the centre of an equilateral triangle, the vertices of which are occupied by X ions. In the case of co-ordination number 4, symmetry considerations require a tetrahedral arrangement of X about A for the greatest stability. For co-ordination number 6, the arrangement is octahedral. For 8 co-ordination, the electro-static repulsion is minimal in a square anti-prismatic arrangement (*Figure 4.20*). This is not found, however, in ionic structures, the alternative cubic arrangement (*Figure 4.21*), with A at the centre and X at the corners of a cube, being preferred. The reason for this is that cubic co-ordination can extend indefinitely in three dimensions but such an extension is not possible for a square anti-prismatic arrangement, despite the fact that this is the more stable for an isolated AX_8 group.

The co-ordination number found in practice depends largely on the require-ments of (b). The relative sizes of two ions is expressed in their radius ratio, r_{A^+}/r_{X^-}, and the magnitude of this will determine the co-ordination number.

For example, in 3 co-ordination, a stable structure obtains when the three X ions are in contact with the central A^+ ion. If A^+ now becomes smaller relative to X^-, i.e. the radius ratio decreases, a stage is reached at which the three X^- ions also touch each other. Below this limiting lower value for the radius ratio, the structure is unstable because all three anions can no longer touch A^+. *Figure 4.22* illustrates the limiting situation. From this, the minimum radius ratio may be calculated for which this structure is stable.

114

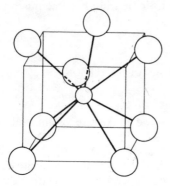

Figure 4.20. Square anti-prismatic co-ordination

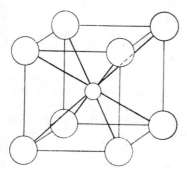

Figure 4.21. Cubic co-ordination

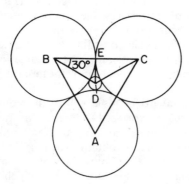

Figure 4.22. The limiting situation for 3 co-ordination

The three anions centred at A, B and C touch the cation centred at D and also touch one another.

In the triangle BDE,

$$\frac{BE}{BD} = \cos 30°$$

$$\frac{r_X}{r_A + r_X} = \frac{\sqrt{3}}{2}$$

$$2r_X = \sqrt{3}r_A + \sqrt{3}r_X$$

$$\frac{r_A}{r_X} = \frac{0 \cdot 268}{1 \cdot 732} = 0 \cdot 1547$$

In a similar way, the radius ratio limits for other co-ordination numbers may be found. These are given in *Table 4.6* for the arrangements which occur most commonly in ionic crystals.

Table 4.6. RADIUS RATIO LIMITS FOR DIFFERENT CO-ORDINATION NUMBERS

Co-ordination number	Shape	Radius ratio limits
3	Plane triangular	0·155 to 0·225
4	Tetrahedral	0·225 to 0·414
6	Octahedral	0·414 to 0·732
8	Cubic	> 0·732

The structures of a number of simple crystals are described in the following pages.

Sodium chloride and caesium chloride

The compounds which should be most nearly ionic in character are those between the most electropositive metals and the most electronegative non-metals — namely the alkali halides. For simple binary compounds of formula AX, two structures are found: sodium chloride (*Figure 4.24*) in which each cation is octahedrally surrounded by 6 anions and each anion by 6 cations; and caesium chloride (*Figure 4.5*) in which the co-ordination number of cation and anion is 8 and the arrangement about each ion is cubic.

Lattice planes in sodium chloride are shown in *Figure 4.23*. The (111) planes of chloride ions have been omitted for clarity.

In the unit cell of sodium chloride, there are eight sodium ions at the corners, each shared with seven other unit cells and giving the equivalent of one ion in each cell. The six sodium ions at the face-centres of the cube are each shared with a second cell and so make the equivalent of three more ions per cell, totalling four sodium ions in all. Similarly, each unit cell contains four chloride ions.

The density of sodium chloride is 2180 kg m^{-3}, its molar mass is 5·845 ×

116

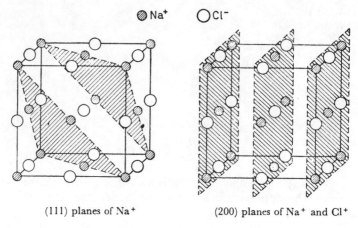

(111) planes of Na$^+$ (200) planes of Na$^+$ and Cl$^+$

Figure 4.23. Lattice planes of sodium chloride

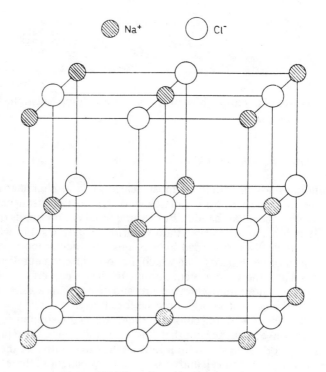

Figure 4.24. The sodium chloride structure

10^{-2} kg mol^{-1} and the Avogadro constant is $6{\cdot}022\,2 \times 10^{23}$ mol^{-1}. The volume of the unit cell is given by

$$v = \frac{4 \times 5{\cdot}845 \times 10^{-2}\ \text{kg mol}^{-1}}{2180\ \text{kg m}^{-3} \times 6{\cdot}022\,2 \times 10^{23}\ \text{mol}^{-1}}$$

$$= 1{\cdot}781 \times 10^{-28}\ \text{m}^3$$

From this, the side of the unit cell is $5{\cdot}626 \times 10^{-10}$ m or 562·6 pm. The internuclear distance, $r_{Na^+} + r_{Cl^-} = 281{\cdot}3$ pm. These are the type of experimental data from which ionic radii have been calculated.

All the alkali metal halides except CsCl, CsBr and CsI (which have the caesium chloride structure*) crystallise with the sodium chloride structure and the radius ratios for many of these compounds do, in fact, lie between the limits 0.414 and 0.732 (*Table 4.7*). According to the radius ratio rule, LiCl,

Table 4.7. RADIUS RATIO VALUES FOR THE ALKALI HALIDES

	Li$^+$	Na$^+$	K$^+$	Rb$^+$	Cs$^+$
F$^-$	0·56	0·77	1·04	1·12	1·41
Cl$^-$	0·41	0·56	0·76	0·82	1·04
Br$^-$	0·38	0·52	0·70	0·77	0·96
I$^-$	0·34	0·46	0·63	0·68	0·85

Table 4.8. SOME COMPOUNDS WITH THE SODIUM CHLORIDE STRUCTURE

Compound	Radius ratio	Compound	Radius ratio
AgF	0·86	AgCl	0·64
AgBr	0·59	MgO	0·51
CaO	0·71	SrO	0·83
BaO	0·97	MnO	0·59
FeO	0·55	CoO	0·52
NiO	0·50	MnS	0·45

LiBr and LiI should have tetrahedrally co-ordinated structures and the simple theory is not able to account for their sodium chloride structure. It is worth noting, however, that of all the alkali halides, the lithium halides are those for which one would expect the greatest deviation from purely ionic bonding.

A number of other ionic crystals having the sodium chloride structure are listed in *Table 4.8* together with the appropriate radius ratios. The caesium chloride unit cell belongs to the primitive cubic system. There are chloride ions at the corners of each cubic unit cell and a caesium ion at the centre. Each caesium ion has eight chloride ions as nearest neighbours and, because the structure is a three-dimensional lattice, each chloride ion is surrounded by eight caesium ions. The arrangement of atoms in the body-centred cubic structure of certain metals (*Figure 4.5*) is the same, each atom showing 8 co-ordination. Because the ion at the centre of the unit cell of caesium chloride

---

* CsCl, CsBr and CsI have the sodium chloride arrangement when deposited from the vapour on to cleavage surfaces of mica and certain other crystals.

is different from those at the eight corners, it is not correct to describe this structure as body-centred cubic but rather as two interpenetrating primitive cubic lattices, one of caesium ions and the other of chloride ions. The caesium chloride structure is much less common than that of sodium chloride. Apart from the three caesium halides mentioned earlier, other compounds which show this structure are CsCN, TlCN, CsSH (the anions here attain spherical symmetry by rotation), TlCl and TlBr. NH_4Cl, NH_4Br and NH_4I crystallise with the sodium chloride structure at temperatures above 457·4, 410·9 and 255·5 K respectively. Transitions occur at these temperatures to the caesium chloride structure.

Zinc blende and wurtzite

The structures of these two forms of ZnS are very closely related; in each case the co-ordination number is 4 with a tetrahedral arrangement of the

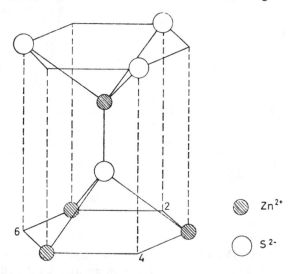

Zn^{2+}

S^{2-}

Figure 4.25. The zinc blende structure. (Wurtzite has the three atoms in the lower hexagon in positions 2, 4 and 6)

four nearest neighbours of any particular ion (*Figure 4.25*). In wurtzite and zinc blende the sulphur atoms may be regarded as arranged in hexagonal and in cubic close-packing respectively. One structure is thus related to the other by movements of part of the structure in a direction parallel to the layers of close-packed atoms. The two arrangements must be very nearly equivalent energetically because many compounds are known which are dimorphous and crystallise with either one or the other structure. *Table 4.9* includes some of these compounds. The radius ratios are, however, frequently outside the limits for 4 co-ordination.

The structure of diamond is also tetrahedral and the disposition of atoms in space (*Figure 4.18*) is the same as in zinc blende. The bonding in diamond is purely covalent, of course, while that in many of the compounds which show the zinc blende structure must have appreciable ionic character otherwise the formation of four covalent bonds by, for example, a Group II metal would result in the accumulation of negative charge on the metal atom, a situation wholly out of accord with its chemical properties. An examination of the compounds listed in *Table 4.9* indicates that considerable polarisation is to be

Table 4.9. COMPOUNDS SHOWING EITHER THE ZINC BLENDE OR WURTZITE STRUCTURE

Compound	Structure	Radius ratio
BeO	W	0·25
BeS	Z-B	0·19
ZnO	W	0·53
ZnS	Z-B, W	0·40
CdS	Z-B, W	0·52
HgS	Z-B	0·55
MnS	Z-B, W	0·45
CuF	Z-B	0·72
CuCl	Z-B	0·53
CuBr	Z-B	0·49
CuI	Z-B	0·44
AgI	Z-B	0·52

Z-B = zince blende; W = wurtzite

expected. It is therefore not surprising that the simple ionic model, which leads on to the concept of limiting radius ratios, cannot account satisfactorily for the structures observed.

Fluorite and rutile

These are ionic structures found for many oxides and fluorides of general formula AX_2. In fluorite, CaF_2, each calcium ion is surrounded by eight fluoride ions arranged at the corners of a cube (*Figure 4.26*) and each fluoride ion is surrounded tetrahedrally by four calcium ions. In rutile, TiO_2, each titanium ion is surrounded octahedrally by six oxide ions and each oxide ion by three titanium ions arranged at the corners of a triangle (*Figure 4.27*). In this type of structure, the octahedron is not necessarily regular and the triangle is not necessarily equilateral. X-ray measurements have shown that the departure from these conditions is usually small.

Geometrical considerations suggest that the fluorite structure should be found for radius ratios greater than 0·732 and the rutile structure for values below this. Some compounds which have these structures are listed in *Table 4.10*. Pauling has shown that for this type of compound the interionic distances are closely approximated by the sum of the crystal radii and the radius ratios quoted in *Table 4.10* have been calculated using the radii given in *Tables 4.4* and *4.5*.

The compounds A_2X, formed between the alkali metals (except Cs) and the elements of Group VI (oxygen, sulphur, etc.), crystallise with the antifluorite

120

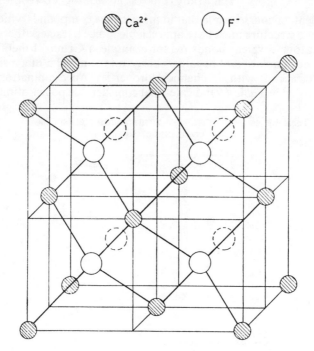

Ca²⁺   F⁻

Figure 4.26. The fluorite structure

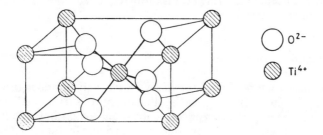

O²⁻

Ti⁴⁺

Figure 4.27. The rutile structure

Table 4.10. COMPOUNDS SHOWING EITHER THE RUTILE OR
FLUORITE STRUCTURE

Fluorite		Rutile	
Compound	Radius ratio	Compound	Radius ratio
CdF_2	0·71	MgF_2	0·54
CaF_2	0·75	ZnF_2	0·56
HgF_2	0·77	MnF_2	0·62
SrF_2	0·87	FeF_2	0·58
BaF_2	1·02	CoF_2	0·55
HfO_2	0·51	NiF_2	0·53

structure. In this, the positions of anions and cations are reversed relative to those in the fluorite structure.

The layer lattices of cadmium chloride and cadmium iodide

These AX_2 structures can be regarded as formed by the superposition of a number of layers each of which is composed of a sheet of cations enclosed between two sheets of strongly polarised anions. Each cation is symmetrically surrounded by six anions at the corners of an octahedron while the three cation neighbours of an anion all lie on one·side of it. The anion is at the apex and the three cations are at the base of triangular pyramid. The forces

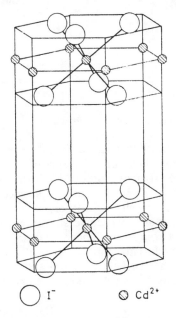

$\bigcirc$ I⁻ $\bigotimes$ Cd²⁺

Figure 4.28. Cadmium iodide structure

between these composite layers cannot be electrostatic in nature because atoms (or ions) of similar kind are next to one another.

These two structures differ only in the way in which the halogen atoms are packed together. In cadmium iodide (*Figure 4.28*), the arrangement is that of hexagonal close-packing: in cadmium chloride, the halogens are cubic close-packed. They may be regarded as intermediate between two limiting types — the ionic fluorite or rutile crystal and a molecular crystal (in which the small finite molecules are held together by weak residual forces). Such layer structures are observed only when there is considerable polarisation of the anion by the cation. This is shown in *Table 4.11* where the transition from a symmetrical ionic structure to a layer lattice is related to the substitution of a more readily polarised ion such as chloride, bromide or iodide for one, like fluoride, which is less easily polarised. It is interesting to note the complementary example of caesium oxide, Cs_2O, which unlike the other alkali metal oxides does not crystallise in the anti-fluorite arrangement but forms instead the anti-cadmium chloride lattice. Here the formation of a layer lattice is attributed to the polarisation of the large caesium cation by the oxide anion.

Table 4.11. TRANSITION OF STRUCTURAL TYPE WITH INCREASING POLARISATION

Symmetrical ionic structures		Layer lattices	
Fluorite	Rutile	Cadmium chloride	Cadmium iodide
CaF_2	$CaCl_2$, $CaBr_2$		CaI_2
	MgF_2	$MgCl_2$	$MgBr_2$, MgI_2
CdF_2		$CdCl_2$, $CdBr_2$	CdI_2
	FeF_2	$FeCl_2$	$FeBr_2$, FeI_2
	CoF_2	$CoCl_2$	$CoBr_2$, CoI_2
	SnO_2		SnS_2
	TiO_2		TiS_2

Increasing polarization→

Rhenium trioxide and perovskite structures

These are two closely-related structures found respectively for metal oxides of formula AO_3 and mixed oxides of formula ABO_3.

The rhenium trioxide structure, ReO_3, is illustrated in *Figure 4.29*. Each rhenium atom is octahedrally co-ordinated by six oxygens and the octahedra are joined to one another by sharing corners.

Rhenium trioxide has an open cubic structure but if the cube centre is occupied by a cation of appropriate charge and size the perovskite structure (*Figure 4.30*) is obtained. This happens typically when one cation, A, is much larger than the other, B. The oxide ions and the larger cations make up a cubic close-packed arrangement and the smaller cations fill the octahedral holes.

Spinel structures

Spinel is the name of magnesium aluminium oxide, $MgAl_2O_4$, and the same structure is found in many compounds of general formula AB_2X_4. The anions, X, are usually oxide but may also be sulphide, selenide or telluride. They form a cubic close-packed arrangement. One-half of the octahedral holes are occupied by B^{3+} ions and one-quarter of all the tetrahedral holes by A^{2+} ions, both kinds of holes being filled in a perfectly regular pattern.

In spinel itself, the total negative charge of four oxide ions is counterbalanced by the positive charges on one A^{2+} and two B^{3+} ions. About 80 per cent of all known spinels have this composition. Most of the A^{2+} ions have radii between 65 and 100 pm and the radii of B^{3+} lie between 60 and 76 pm. Alternative spinel structures are found when there is one A^{4+} and two B^{2+} ions or one A^{6+} and two B^+ ions.

An inverse spinel structure is sometimes found for compounds of formula AB_2X_4. By this is meant a structure in which the tetrahedral sites are occupied by half the B cations and the octahedral sites by the other half of the B cations and all the A cations. Several factors determine whether a normal or an inverse spinel structure is found. These include the relative sizes of A and B,

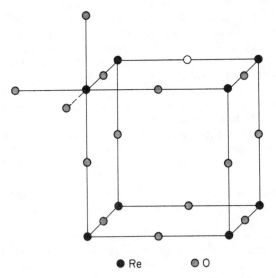

● Re   ◉ O

Figure 4.29. The rhenium trioxide structure

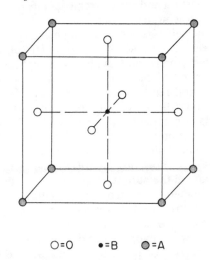

○=O   •=B   ◉=A

Figure 4.30. The perovskite structure, ABO_3

the extents to which polarisation of the anion by A and B occur and the Madelung constants for the two structures. It is, of course, the total lattice energy which measures the stability of a particular structure and these factors all contribute towards this. Examples of spinel and inverse spinel structures are found in the compounds of transition metals in particular and these are further discussed in Chapter 10.

NON-STOICHIOMETRIC COMPOUNDS

One of the basic tenets of chemistry which emerged from the quantitative study of the formation of compounds was that a chemical compound, however prepared, had a constant composition and so contained its various elements in fixed proportions by weight. In the nineteenth century much research was directed towards devising new and improving existing methods for the measurement of combining, or equivalent, weights of the elements. An equivalent weight was assigned to each element that represented its combining weight in grammes relative to a fixed weight of another element chosen as a standard (1 g of hydrogen or 8 g of oxygen were commonly used). The atomic weight of an element (again a relative quantity) could also be determined by taking the valency into account. This resulted in the fundamental law that when atoms combine they do so in simple numerical ratios. This law is implicit in the formulation of a chemical compound and in the use of its formula in writing a chemical equation. It is also basic to the whole of quantitative analysis.

As analytical methods became further refined it became evident that although the vast majority of compounds obeyed the fundamental laws of chemical composition there were some substances which quite definitely did not. These cannot be represented by a simple chemical formula because their composition is not constant and varies over a definite range. Such compounds are examples of non-stoichiometry.

Some oxides and sulphides of the transition metals provide good examples of non-stoichiometric compounds. Thus iron (II) sulphide never has the exact composition FeS and sulphur is always in slight excess. This is due to the fact that a proportion of the positions which would be occupied by iron atoms in an ideal crystal remain vacant. Iron (II) oxide does not have the ideal formula FeO but shows a range of composition from $Fe_{0.91}O$ to $Fe_{0.95}O$. Other transition metal oxides such as VO, NiO, and sulphides such as CrS, show similar non-stoichiometry. Titanium (II) oxide, with the sodium chloride structure, shows a particularly wide range of composition from $TiO_{0.64}$ to $TiO_{1.27}$. Praseodymium (IV) and terbium (IV) oxides are very difficult to prepare in the pure state. If the carbonate or hydroxide of the tripositive state of the metal is heated then the coloured oxides Pr_6O_{11} and Tb_4O_7 respectively are produced. These can be regarded as defect structures in which some of the oxygen positions are empty.

The compounds between hydrogen and transition metals show characteristic non-stoichiometry. For instance, the compounds represented by the formulae $PdH_{0.7}$, $ZrH_{1.92}$, $TiH_{1.73}$, $LaH_{2.76}$ and $CeH_{2.7}$.

Another type of variable composition is shown in the compounds known as the tungsten bronzes, which are formed when alkali metal tungstates are reduced by heating with tungsten. They are deep-coloured unreactive substances with semi-metallic properties. The formula varies between $Na_{0.3}WO_3$ and $NaWO_3$. In $NaWO_3$, the tungsten is present entirely in the

5-valent state but when the compound is deficient in sodium ion, the tungsten is partly in the 6-valent state. Tungsten trioxide has a distorted rhenium trioxide structure. Non-stoichiometry in the tungsten bronzes is due to the uptake of variable amounts of alkali metal ion to give, what is in effect, a defective perovskite structure.

Non-stoichiometry is most commonly found in compounds of the transition metals where the metal can exist in more than one oxidation state and is in combination with polarisable anions. It is a property which is, however, not confined to these metals. For example, the colour change of zinc oxide from white to yellow on heating is associated with a non-stoichiometric composition with excess of metal incorporated as interstitial cations. Even a compound like sodium chloride can show non-stoichiometry. Thus, on heating in sodium vapour, metal is taken up by the lattice to form a non-stoichiometric compound.

The number of such compounds known will undoubtedly increase as more searching studies are made of crystalline solids but the exact and simple stoichiometry of most substances is likely to remain a cornerstone of inorganic chemistry.

INTERSTITIAL COMPOUNDS

These are compounds between transition metals and the non-metallic elements, nitrogen, carbon, boron and hydrogen.

Interstitial nitrides are prepared by heating the finely-divided metal in ammonia to 1400 to 1500 K and the carbides are made similarly by heating with carbon to *ca* 2500 K. Borides can also be synthesised directly by high-temperature reactions between the elements or by co-reduction of the metal oxide and boric oxide with carbon or aluminium at a high temperature. Interstitial hydrides are formed by some metals which have the property of taking up many times their own volume of hydrogen at moderate temperatures (between room temperature and *ca* 600 K) and pressures.

Interstitial nitrides, carbides and borides are technically important for a number of reasons. They have high melting points (examples are: TiC, 3410; NbC, 3770; TiN, 3220; ZrN, 3225 K), they are very inert chemically except towards oxidising agents, they are extremely hard and they have metallic appearances and conductivities. Iron–carbon compounds are particularly important in steels. Interstitial hydrides also have a metallic appearance and high electrical conductivity and are good catalysts for hydrogenation.

Interstitial compounds can be regarded as being derived structurally from the metals themselves. For example, in a face-centred cubic close-packed structure, there are a number of 'octahedral' holes (located at the centres of the sides of the unit cell and marked 'a' in *Figure 4.31*) into which small atoms such as those of nitrogen, carbon, boron or hydrogen can fit. When all these octahedral holes are occupied, the structure obtained is that of sodium chloride although the properties of the solids show that the type of bonding is very different. Compounds which have the sodium chloride structure include TiC, TiN, ZrN, VC and TaC. Other interstitial compounds arise when some of the interstitial sites remain vacant: when half of all possible octahedral sites are filled, compounds of general formula A_2X are obtained (for example, W_2N and Mo_2N); if only a quarter are filled, the formula is A_4X (as in Mn_4N, Mn_4B and Fe_4N).

In a face-centred cubic structure there are also smaller tetrahedral holes located at the centres of the eight small cubes into which the unit cell may be divided. When all of these are occupied by the non-metallic element, the calcium fluoride type of structure results (as in TiH_2). When half are filled, the zinc blende structure is found (ZrH and TiH); when a quarter are filled, compounds of formula A_2X are obtained (Pd_2H) and when only one-eighth, the general formula is A_4X (Zr_4H).

Compounds with interstitial structures are frequently non-stoichiometric. For example, in the palladium–hydrogen system, two non-stoichiometric face-centred cubic phases have been identified. One, the α-phase, occurs at very low concentrations of hydrogen, and the second, β-phase, is found at

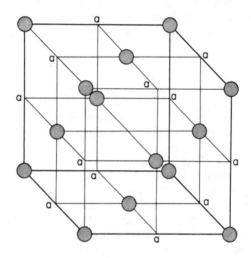

Figure 4.31. Octahedral holes in a face-centred close-packed structure

higher concentrations and this reaches a limiting composition of $PdH_{0.7}$ at normal temperatures and pressures.

Some metallic borides can be classified as interstitial because their structure consists of isolated boron atoms in a network of metal atoms. Examples include Mn_4B, Co_3B and Ni_3B. These form only a small part of the range of metal borides, for, as the proportion of boron to metal increases, there is a strong tendency for the boron atoms to begin to form bonds with each other with a consequent departure from a true interstitial structure.

The physical properties and chemical stability of interstitial carbides, nitrides and borides suggest the presence of metal to non-metal bonds of considerable strength. Their electrical conductivity indicates that some, at least, of the valency electrons are non-localised. In the case of metallic hydrides, it appears that the hydrogen is present as hydride ion, being formed by the transfer of an electron from the valency shell of the metal to a hydrogen atom. The remainder of the electrons from the valency shell of the metal are used in bonding between the metal atoms and, as in pure metals, occupy bands of permitted energy states. This explanation of the bonding accounts satisfactorily for the metallic character and electrical conductivity of the hydrides and for their non-stoichiometry.

THE MINERAL SILICATES

The Earth's crust consists almost entirely of silicates and silica. The glass, ceramic and cement industries are based on silicate chemistry and many metallurgical processes are concerned with the removal of silicates as a necessary stage in the extraction of pure metals. The elucidation of the structures has been difficult because of the chemical complexity of silicates but the basic principles are now well understood. Some of the more important aspects of silicate structures are dealt with below: for a more detailed survey, reference should be made elsewhere.

Silica itself is known in three different crystalline modifications, each of which has a high and low temperature form. In one of these, *cristobalite*, the silicon atoms are arranged in a diamond-type lattice with an oxygen atom between each pair of silicons. The co-ordination number of silicon is thus 4 and that of oxygen is 2. Alternatively, the three-dimensional lattice may be described as composed of SiO_4 tetrahedra, in which each oxygen is shared between two adjacent tetrahedra. *Quartz* and *tridymite*, the other two modifications of silica, have the same tetrahedral unit but in quartz there is a helical arrangement of atoms which gives rise to optical activity.

According to Pauling, the silicon–oxygen bond is 50 per cent ionic. The appreciable difference in electronegativity of the two elements leads to this conclusion. The fundamental unit for all silicate structures is the SiO_4 tetrahedron and various arrangements are possible because of the different ways in which the tetrahedra are joined together by the sharing of oxygens:

(a) *Discrete anions* — these include the orthosilicates which contain SiO_4^{4-} (for example, *zircon*, $ZrSiO_4$, and *phenacite*, Be_2SiO_4); the pyrosilicates which contain $[Si_2O_7]^{6-}$, wherein one oxygen is shared between two tetrahedra (*thortveitite*, $Sc_2Si_2O_7$); and closed rings such as $[Si_3O_9)^{6-}$ and $[SiO_{18}]^{12-}$, wherein each tetrahedron shares two oxygens (*benitoite*, $BaTiSi_3O_9$, and *beryl*, $Be_3Al_2Si_6O_{18}$).

(b) *Extended anions* — these may either be infinite chains, such as $[(SiO_3)_n]^{2n-}$ and $[(Si_4O_{11})_n]^{6n-}$, in which each tetrahedron again shares two oxygens or sheets of composition $[(Si_2O_5)_n]^{2n-}$, in which each tetrahedron shares three oxygens. Examples of the first type are the *pyroxenes* like *diopside*, $CaMg(SiO_3)_2$, and *spodumene*, $LiAl(SiO_3)_2$ and the *amphiboles* like *tremolite*, $(OH)_2Ca_2Mg_5(Si_4O_{11})_2$. Ions of the second type are found in micas and clays.

(c) *Three-dimensional networks* — the tetrahedra are completely linked by all four oxygens. These are the various forms of *silica*, SiO_2.

The anions described in the preceding section are illustrated in *Figure 4.32*.

Another factor responsible for the complexity of mineral silicates is the freedom with which isomorphous replacement of one ion by another of similar charge or size has occurred during their formation. For instance, the mineral *olivine* has a variable composition depending on its source because of the replacement of magnesium by iron to differing extents. Although this mineral is essentially magnesium orthosilicate, Mg_2SiO_4, at least some of the magnesium ions have been replaced by iron (II). Calcium is another element whose ions can undergo isomorphous replacement with magnesium.

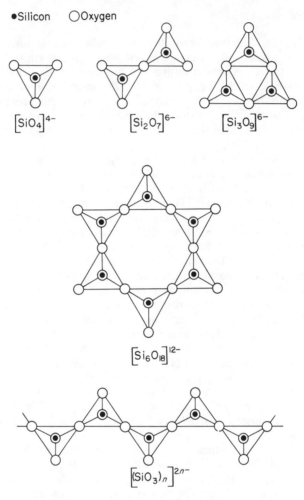

Figure 4.32. The structure of some silicate ions

The replacement which has the greatest structural significance is the substitution of some of the silicon ions, Si^{4+} ($= 41$ pm), by aluminium, Al^{3+} ($= 50$ pm). Thus many alumino-silicates are based on a three-dimensional network of linked tetrahedra derived from silica itself by the replacement of tetrapositive silicon with tripositive aluminium. This substitution requires that a univalent cation like Na^+ or K^+ be incorporated into the structure for every aluminium ion introduced or, in silicates where cations are already present, the replacement of one cation by another of higher charge. A third possibility is the addition of an anion such as fluoride or chloride.

The formulae of the *feldspar* and *zeolite* alumino-silicates illustrate the substitutions which have taken place:

Feldspars:	*orthoclase*	$KAlSi_3O_8$
	celsian	$BaAl_2Si_2O_8$
	albite	$NaAlSi_3O_8$
	anorthite	$CaAl_2Si_2O_8$
Zeolites:	*analcite*	$NaAlSi_2O_6.H_2O$
	heulandite	$CaAl_2Si_7O_{18}.6H_2O$
	natrolite	$Na_2Al_2Si_3O_{10}.2H_2O$
	thomsonite	$NaCa_2Al_5Si_5O_{20}.6H_2O$

In these crystals, the aluminium, like silicon, is tetrahedrally co-ordinated and the ratio of the number of oxygens to the combined total of aluminium and silicon is $2:1$. Sufficient alkali and alkaline-earth metal ions are also present to maintain electrical neutrality. Zeolites are characterised by a particularly open structure and this imparts the property of ion exchange to these minerals. The cations are replaced by other positive ions when the zeolite is brought into contact with an aqueous solution containing these other ions. These naturally occurring materials and their synthetic analogues have found extensive use in water-softening.

Zeolites have also been used for the absorption of gases and vapours. Throughout their rigid structure run channels of a size suitable for the retention of small molecules. The diameters of these interstitial channels and hence the absorptive behaviour towards particular gases can be varied by changing the ionic form of the zeolites. The Linde Molecular Sieves are commercially available synthetic zeolites. Linde Sieve 4A (channel diameter $= 400$ pm) has the formula $Na_{12}[12\,AlO_2 . 12\,SiO_2]\,NaAlO_2 . 29H_2O$; Linde Sieve 5A (channel diameter $= 500$ pm) is the corresponding calcium compound. Type 4A will absorb permanent gases (N_2, O_2, CH_4), polar molecules (H_2O, NH_3), C_2 hydrocarbons and the lower alcohols. Type 5A will absorb *n*-paraffins up to C_{14}, *n*-olefins and *n*-alcohols but not branched-chain organic molecules. Many chromatographic separations have been effected with these materials and they are widely used on an industrial scale for drying gases.

Another factor which further complicates the structures is that aluminium can be 6 as well as 4 co-ordinate with respect to oxygen. For example, in *hydrargillite*, $Al(OH)_3$, the structure is built up from sheets of $Al(OH)_6$ octahedra, each octahedron sharing three of its edges with three adjacent octahedra. When such a sheet is combined with a silicon/oxygen sheet like $(H_2Si_2O_5)_n$, elimination of water occurs and composite layers are built up. *Kaolin*, $Al_2Si_2O_5(OH)_4$, is an example of this (*Figure 4.33*). When silicon/oxygen layers are condensed on both sides of a hydrargillite layer, the clay mineral *pyrophyllite*, $Al_2Si_4O_{10}(OH)_2$, is obtained.

Similarly, *brucite*, $Mg(OH)_2$, contains 6 co-ordinated metal ions and the mineral *talc*, $Mg_3Si_4O_{11} . H_2O$, is composed of a layer of brucite between two silicon/oxygen layers. These composite layers, in this case, are electrically neutral and they are loosely packed in the crystal. Hence the softness and pronounced cleavage of talc.

When one-quarter of the silicon ions in a pyrophyllite layer are replaced by aluminium ions, a layer of composition $Al_2AlSi_3O_{10}(OH)_2$ is obtained. In the crystal, layers of this type alternate with layers of positive ions. This structure is typical of that found in micas such as *muscovite*, $KAl_3Si_3O_{10}(OH)_2$.

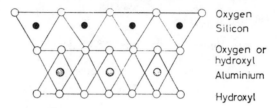

Oxygen
Silicon

Oxygen or
hydroxyl
Aluminium

Hydroxyl

Figure 4.33. Diagrammatic representation of the composite layers of kaolin

A similar replacement in talc produces a mica having the composition $KMg_3AlSi_3O_{10}(OH)_2$, as in *phlogopite*. Mica crystals are held together by electrostatic bonds and so are not as soft as talc and pyrophyllite where the composite layers are held together by residual bonds alone. Cleavage in one plane, however, is still a very characteristic property.

SUGGESTED REFERENCES FOR FURTHER READING

ADDISON, W. E. *Structural Principles in Inorganic Compounds*, 1st edn, Longmans Green, London, 1961.

ADDISON, W. E. *Allotropy of the Elements*, Oldbourne, London, 1964.

CROFT, R. C. 'Lamellar compounds of graphite', *Quart. Rev. chem. Soc., Lond.*, 14 (1961) 1.

DORAIN, P. B. *Symmetry in Organic Chemistry*, Addison-Wesley, Reading, Mass., 1965.

EVANS, R. C. *An Introduction to Crystal Chemistry*, 2nd edn, Cambridge, 1964.

GREENWOOD, N. N. *Ionic Crystals, Lattice Defects and Non-Stoichiometry*, Butterworths, London, 1968.

HUME-ROTHERY, W. *Atomic Theory for Students of Metallurgy*, 3rd edn, Institute of Metals, London, 1962.

WELLS, A. F. *Structural Inorganic Chemistry*, 3rd edn, Oxford, 1962.

WYCKOFF, R. W. G. *Crystal Structures*, Interscience, New York, 1948 onward. (Several volumes of this comprehensive reference work have been published.)

5   Reactions in water and in non-aqueous solvents

OXIDATION AND REDUCTION

A very large number of inorganic reactions take place in aqueous solution and, of these, many may be classified under the heading of *oxidation–reduction* or *redox* reactions. The process of oxidation may be considered as the loss of electrons by an atom, ion or molecule and reduction as the reverse of this. It is, however, confusing to many students that some substances behave as reducing agents when reacted with certain compounds but as oxidising agents with others. Thus, the nitrite ion, NO_2^-, reacts as an oxidising agent with iodide ion in acid solution, converting it to iodine, yet reduces permanganate in acid solution to Mn^{2+}. Such behaviour can only be understood by the quantitative consideration of the electronic changes involved in oxidation and reduction.

The two processes of oxidation and reduction are inseparable and take place simultaneously. Thus, in the case of the reduction of the mercury (II) ion by the tin (II) ion, the overall reaction is

$$2Hg^{2+} + Sn^{2+} = Hg_2^{2+} + Sn^{4+}$$

and this may be split up into two ion half-reactions:

$$2Hg^{2+} + 2e^- = Hg_2^{2+}$$
$$Sn^{2+} - 2e^- = Sn^{4+}$$

the first of which is a reduction and the second an oxidation reaction. Similarly, the deposition of metallic copper from an aqueous solution of copper (II) sulphate by zinc may be written as

$$Zn + Cu^{2+} = Zn^{2+} + Cu$$

for which the two ion half-reactions are

$$Cu^{2+} + 2e^- = Cu \quad \text{(reduction)}$$
$$Zn - 2e^- = Zn^{2+} \quad \text{(oxidation)}$$

Reactions such as those above will only take place if there is a decrease in *free energy* of the system. This means that the total free energy of the products must be less than that of the reactants. Absolute free energies for ions etc., are not known, but relative values, or *standard free energies*, measured at 298 K have been determined and these may be used instead to predict the direction of chemical change.

If as an example we choose the system of copper (II) ions and zinc metal, the standard free energies of the elements are, by convention, zero and those of the two ions Zn^{2+} and Cu^{2+} are -146.86 and $+66.53\,kJ\,mol^{-1}$ respectively. Hence the free energy change (ΔG°) for the reaction

$$Zn + Cu^{2+} = Zn^{2+} + Cu$$

is $213.39\,kJ\,mol^{-1}$ and the reaction therefore proceeds spontaneously from left to right. Had the free energy change been positive, the reaction would have taken place from right to left.

Besides using free energy change, as such, for an estimate of the driving force behind a particular reaction it is possible to approach the subject from an electrochemical standpoint. The tendency for an element to lose electrons to a solution is measured by its *electrode potential*. When a metal such as zinc is placed in a solution of one of its salts in water, a potential difference is set up between the metal and the solution that is dependent upon the metal, the *relative activity* $(a = (m\gamma/m^\circ)$ where m = molality and γ the activity coefficient of the species; m° = standard molality) of the metal ion in the aqueous solution and the temperature. When the ions are at unit relative activity and the temperature is 298 K, the potential difference is termed the *standard electrode potential* (E^0). Standard electrode potentials are measured by comparison with the standard hydrogen electrode (S.H.E.) or any standard electrode such as the calomel electrode; the units used are volts (V).

The sign attributed to an electrode potential differs according to the particular sign convention being used; that used here will be the one recommended by the International Union of Pure and Applied Chemistry (IUPAC). If we take the system $Cu^{2+}(aq)/Zn$ and consider the two half-reactions, we may set up a cell where the two systems undergoing oxidation and reduction are separated by a porous diaphragm (*Figure 5.1*) which acts as a salt-bridge in permitting the passage of ions but preventing the mixing of the two aqueous solutions. On the IUPAC convention, the sign and magnitude of the e.m.f. of this cell are defined by the cell diagram and are identical with the sign and magnitude of the potential of a lead to the electrode on the right-hand side with respect to a similar lead to the electrode on the left-hand side whose potential is zero, the cell being in open circuit. The e.m.f. of the cell is thus positive if the negative electrode is on the left-hand side and, on discharging the cell, oxidation takes place at this electrode and reduction occurs at the right-hand electrode. The corresponding redox reaction must be written down in a similar manner with the metal atoms on the same sides of the redox equation as they are in the cell. Hence we write zinc metal on the left-hand side and copper on the right-hand side of the equation as above.

The relation between ΔG° and E° is

$$\Delta G^\circ = -zFE^\circ$$

where z is the number of electrons transferred in the reduction process and F is the Faraday. Hence, if E° is positive, the reaction proceeds as written.

The potential of the electrode is measured with reference to the S.H.E. in which hydrogen gas at 101·325 kPa (1 atm) is bubbled over an inert metal electrode (e.g. platinum) immersed in a solution of HCl that is of unit relative activity with respect to H^+. The electrode potential at 298 K, under these conditions, for the system

$$H_2 = 2H^+ + 2e^-$$

is defined as zero volts. The potential of an ion half-reaction refers to the cell in which the S.H.E. is on the left-hand side and the electrode in question is on

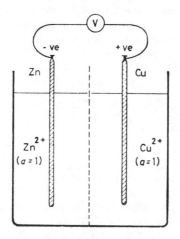

Zn | Zn²⁺ $(a=1)$ ‖ Cu²⁺ $(a=1)$ | Cu   E = + 1·10 V
− ve   + ve

Figure 5.1. Cell composed of the system Zn^{2+}/Zn and Cu^{2+}/Cu (unit relative activity)

the right-hand side. The half-reaction thereby examined is always formulated as a reduction:

$$\text{Oxidised state} + ze^- = \text{Reduced state}$$
$$(Ox) \qquad\qquad\qquad (Red)$$

The electrode potential of the ion half-reaction is given by the Nernst equation

$$E = E^\circ_{Ox/Red} + \frac{RT}{zF}\ln\frac{a_{Ox}}{a_{Red}}$$

where R is the gas constant (8·314 J K^{-1} mol^{-1}), T is the absolute temperature (K), z is the number of electrons transferred, F is the Faraday and a denotes the relative activity of the species. When a_{Ox} and a_{Red} are both unity, the logarithmic term is zero and the electrode potential is the standard potential. The activities of elements under standard conditions are defined as unity.

If the system M^{z+}/M is a better reducing agent than hydrogen, under standard conditions, the electrode potential is negative and if it is a poorer reducing agent, the potential is positive. The standard electrode potentials for the systems Zn^{2+}/Zn and Cu^{2+}/Cu are -0.763 V and $+0.337$ V respectively, giving for the overall cell e.m.f., which is defined as

$$E = (E^{\circ}_{R.H.S.} - E^{\circ}_{L.H.S.}) = (E^{\circ}_{reduced} - E^{\circ}_{oxidised})$$

a value of $+1.10$ V.

The standard electrode potentials of the elements arranged in order of the most negative to the most positive constitute the electrochemical series of the elements. This series may be used to interpret and also predict the reactivity of the elements. A cation-forming element will displace from aqueous solution another element which lies below it in this series and an anion-forming element will displace from aqueous solution another element which is above it in the series.

The order in the series is, in a shortened form:

$Li^+/Li < K^+/K < Ca^{2+}/Ca < Na^+/Na < Mg^{2+}/Mg < Al^{3+}/Al < Zn^{2+}/Zn$
$< Fe^{2+}/Fe < Sn^{2+}/Sn < H^+/\frac{1}{2}H_2 < Cu^{2+}/Cu < \frac{1}{2}I_2/I^- < Ag^+/Ag <$
$\frac{1}{2}Br_2/Br^- < \frac{1}{2}Cl_2/Cl^- < \frac{1}{2}F_2/F^-$

VARIATION OF THE MAGNITUDE OF STANDARD ELECTRODE POTENTIALS

The factors affecting the magnitude of standard electrode potentials, both for metals and non-metals, may be discussed in terms of an enthalpy cycle (*Figure 5.2*). [Approximate comparisons may be made by using *enthalpy* changes (ΔH°) instead of *free energy* changes (ΔG°).]

$$M (s) \longrightarrow M^+ (aq) + e^-$$
$$\downarrow \qquad\qquad \uparrow$$
$$M (g) \longrightarrow M^+ (g) + e^-$$

$$\tfrac{1}{2}X_2 (g) + e^- \longrightarrow X^- (aq)$$
$$\downarrow \qquad\qquad \uparrow$$
$$X (g) + e^- \longrightarrow X^- (g)$$

Figure 5.2. Enthalpy cycles for the formation of M^+ (aq) *and* X^- (aq)

For metals, three steps may be considered in the formation of the metal ion M^+ in aqueous solution from the metal M. These are:
(a) atomisation of the metal (ΔH°_{at});
(b) ionisation of the gaseous metal atoms (ΔH°_{I});
(c) hydration of the gaseous ions (ΔH°_{h}).
The enthalpy change (ΔH°) for the process $M (s) \longrightarrow M^+(aq)$ is therefore given by

$$\Delta H^{\circ} = \Delta H^{\circ}_{at} + \Delta H^{\circ}_{I} + \Delta H^{\circ}_{h}$$

If the sign of ΔH° be reversed, the enthalpy change refers to the process

$$M^+(aq) + e(g) = M (s)$$

and may therefore be used as a measure of the electrode potential for this system. A similar consideration applies to the halogens.

A similar cycle may also be drawn for hydrogen by replacing ΔH_{at}° by ΔH_d° (one-half the bond dissociation energy of the hydrogen molecule).

For the alkali metals the following enthalpy changes apply

	Li	Na	K	Rb	Cs
$\Delta H^{\circ}_{at}/kJ\ mol^{-1}$	+159·0	+108·8	+90·0	+85·8	+78·7
$\Delta H_I^{\circ}/kJ\ mol^{-1}$	+520·1	+495·4	+418·4	+402·9	+373·6
$\Delta H_h^{\circ}/kJ\ mol^{-1}$	−507·1	−395·8	−317·1	−289·5	−259·4
$\Delta H^{\circ}/kJ\ mol^{-1}$	+172·0	+208·4	+191·3	+199·2	+192·9

For hydrogen the enthalpy changes are

$\Delta H_d^{\circ}/kJ\ mol^{-1}$	+218·0
$\Delta H_I^{\circ}/kJ\ mol^{-1}$	+1312·1
$\Delta H_h^{\circ}/kJ\ mol^{-1}$	−1071·1
$\Delta H^{\circ}/kJ\ mol^{-1}$	+459·0

Referring the values for the alkali metals to hydrogen equal to zero, gives values for the system $M^+(aq)+e(g) = M(s)$ of

	Li	Na	K	Rb	Cs
$-\Delta H^{\circ}/kJ\ mol^{-1}$	+287	+250.6	+267·7	+259·8	+266.1

Hence E° becomes less negative from the Li^+/Li system to the Na^+/Na system in the order above.

For the halogens, the values of $(\Delta H_d^{\circ} - \Delta H_E^{\circ})$, where ΔH_E° is the electron affinity of the halogen atom, lie between -226 and $-260\ kJ\ mol^{-1}$; here, therefore, the heat of hydration is the determining factor in making the electrode potential less positive as the atomic number of the halogen increases.

	F	Cl	Br	I
$\Delta H_d^{\circ}/kJ\ mol^{-1}$	+77·4	+121·3	+96·2	+75·3
$\Delta H_E^{\circ}/kJ\ mol^{-1}$	−335·6	−356·1	−332·6	−303·8
$\Delta H_h^{\circ}/kJ\ mol^{-1}$	−513·0	−371·1	−340·6	−301·7
$\Delta H^{\circ}/kJ\ mol^{-1}$	−771·2	−605·9	−577·0	−530·2
$\Delta H^{\circ}/kJ\ mol^{-1}$ (H=0)	−1230·2	−1064·9	−1036·0	−989·2

Summarising, we may say that a reducing metal is characterised by (i) a low ionisation enthalpy, (ii) a low sublimation enthalpy and (iii) a high enthalpy of hydration. Oxidising potentials of high magnitude are favoured by (i) a large electron affinity, (ii) a small dissociation enthalpy and (iii) a high enthalpy of hydration.

So far, the discussion of electrode reactions has been limited to the system $M^{z+}(aq)/M(s)$. It is possible, however, that the oxidised and reduced forms are both ionic and soluble, for example Fe^{3+}/Fe^{2+} and MnO_4^-/Mn^{2+}. These systems also have standard electrode potentials which may be determined by constructing an electrode in which a platinum wire is immersed in a solution of unit relative activity with respect to both the oxidised and reduced forms and comparing its potential against a standard hydrogen electrode.

$$Pt\,|\,H_2(1\ atm)\,|\,H^+(a = 1)\,\left\|\,\begin{array}{c} Fe^{3+}(a = 1) \\ Fe^{2+}(a = 1) \end{array}\right|\,Pt$$

The couple Fe^{3+}/Fe^{2+} under these conditions has an electrode potential of $+0.771$ V.

A reaction such as that of the reduction of permanganate to Mn^{2+}, besides involving a transfer of electrons in the reduction process, also leads to the transfer of atoms. Oxidation of compounds by MnO_4^- are pH dependent, the overall reaction being

$$MnO_4^- + 8H^+ + 5e^- = Mn^{2+} + 4H_2O$$

It is necessary that the activity of the hydrogen ion shall also be unity for the determination of E° for this system, since

$$E = E^\circ + \frac{RT}{zF}\ln\frac{a_{MnO_4^-} \times (a_{4+})^8}{a_{Mn^{2+}}}$$

The value of E° is $+1.51$ V.

Values of other standard electrode potentials similarly obtained are given in *Table 5.1*.

Standard electrode potentials may be used to indicate if a particular reaction is likely to take place. They can, however, give no information on the rapidity of the reaction nor will they indicate if the reaction is likely to take place in the absence of suitable catalysts. Thus, the reaction between arsenite and dichromate may be written as

$$Cr_2O_7^{2-} + 3HAsO_2 + 8H^+ = 2Cr^{3+} + 3H_3AsO_4 + H_2O$$

for which the half-reactions are

$$Cr_2O_7^{2-} + 6e^- + 14H^+ = 2Cr^{3+} + 7H_2O \quad E^\circ = +1.33\ V$$
and $\quad H_3AsO_4 + 2e^- + 2H^+ = HAsO_2 + 2H_2O \quad E^\circ = +0.559\ V$

The overall potential of a redox cell composed of these two systems would be $(+1.33 - 0.559)$ V and since E° is positive, dichromate should oxidise arsenite. Unless the reaction is catalysed by the addition of a trace of KI or by OsO_4 it proceeds extremely slowly.

Similarly the halogens Cl_2, Br_2 and I_2 are all capable of oxidising $S_2O_3^{2-}$ to SO_4^{2-} yet iodine only oxidises as far as the intermediate tetrathionate $S_4O_6^{2-}$.

THE EFFECT OF NON-STANDARD CONDITIONS UPON ELECTRODE POTENTIALS

The Nernst equation

$$E = E^\circ + \frac{RT}{zF}\ln\frac{a_{Ox}}{a_{Red}}$$

indicates that the potential of a particular solution containing both the oxidised and reduced forms will vary with the concentrations of the various forms. At 298 K the Nernst equation becomes

$$E = E^\circ + \frac{0{\cdot}059}{z}\log_{10}\frac{a_{Ox}}{a_{Red}}$$

Table 5.1. STANDARD ELECTRODE POTENTIALS

Couple	E°/V	Couple	E°/V
$Li^+ + e^- = Li$	$-3{\cdot}045$	$Fe(CN)_6^{3-} + e^- = Fe(CN)_6^{4-}$	$+0{\cdot}36$
$K^+ + e^- = K$	$-2{\cdot}925$	$2H_2SO_3 + 2H^+ + !e^-$	
$Rb^+ + e^- = Rb$	$-2{\cdot}925$	$\quad = S_2O_3^{2-} + 3H_2O$	$+0{\cdot}40$
$Cs^+ + e^- = Cs$	$-2{\cdot}923$	$4H_2SO_3 + 4H^+ + 6e^-$	
$Ba^{2+} + 2e^- = Ba$	$-2{\cdot}90$	$\quad = S_4O_6^{2-} + 6H_2O$	$+0{\cdot}51$
$Sr^{2+} + 2e^- = Sr$	$-2{\cdot}89$	$Cu^+ + e^- = Cu$	$+0{\cdot}521$
$Ca^{2+} + 2e^- = Ca$	$-2{\cdot}87$	$I_2 + 2e^- = 2I^-$	$+0{\cdot}5355$
$Na^+ + e^- = Na$	$-2{\cdot}714$	$Cu^{2+} + Cl^- + e^- = CuCl$	$+0{\cdot}538$
$Mg^{2+} + 2e^- = Mg$	$-2{\cdot}37$	$H_3AsO_4 + 2H^+ + 2e^-$	
$Lu^{3+} + 3e^- = Lu$	$-2{\cdot}25$	$\quad = HAsO_2 + 2H_2O$	$+0{\cdot}559$
$\frac{1}{2}H_2 + e^- = H^-$	$-2{\cdot}25$	$MnO_4^- + e^- = MnO_4^{2-}$	$+0{\cdot}564$
$Sc^{3+} + 3e^- = Sc$	$-2{\cdot}08$	$Cu^{2+} + Br^- + e^- = CuBr$	$+0{\cdot}64$
$Be^{2+} + 2e^- = Be$	$-1{\cdot}85$	$O_2 + 2H^+ + 2e^- = H_2O_2$	$+0{\cdot}682$
$Al^{3+} + 3e^- = Al$	$-1{\cdot}66$	$Fe^{3+} + e^- = Fe^{2+}$	$+0{\cdot}771$
$Ti^{2+} + 2e^- = Ti$	$-1{\cdot}63$	$Hg_2^{2+} + 2e^- = 2Hg$	$+0{\cdot}789$
$Mn^{2+} + 2e^- = Mn$	$-1{\cdot}18$	$Ag^+ + e^- = Ag$	$+0{\cdot}7991$
$V^{2+} + 2e^- = V$	$\approx -1{\cdot}18$	$Cu^{2+} + I^- + e^- = CuI$	$+0{\cdot}86$
$Zn^{2+} + 2e^- = Zn$	$-0{\cdot}763$	$2Hg^{2+} + 2e^- = Hg_2^{2+}$	$+0{\cdot}920$
$Cr^{3+} + 3e^- = Cr$	$-0{\cdot}74$	$NO_3^- + 3H^+ + 2e^-$	
$Ga^{3+} + 3e^- = Ga$	$-0{\cdot}53$	$\quad = HNO_2 + H_2O$	$+0{\cdot}94$
$H_3PO_3 + 2H^+ + 2e^-$		$NO_3^- + 4H^+ + 3e^-$	
$\quad = H_3PO_2 + H_2O$	$-0{\cdot}50$	$\quad = NO + 2H_2O$	$+0{\cdot}96$
$Fe^{2+} + 2e^- = Fe$	-0.440	$Br_2(liq) + 2e^- = 2Br^-$	$+1{\cdot}0652$
$Cr^{3+} + e^- = Cr^{2+}$	$-0{\cdot}41$	$SeO_4^{2-} + 4H^+ + 2e^-$	
$Ti^{3+} + e^- = Ti^{2+}$	$\approx -0{\cdot}37$	$\quad = H_2SeO_3 + H_2O$	$+1{\cdot}15$
$In^{3+} + 3e^- = In$	$-0{\cdot}342$	$IO_3^- + 6H^+ + 5e^-$	
$Tl^+ + e^- = Tl$	$-0{\cdot}3363$	$\quad = \frac{1}{2}I_2 + 3H_2O$	$+1{\cdot}195$
$Co^{2+} + 2e^- = Co$	$-0{\cdot}277$	$O_2 + 4H^+ + 4e^- = 2H_2O$	$+1{\cdot}229$
$H_3PO_4 + 2H^+ + 2e^-$		$Tl^{3+} + 2e^- = Tl^+$	$+1{\cdot}25$
$\quad = H_3PO_3 + H_2O$	$-0{\cdot}276$	$Cr_2O_7^{2-} + 14H^+ + 6e^-$	
$V^{3+} + e^- = V^{2+}$	$-0{\cdot}255$	$\quad = 2Cr^{3+} + 7H_2O$	$+1{\cdot}33$
$Ni^{2+} + 2e^- = Ni$	$-0{\cdot}250$	$Cl_2 + 2e^- = 2Cl^-$	$+1{\cdot}3595$
$CuI + e^- = Cu + I^-$	$-0{\cdot}185$	$Au^{3+} + 3e^- = Au$	$+1{\cdot}50$
$AgI + e^- = Ag + I^-$	$-0{\cdot}151$	$Mn^{3+} + e^- = Mn^{2+}$	$+1{\cdot}51$
$Sn^{2+} + 2e^- = Sn$	$-0{\cdot}136$	$MnO_4^- + 8H^+ + 5e^-$	
$Pb^{2+} + 2e^- = Pb$	$-0{\cdot}126$	$\quad = Mn^{2+} + 4H_2O$	$+1{\cdot}51$
$2H^+ + 2e^- = H_2$	$0{\cdot}00$	$BrO_3^- + 6H^+ + 5e^-$	
$Ag(S_2O_3)_2^{3-} + e^-$		$\quad = \frac{1}{2}Br_2 + 3H_2O$	$+1{\cdot}52$
$\quad = Ag + 2S_2O_3^{2-}$	$+0{\cdot}01$	$Ce^{4+} + e^- = Ce^{3+}$	$+1{\cdot}61$
$CuBr + e^- = Cu + Br^-$	$+0{\cdot}033$	$Au^+ + e^- = Au$	$\approx +1{\cdot}68$
$S_4O_6^{2-} + 2e^- = 2S_2O_3^{2-}$	$+0{\cdot}08$	$MnO_4^- + 4H^+ + 3e^-$	
$CuCl + e^- = Cu + Cl^-$	$+0{\cdot}137$	$\quad = MnO_2 + 2H_2O$	$+1{\cdot}695$
$Sn^{4+} + 2e^- = Sn^{2+}$	$+0{\cdot}15$	$H_2O_2 + 2H^+ + 2e^- = 2H_2O$	$+1{\cdot}77$
$Cu^{2+} + e^- = Cu^+$	$+0{\cdot}153$	$Co^{3+} + e^- = Co^{2+}$	$+1{\cdot}82$
$SO_4^{2-} + 4H^+ + 2e^-$		$Ag^{2+} + e^- = Ag^+$	$+1{\cdot}98$
$\quad = H_2SO_3 + H_2O$	$+0{\cdot}17$	$S_2O_8^{2-} + 2e^- = 2SO_4^{2-}$	$+2{\cdot}01$
$Cu^{2+} + 2e^- = Cu$	$+0{\cdot}337$	$F_2 + 2e^- = 2F^-$	$+2{\cdot}87$

Data from Latimer, W. M. *The Oxidation States of the Elements and their Potentials in Aqueous Solutions*, 2nd edn, Prentice-Hall, New York, 1952.

Thus, in the case of the system Fe^{3+}/Fe^{2+}, with a ratio of $a_{Fe^{3+}}/a_{Fe^{2+}}$ of 10^2, the potential is given by

$$E = E^\circ + \frac{0.059}{1} \log_{10} 10^2$$
$$= 0.771 + 0.118$$
$$= 0.889 \text{ V}$$

From this value it can be seen that in the titration of Fe^{2+} by $Cr_2O_7^{2-}$, an indicator such as diphenylamine sulphonate, for which the electrode potential is $+0.83$ V, is suitable provided the concentration of Fe^{3+} is low. Otherwise the end point will be inaccurate since, for large concentrations of Fe^{3+}, the Fe^{3+}/Fe^{2+} system will have a potential greater than that required for oxidising the indicator. This will occur before the equivalent amount of dichromate has been added and give a premature end point. This is the reason for the addition of phosphoric acid which suppresses the concentration of Fe^{3+} ions in the titration solution by the formation of complex ions.

A cell may be constructed from two electrodes of the same metal dipping into two solutions of the same compound but of different ionic strengths. This is termed a concentration cell and may be represented as

$$M \mid M^{z+}(c_1) \parallel M^{z+}(c_2) \mid M$$

hence

$$E_1 - E_2 = \frac{0.059}{z} \log_{10} \frac{c_1}{c_2} \quad (c_1 > c_2)$$

This type of cell has been used to determine the nature of the mercurous, Hg^I, ion.

CALCULATION OF EQUILIBRIUM CONSTANTS

Because redox reactions are of the reversible type, they are governed by the laws of chemical equilibrium. If a cell is set up and allowed to discharge until its e.m.f. is zero, the products and reactants are then at equilibrium since ΔG^0 is zero. Hence, the difference in potentials for the two component half-reactions must be zero. A general form of a redox reaction may be written as

$$a\, Ox_A + b\, Red_B \rightleftharpoons a\, Red_A + b\, Ox_B$$

$$E_A = E_A^\circ + \frac{RT}{zF} \ln \frac{(a_{Ox_A})^a}{(a_{Red_A})^a}$$

$$E_B = E_B^\circ + \frac{RT}{zF} \ln \frac{(a_{Ox_B})^b}{(a_{Red_B})^b}$$

at equilibrium and at 298 K

$$E_A - E_B = 0$$

or

$$E_A^\circ - E_B^\circ = \frac{0.059}{z} \log_{10} \frac{(a_{Ox_B})^b (a_{Red_A})^a}{(a_{Ox_A})^a (a_{Red_B})^b}$$

$$= \frac{0.059}{z} \log_{10} K$$

where K is the equilibrium constant.

Hence

$$\log_{10} K = \frac{(E_A^\circ - E_B^\circ)z}{0.059}$$

The value of K so derived is a measure of the completeness of the reaction under the conditions of the experiment. If K is large, then the reaction proceeds from left to right. Very small values of K mean the reaction takes place from right to left. If K is nearly unity the reaction will not be complete in either direction.

The following systems, under standard conditions, are examples:

(i) $\qquad MnO_4^- + 8H^+ + 5Fe^{2+} \rightleftharpoons Mn^{2+} + 5Fe^{3+} + 4H_2O$
(ii) $\qquad Sn^{4+} + 2Ce^{3+} \rightleftharpoons Sn^{2+} + 2Ce^{4+}$
(iii) $\qquad Ag + \frac{1}{2}Hg_2^{2+} \rightleftharpoons Hg + Ag^+$

The values of the equilibrium constants are

(i) $\log_{10} K_A = \log_{10} \dfrac{a_{Mn^{2+}} \times (a_{Fe^{3+}})^5}{a_{MnO_4^-} \times (a_{H^+})^8 \times (a_{Fe^{2+}})^5} = \dfrac{(1.51 - 0.77) \times 5}{0.059}$

$\qquad\qquad\qquad \approx +62$

(ii) $\log_{10} K_B = \log_{10} \dfrac{(a_{Ce^{4+}})^2 \times a_{Sn^{2+}}}{(a_{Ce^{3+}})^2 \times a_{Sn^{4+}}} = \dfrac{(+0.15 - 1.61) \times 2}{0.059}$

$\qquad\qquad\qquad \approx -50$

(iii) $\log_{10} K_C = \log_{10} \dfrac{a_{Ag^+} \times a_{Hg}}{a_{Ag} \times (a_{Hg_2^{2+}})^{\frac{1}{2}}} = \dfrac{(+0.789 - 0.799) \times 1}{0.059}$

$\qquad\qquad\qquad \approx -0.17$

DISPROPORTIONATION AND STABILIZATION OF VALENCY STATES

Many elements exhibit more than one oxidation state and for a particular element certain states are more stable than others. Electrode potential data can be used to illustrate the relative stabilities of oxidation states.

If we take the case of the Fe^{II}/Fe^{III} system, the standard electrode potentials for Fe^{3+}/Fe^{2+} and Fe^{2+}/Fe are $+0.771$ V and -0.44 V respectively. The possibility of the following disproportionation occurring may now be considered:

$$3Fe^{2+} = 2Fe^{3+} + Fe$$
$$E_{system}^0 = -0.44 - 0.771$$
$$= -1.211 \text{ V}$$

Under these conditions ΔG° is positive and hence disproportionation does not occur.

With copper, the two main oxidation states are Cu^{2+} and Cu^+. The possibility of the disproportionation

$$2Cu^+ = Cu^{2+} + Cu$$

may be considered from a knowledge of the two standard electrode potentials for Cu^{2+}/Cu^+ and Cu^+/Cu which are $+0.153$ V and $+0.521$ V respectively.

$$E = +0.521 - 0.153$$
$$= +0.368 \text{ V}$$

Since ΔG° is therefore negative, the disproportionation takes place as shown.

Stabilisation of Cu^+ may be effected by the presence of cyanide or iodide ions. The disproportionation

$$2CuI = Cu^{2+} + Cu + 2I^-$$

may be split into the two half-reactions

$$CuI + e^- = Cu + I^- \quad E^{\circ} = -0.185 \text{ V}$$
$$Cu^{2+} + I^- + e^- = CuI \quad E^{\circ} = +0.86 \text{ V}$$

Hence,

$$E^{\circ}_{system} = -0.185 - 0.86$$
$$= -1.045 \text{ V}$$

This negative value of E° means that the free energy change of the reaction is positive and hence CuI is stable. Examples of stabilisation of other transition metal ions by complex-formation are given in Chapter 10.

The potentials existing between various oxidation states of an element in aqueous solution are frequently written in diagrammatic form and this serves as a convenient summary of the chemistry of the element. Thus, for manganese, the potential diagram for the oxidation states in acid solution is:

$$MnO_4^- \xrightarrow{+0.564 \text{ V}} MnO_4^{2-} \xrightarrow{+2.26 \text{ V}} MnO_2 \xrightarrow{+0.95 \text{ V}} Mn^{3+} \xrightarrow{+1.51 \text{ V}} Mn^{2+} \xrightarrow{-1.18 \text{ V}} Mn$$

Recently a graphical method has been suggested by Ebsworth for summarizing the oxidation state interrelationship. The approach is to plot the free energy of the process (half-reaction).

$$M \longrightarrow M^{z+} + ze$$

against the number of electrons involved (z), and the various points are joined up. Such plots have interesting features and examples are given in *Figures 5.3 to 5.5*.

(a) A maximum, other than the lowest or highest oxidation state point, is one that is unstable (thermodynamically) with respect to disproportionation.

(b) A minimum in the plot represents an oxidation state that is thermodynamically stable with respect to its neighbours.

(c) A state represented by a convex point will be unstable with respect to disproportionation, i.e. at equilibrium there will be less of that state present than of the products of disproportionation. The reverse applies to a concave point.

(d) A state which falls on a line between a higher and lower oxidation state will have an equilibrium constant of one for the disproportionation.

(e) In aqueous solution at pH = 0, the slope of a line joining two states gives $E°$ for the change since

$$E° = -\frac{d(\Delta G°)}{dz}$$

Thus, in the case of the oxidation states of chlorine (*Figure 5.3*), where all points lie close to the line that joins chloride ion to perchlorate ion, there are

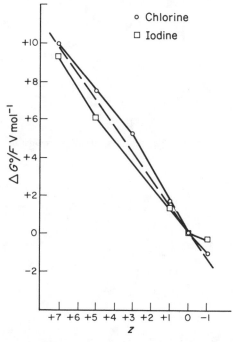

Figure 5.3. Oxidation state diagram for chlorine and iodine

no maxima or minima. This indicates the stability of the -1 and $+7$ oxidation states to disproportionation; of the remaining states elementary chlorine is the most stable and chlorite ion the least stable. Also all states except -1 are oxidising. A similar diagram for iodine in the same figure shows the stability of iodate in contrast to chlorate ion.

With copper (*Figure 5.4*) the $+1$ oxidation state falls on a convex point and is thus unstable with respect to disproportionation whereas the same state for silver is stable.

A plot for the first period transition metals (*Figure 5.5*) shows the increased oxidising power of the higher oxidation states across the series and the decreasing stability of the lower states relative to the metal.

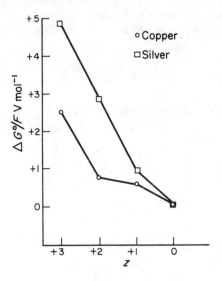

Figure 5.4. Oxidation state diagram for copper and silver

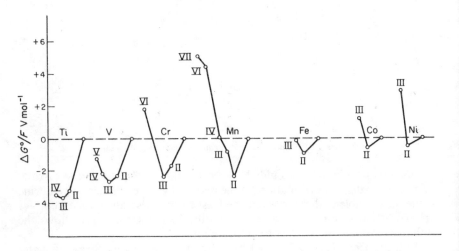

Figure 5.5. Oxidation state diagram for the first row transition metals

Such plots are useful in the qualitative approach to oxidation states but have their limitations. Thus, unless plotted on large scale diagrams there is sometimes difficulty in distinguishing between concave and convex points. Also the fact that the energy data plotted apply to particular environments must be remembered. Since they refer to energy, the plots must be treated with caution in their interpretation as in fact must all energy data; thus, in the case of ClO_3^-, although the energy data predict the instability of this ion with respect to disproportionation, the reaction is very slow.

ACIDS AND BASES

Many substances were formerly classified as acids or alkalis according to their chemical nature. Typical properties associated with acids were sour taste, high solvent power and ability to change the colour of vegetable dyes like litmus. Alkalis also had a number of positive properties such as detergency, the ability to dissolve sulphur and the distinctive soapiness of their solutions, but their chief characteristic was reaction with an acid to form a salt, a substance which had none of the properties of an acid or an alkali. This reaction of an alkali with an acid represented a neutralisation of the characteristics of both.

Many other substances, including the oxides and hydroxides of most metals, neutralised acids but they did not have the characteristic properties associated with alkalis. Therefore the more general term *base* was introduced to denote any substance which would react with an acid to form a salt.

The historical development of ideas on the nature of acids and bases has been discussed elsewhere and we shall be concerned with the more modern concepts, particularly those which are useful in the interpretation of inorganic reactions in water and other solvents.

Water is the most familiar solvent in which chemical reactions have been studied. It is, however, quite exceptional in many of its properties. The water molecule is strongly polar, intermolecular hydrogen bonding occurs and therefore water behaves as an associated liquid. Moreover, the high dielectric constant (*Table 5.4*) promotes the ionisation of solutes and so reactions in aqueous solutions generally take place between ions. Towards the end of the nineteenth century investigations were carried out on the properties of other ionising solvents with the use of these as media for chemical reactions as a major objective. Acid–base concepts primarily defined for aqueous systems were not comprehensive enough to include all the non-aqueous solvents studied so more general definitions were made. Non-aqueous solvents are now increasingly used in synthetic inorganic chemistry. A more fundamental understanding of the chemical properties of the elements and their compounds is likely to emerge from these studies. Hitherto, the special nature of water as a solvent has tended to obscure this by over-emphasis on those chemical substances which exist in the presence of water.

THE BRONSTED–LOWRY CONCEPT OF PROTONIC ACIDS

Arising from the theory of electrolytic dissociation proposed in 1884 by Arrhenius and developed later by Ostwald, a general definition of acids and bases was proposed in 1923 independently and almost simultaneously by J. N. Bronsted and T. M. Lowry. According to this, an acid is regarded as a donor of protons and a base as a proton-acceptor. Every acid A has its *conjugate base* B$^-$ to which it is related by

$$A \rightleftharpoons B^- + H^+$$

The acid A can only release its proton when brought into contact with a substance of higher proton affinity than B$^-$.

In water, self-ionisation occurs to a very limited extent:

$$H_2O + H_2O \rightleftharpoons H_3O^+ + OH^-$$

H_3O^+ is the solvated proton or oxonium ion. The bare proton has an enormous affinity for electrons and is invariably solvated in water. In the self-ionisation of water some molecules behave as Bronsted–Lowry acids and an equal number as bases.

When an acid dissolves in water it releases protons which become solvated to produce a higher concentration of oxonium ions. For example, hydrogen chloride ionises in water almost completely according to

$$H_2O + HCl \rightleftharpoons H_3O^+ + Cl^-$$

By definition, the ions HSO_4^- and NH_4^+ are also acids in aqueous media because of the dissociations

$$HSO_4^- + H_2O \rightleftharpoons SO_4^{2-} + H_3O^+$$

and $$NH_4^+ + H_2O \rightleftharpoons NH_3 + H_3O^+$$

The sulphate ion and ammonia molecule are the conjugate bases of the acids HSO_4^- and NH_4^+ respectively. Other Bronsted–Lowry acids are listed in *Table 5.2.*

Table 5.2. BRONSTED–LOWRY ACIDS AND BASES IN AQUEOUS SOLUTION

Species	*Acids*	*Bases*
Molecules	HI, HBr, HCl, HF, HNO$_3$, H$_2$SO$_4$, HClO$_4$, H$_3$PO$_4$, H$_2$S, H$_2$O	NH$_3$, N$_2$H$_4$, NH$_4$OH, amines, H$_2$O
Anions	HSO$_4^-$, H$_2$PO$_4^-$, HPO$_4^{2-}$, HS$^-$	I$^-$, Br$^-$, Cl$^-$, F$^-$, HSO$_4^-$, SO$_4^{2-}$, O^{2-}
Cations	NH$_4^+$, Al(H$_2$O)$_6^{3+}$, Fe(H$_2$O)$_6^{3+}$	HPO$_4^{2-}$, CO$_3^{2-}$, OH$^-$, Al(H$_2$O)$_5$(OH)$^{2+}$, Fe(H$_2$O)$_5$(OH)$^{2+}$

A base, B$'$, accepts a proton from the water molecule:

$$B' + H_2O \rightleftharpoons B'H^+ + OH^-$$

Many anions are bases and take up a proton to form a neutral molecule. Basic behaviour in an aqueous medium results in an increase in hydroxide ion concentration and therefore a decrease in the concentration of H_3O^+. The oxides of electropositive metals are well-known bases. They contain the oxide ion which has an enormous affinity for protons:

$$O^{2-} + H_3O^+ \rightleftharpoons OH^- + H_2O$$

Metal hydroxides which ionise in water to produce hydroxide ions directly are also classified as bases. Other typical bases are listed in *Table 5.2*.

An acid which has a great tendency to donate its proton is described as strong, one with only a slight tendency as weak. A base may be similarly described in terms of its proton affinity. A strong acid, e.g. HCl, has a weak conjugate base (Cl^-): a strong base, e.g. the carbonate ion, has a weak conjugate acid (HCO_3^-).

Hydrolysis

Neutralisation of an acid by a base takes place in aqueous solution to form a salt and water. The process may go to completion as in the reaction between sodium hydroxide and hydrochloric acid which may be represented as follows in terms of the participating ions:

$$H_3O^+ + Cl^- + Na^+ + OH^- = 2H_2O + Na^+ + Cl^-$$

or it may be reversible as in the reaction between aluminium hydroxide and hydrochloric acid:

$$3HCl + Al(OH)_3 = AlCl_3 + 3H_2O$$

The reaction of aluminium chloride with water represents the hydrolysis of this compound. (Hydrolysis is used as a more general term to describe the reaction between a compound and water and includes, for instance, the decomposition of the halides of non-metals to produce two acids, e.g. the hydrolysis of phosphorus trichloride according to the equation

$$PCl_3 + 3H_2O = 3HCl + H_3PO_3)$$

Concerning the hydrolysis of salts, three types of reaction may be distinguished.
- (a) When the salt of a strong acid and a strong base dissolves in water, the constituent ions become solvated. This is the only change which occurs and the pH remains constant. Sodium chloride is one of many examples.
- (b) When a salt of a strong base and a weak acid dissolves, the cations become solvated as in (a). The anions are strongly basic, however, having a high proton affinity, and they remove protons from the solvent causing the pH to rise. Sodium carbonate is a case in point, the pH increase being due to the reaction

$$CO_3^{2-} + H_2O \rightleftharpoons HCO_3^- + OH^-$$

- (c) When the salt of a weak base and a strong acid dissolves, the pH falls. The hydrolysis of aluminium chloride is an example of this.

It is interesting to consider the hydrolytic reactions (b) and (c) from the viewpoint of the Bronsted–Lowry concept. In (b) the carbonate ion may be regarded as having markedly basic properties compared with the extremely weak acidic properties of the hydrated sodium ion. This has very little tendency to donate a proton. The net removal of protons from solvent molecules results in a pH increase. In (c), the hydrated metal ion is $[Al(H_2O)_6]^{3+}$

and this is a Bronsted–Lowry acid. Its tendency to lose a proton is stronger than that of the chloride ion to gain one and hence there is a decrease in pH.

THE RELATIVE STRENGTHS OF ACIDS AND BASES

It is very useful to have a quantitative measure of the tendency of an acid to lose a proton and of a base to gain one, i.e. of the acid and base strengths.

For a protonic acid HX, the following equilibrium is established in aqueous solution:

$$HX + H_2O \rightleftharpoons X^- + H_3O^+$$

The equilibrium constant of this system, K, expresses the relative tendencies of HX and H_3O^+ to donate a proton:

$$K = \frac{a_{X^-} \times a_{H_3O^+}}{a_{HX} \times a_{H_2O}} \tag{1}$$

a_{H_2O} is virtually constant for dilute solutions and it is usual to define a second constant, K_a, as

$$K_a = \frac{a_{X^-} \times a_{H_3O^+}}{a_{HX}} \tag{2}$$

K_a denotes the acid strength, that is, the extent to which HX is ionised. In water the conjugate base X^- can undergo the reaction

$$X^- + H_2O \rightleftharpoons HX + OH^-$$

for which

$$K_b = \frac{a_{HX} \times a_{OH^-}}{a_{X^-}}$$

$K_a K_b = K_w (= a_{H_3O^+} \times a_{OH^-})$ the ionic product for water which has the value

$$K_w = 1{\cdot}0 \times 10^{-14} \text{ at } 298{\cdot}15 \text{ K}$$

A number of experimental methods are available for the measurement of acid and base strengths; reference should be made to a text-book of physical chemistry for details of these. The values of K vary through many powers of ten for different acids and bases. It is generally more convenient to express acid strengths as dissociation exponents, $pK_a (= -\log_{10} K_a)$. A large value of pK_a means that the acid is little dissociated and therefore weak: a small value is found for pK_a when the acid is strong.

THE ACIDIC AND BASIC PROPERTIES OF HYDRIDES

Any binary compound containing hydrogen which is chemically bound to a more electronegative element than itself can, in principle, behave as an acid. In aqueous solution, acid properties are associated chiefly with the hydrides of the non-metallic elements of Groups VI and VII. Values of pK_a are listed in *Table 5.3*. The values for HCl, HBr and HI are known only approximately because of virtually complete ionisation in dilute solutions. These compounds are all polar molecules with the positive end of the dipole located on the hydrogen atom. No simple relationship exists between polarity and acid strength, for the compounds of greatest polarity are the weakest acids (H_2O and HF). In the two series H_2O to H_2Te and HF to HI, the dipole moment decreases but the acid strength increases as the molecular weight increases. The observed trends in acid strength are contrary to expectation if the

Table 5.3. EXPERIMENTAL pK_a VALUES FOR BINARY HYDRIDES

Compound	pK_a	Compound	pK_a
H_2O	15·89	HF	3·17
H_2S	7·06	HCl	-7
H_2Se	3·72	HBr	< -7
H_2Te	2·64	HI	< -7

tendency to dissociate were simply related to the electronegativity of the non-metal present: the trend for the hydrogen halides is more profitably considered from the point of view of the energy changes when dissociation in aqueous solution takes place.

It is interesting to note that ammonia is the only simple hydride which has basic properties in aqueous solution. Both the NH_3 and H_2O molecules have, as we have seen, a tendency to accept protons from donor molecules. Evidently, this tendency is the greater with ammonia.

THE PROPERTIES OF HYDROXIDES AND OXO-ACIDS

In an ionising solvent like water, the group $M-O-H$, where M represents one atom of a particular element which may or may not be bound to other atoms or groups beside the hydroxyl group, can dissociate in one of two ways:

(i) $\qquad M-O-H = (M-O)^- + H^+$
(ii) $\qquad M-O-H = M^+ + (O-H)^-$

If M is an electronegative element, it exerts a strong attraction on the electrons associated with the oxygen. This withdrawal of electrons weakens the $O-H$ bond and ionisation according to (i) is favoured. MOH is then an oxo-acid. On the other hand, if M is an electropositive metal, dissociation (ii) occurs and MOH is a basic hydroxide. Certain MOH groups are amphoteric: that is, dissociation (i) is promoted by the presence of a base to remove hydrogen ions, (ii) is favoured when an acid is present. Amphoteric properties arise when M is a weakly electropositive metal.

Two empirical rules, enunciated by Pauling, are useful approximate relationships between the strengths of oxo-acids.

Rule 1: The successive dissociation constants, K_1, K_2, K_3, etc., for a polybasic acid are in the ratios

$$1 : 10^{-5} : 10^{-10} : \ldots$$

For example, for phosphoric acid, H_3PO_4,

$$K_1 = 0.75 \times 10^{-2}$$
$$K_2 = 0.62 \times 10^{-7}$$
$$K_3 = 1.0 \ \times 10^{-12}$$

Rule 2: The first dissociation constant of the oxo-acid $XO_m(OH)_n$ is determined by the value of m.

When $m = 0$, the acid is very weak and $K \leqslant 10^{-7}$.
Examples:

        boric acid, $B(OH)_3$; $K_1 = 5.8 \times 10^{-10}$
        hypobromous acid, $Br(OH)$; $K_1 = 2.0 \times 10^{-9}$.

When $m = 1$, the acid is moderately weak and $K_1 = \approx 10^{-2}$.
Examples:

        sulphurous acid, $SO(OH)_2$; $K_1 = 1.2 \times 10^{-2}$
        arsenic acid, $AsO(OH)_3$; $K_1 = 0.5 \times 10^{-2}$
        nitrous acid, $NO(OH)$; $K_1 = 4.5 \times 10^{-4}$

When $m = 2$, the acid is strong and K_1 is large.
Examples:

        nitric acid, $NO_2(OH)$; $K_1 = \approx 10^1$
        chloric acid, $ClO_2(OH)$; $K_1 = \approx 10^1$
        sulphuric acid, $SO_2(OH)_2$; $K_1 = \approx 10^3$

When $m = 3$, the acid is very strong and K_1 is very large.
Example: perchloric acid, $ClO_3(OH)$; $K_1 = \approx 10^8$

The first rule reflects the increase in electrostatic attraction of the negative ion for a proton as the extent of ionisation increases. The second rule expresses the well-known fact that the acid strength increases with the proportion of oxygen in the molecule. For exa mple, the four oxo-acids of chlorine ionise to give the anions ClO^- (hypochlorite), ClO_2^- (chlorite), ClO_3^- (chlorate) and ClO_4^- (perchlorate). The single negative charge may be regarded as located on one oxygen in ClO^-, 'shared' between two oxygens in ClO_2^-, and so on. The attraction of the anion for a proton therefore decreases as the number of oxygen atoms present increases and the acid strength increases in the order

$$HClO < HClO_2 < HClO_3 < HClO_4$$

It is interesting to note that phosphonic acid, H_3PO_3, and phosphinic acid, H_3PO_2, are both moderately weak with K_1 values of 1.6×10^{-2} and 1×10^{-2} respectively. At first :ight it does not appear possible to classify them according to *Rule 2*. However, their chemical properties and structures show that P—H bonds are present and that their formulae are more accurately written as $HPO(OH)_2$ and $H_2PO(OH)$. Their K_1 values are then typical of the oxo-acids for which $m = 1$.

EXTENSION OF THE PROTONIC CONCEPT TO NON-AQUEOUS SOLVENTS

Non-aqueous solvents which have been most extensively studied include NH_3, HF, HCN; anhydrous acids such as H_2SO_4, HNO_3 and CH_3COOH;

SO_2, N_2O_4; and covalent halides like BrF_3, ICl and $AsCl_3$. The relevant physical properties are summarised in *Table 5.4*.

With the necessary exception of those solvents which do not contain hydrogen, the protonic concept can be applied to reactions in non-aqueous solvents. Four types of solvent can be recognised:

Acidic. These have a marked tendency to release protons.

Examples: HF, H_2SO_4, CH_3COOH, HCN.

Basic. These accept protons strongly.

Examples: NH_3, N_2H_4.

Amphiprotic. These can either accept or release protons.

Examples: Water and the alcohols.

Aprotic. No tendency to release or accept protons.

Examples: Benzene, chloroform.

A typical acidic solvent such as acetic acid undergoes a limited amount of self-ionisation:

$$2CH_3COOH \rightleftharpoons CH_3COO^- + CH_3COOH_2^+$$

This accounts for the electrical conductivity shown by the pure solvent.

Table 5.4. SOME PHYSICAL CONSTANTS OF WATER AND OTHER SOLVENTS

Solvent	m.p./K	b.p./K	Dielectric constant (ε_r)	Electrical conductivity κ/S m^{-1}
H_2O	273·15	373·15	81·8 (291·15 K)	6×10^{-6} (298·15 K)
NH_3	195·45	239·80	22 (240·15 K)	4×10^{-9} (195·45 K)
HF	188·15	292·65	83·6 (273·15 K)	1×10^{-3} (236·15 K)
HCN	259·75	298·75	123·0 (289·15 K)	5×10^{-5} (273·15 K)
H_2SO_4	283·55	547·15	≈ 85 (293·15 K)	2×10^0 (291·15 K)
HNO_3	232·15	359·15		9×10^{-1} (273·15 K)
CH_3COOH	289·75	391·15	9·7 (291·15 K)	4×10^{-7} (298·15 K)
SO_2	197·45	263·15	13·8 (288·15 K)	1×10^{-5} (273·15 K)
N_2O_4	261·85	294·15	2·4 (291·15 K)	1×10^{-10} (290·15 K)
BrF_3	264·15	400·15		8×10^{-1} (298·15 K)
ICl	300·35	370·55		5×10^{-1} (308·15 K)
$AsCl_3$	255·15	403·35	12·8 (293·15 K)	1×10^{-5} (293·15 K)
$POCl_3$	274·35	378·95	13·9 (295·15 K)	2×10^{-6} (293·15 K)
$SOCl_2$	168·65	348·85	9·1 (293·15 K)	3×10^{-7} (293·15 K)

The molecule is a poor proton acceptor but a somewhat better proton donor. Only those compounds which behave as strong acids in water exhibit markedly acid character in acetic acid solution. This solvent therefore acts as a *differentiating* solvent towards acidic solutes, and many acids which are completely dissociated in water, that is, appear equally strong in water, are only partially ionised in acetic acid. Conductivity measurements on their solutions in acetic acid have shown that the strengths of a number of common mineral acids decrease in the order

$$HClO_4 > HBr > H_2SO_4 > HCl > HNO_3$$

The proton-donor property of acetic acid means that a solute which acts as a weak base in aqueous solution usually becomes a strong base in acetic acid.

The solvent thus exerts a *levelling effect* on bases because in it they tend to become equally strong. By analogy with the behaviour of metal hydroxides in water, metal acetates in acetic acid act as bases. They can, for instance, be neutralised by a strong acid such as perchloric.

Liquid ammonia typifies a solvent with predominant basic properties. Self-ionisation occurs thus:

$$2NH_3 \rightleftharpoons NH_4^+ + NH_2^-$$

Ammonia co-ordinates the proton strongly and enhances the acidic character of hydrogen-containing solutes. A solute which is a weak acid in water becomes much stronger when dissolved in ammonia. All acids tend to become equally strong, that is, a basic solvent exerts a levelling effect towards them. The ammonium ion, NH_4^+, in liquid ammonia corresponds to the oxonium ion, H_3O^+, in water. An ammonium salt dissolved in liquid ammonia acts as an acid. Conversely, a solute such as a metal amide which increases the concentration of amide ions, NH_2^-, is a base.

Many reactions in non-aqueous solvents have important analytical applications. For instance, it is often feasible to carry out rapid and reliable volumetric analyses for weak bases or acids in a suitable non-aqueous solvent in cases where the use of aqueous solutions is precluded because of the limited ionisation therein. For details, reference should be made to appropriate textbooks on analytical chemistry.

Lewis acids and bases

According to the electronic theory of G. N. Lewis, proposed in 1923, an acid is a substance which accepts an electron pair, whereas a base is a substance which can donate an electron pair. The process of neutralisation involves the formation of a co-ordinate bond. The compound formed may subsequently ionise.

For example, ammonia is a Lewis base because the nitrogen atom carries a lone pair of electrons which it can donate to a proton:

$$H^+ + {}^x_x NH_3 \rightleftharpoons (H \leftarrow NH_3)^+$$

Hydrogen chloride is a Lewis acid because it can co-ordinate with a base and then ionisation takes place. The reaction with water can be represented as

$$H_2O + HCl = H_2O \rightarrow HCl = H_3O^+ + Cl^-$$

and that with a tertiary amine as

$$R_3N + HCl = R_3N \rightarrow HCl = R_3NH^+ + Cl^-$$

This is a more general approach than the protonic concept in so far as acid–base behaviour is not dependent on the presence of one particular element or on the presence or absence of a solvent. However, it does mean that many substances must be regarded as Lewis acids which are not acids at all in the Bronsted–Lowry sense. For example, the Lewis definition requires that the reaction between a metal ion and a ligand shall be regarded as the neutralisation of an acid (the metal ion) by a base (the ligand), i.e. a case such as

$$Ag^+ + {}_x^xNH_3 \rightleftharpoons (Ag \leftarrow NH_3)^+$$

is exactly comparable with the formation of the ammonium ion. The analogy here is of great utility in the correlation of the behaviour of an electron-pair donor (i) as a base which reacts with a proton-donating acid and (ii) as a ligand which forms complexes with metals.

One disadvantage of the Lewis approach is the lack of a uniform scale of acid or base strength. Instead, acid and base strengths are variable and dependent on the reaction chosen. It is, however, of great value in circumstances where the protonic concept is inapplicable, as in reactions between acidic and basic oxides in the fused state.

A convenient general definition which can be used for non-aqueous solvents is that due to Cady and Elsey, who defined acids as solutes which increase the concentration of cations characteristic of the pure solvent and bases as solutes which increase the concentration of anions characteristic of the pure solvent. This applies equally well to protonic systems, such as

$$2NH_3 \rightleftharpoons NH_4^+ + NH_2^-$$

and to non-protonic systems like

$$2BrF_3 \rightleftharpoons BrF_2^+ + BrF_4^-$$

In bromine trifluoride, potassium fluoride behaves as a base because it increases the concentration of BrF_4^- ions:

$$KF + BrF_3 \rightleftharpoons K^+ + BrF_4^-$$

Antimony pentafluoride acts as an acid in this solvent because of the reaction

$$SbF_5 + BrF_3 = BrF_2^+ + SbF_6^-$$

A neutralisation reaction in this solvent proceeds according to

$$KBrF_4 + BrF_2SbF_6 = KSbF_6 + 2BrF_3$$

REACTIONS IN NON-AQUEOUS SOLVENTS

Many inorganic substances dissolve in a solvent with chemical change, that is, they undergo solvolysis. A particular example is the hydrolysis of salts in aqueous solution. Alternatively, the solvent itself may not be involved but is merely used as a medium in which to carry out the reaction. Five non-aqueous solvents which are of special interest are now discussed in more detail.

Liquid ammonia

Ammonia has been extensively used as a solvent. Although it shows certain resemblances to water in its solvent properties, it is less strongly associated through hydrogen bonding and its melting point and boiling point are correspondingly lower than those of water. Another consequence is that all salts are weakly dissociated in ammonia because considerable ion-pair formation occurs; for instance, ammonium halides exist largely as ion-pairs, $NH_4^+X^-$. Nevertheless, ammonia is a good solvent for many nitrogen-containing solutes like nitrates and nitrites and for organic compounds such as amines, phenols and carboxylic acids which can form hydrogen bonds with solvent molecules. The salts of transition metals such as nickel, zinc, silver and copper often have much greater solubilities in ammonia than in water because of the tendency of ammonia to co-ordinate strongly with these metals. Silver halides show appreciable solubility in ammonia for this reason and the metathetical reaction between silver chloride and barium nitrate to precipitate sparingly soluble barium chloride illustrates one application of this difference in solubilities:

$$2AgCl + Ba(NO_3)_2 = BaCl_2 + 2AgNO_3$$

Ammonolysis occurs with many solutes. The tetrahalides of silicon and germanium (except the fluorides) are readily and completely converted to the amides $M(NH_2)_4$ at low temperatures.

$$223\ K$$
$$SiCl_4 + 8\ NH_3 = Si(NH_2)_4 + 4NH_4Cl$$

The comparable solvolysis with water is the conversion to hydrated silica.

Sn(IV) halides and transition metal halides like $TiCl_4$, $ZrCl_4$, VCl_4 and $MoCl_5$ are only partially ammonolysed.

$$240\ K$$
$$SnCl_4 + 6NH_3 = SnCl(NH_2)_3 + 3NH_4Cl$$

Reactions of chlorine, bromine and iodine with ammonia are typified by:

$$Cl_2 + 2NH_3 = NH_4Cl + NH_2Cl$$

and chloramine reacts further to give hydrazine and nitrogen:

$$NH_2Cl + 2NH_3 = NH_2.NH_2 + NH_4Cl$$

Neutralisation reactions take place between an acid and a base. Representative of these are the reactions between an ammonium halide (acid) and a metal amide, imide or nitride (base):

$$NH_4Cl + KNH_2 = KCl + 2NH_3$$
$$2NH_4I + PbNH = PbI_2 + 3NH_3$$
$$3NH_4I + BiN = BiI_3 + 4NH_3$$

The course of such reactions may be followed by conductometric titration methods and sometimes by the use of indicators. For example, phenolphthalein is pale red in liquid ammonia but becomes intensely red in the presence of amides. As a result, the alkali amides can be titrated with ammonium salts in liquid ammonia with phenolphthalein as indicator in exactly the same way as acids can be titrated with alkalis in aqueous solution.

In some cases it is preferable to carry out a reaction in liquid ammonia rather than in water. For instance, ammonium bromide in ammonia gives a better yield than aqueous hydrochloric acid in the preparation of silanes from magnesium silicide:

$$Mg_2Si + HCl\,(in\ H_2O) = Silanes\,(Si_2H_6,\ etc.)\ 25\%\ yield$$
$$Mg_2Si + NH_4Br\,(in\ NH_3) = Silanes\,(chiefly\ SiH_4,\ Si_2H_6)\ 70–80\%\ yield$$

Extensive hydrolysis of the products reduces the yield in aqueous solution, but ammonolysis is less extensive than hydrolysis because the N—H bonds of ammonia are less easily broken than the O—H bonds of water.

Metal–ammonia solutions

One characteristic property of liquid ammonia is its ability to dissolve the alkali and alkaline-earth metals. These metal–ammonia solutions are meta-stable and are catalytically decomposed by platinum black to hydrogen and the metal amide:

$$Na + NH_3 = \tfrac{1}{2}H_2 + NaNH_2$$

The alkali metals can be recovered unchanged by evaporation of their am-monia solutions; an alkaline-earth metal is obtained as its hexammoniate, $M.6NH_3$. In this respect, ammonia and water differ sharply in their be-haviour.

Dilute solutions of alkali metals in ammonia have a deep-blue colour, they are less dense than the parent solvent, excellent electrical conductors and strongly paramagnetic. At high concentrations, the conductivity approaches that of the metals themselves; at low concentrations it is similar to that of electrolytic solutions in water. Metal–ammonia solutions are noted for their powerful reducing properties. To account for their behaviour it has been suggested that the metal dissolves in liquid ammonia to produce solvated cations and valency electrons moving in expanded orbitals around the cations,

$$Na(solvent)^+ + e^-$$

The isolation of solids such as $Ca(NH_3)_6$ suggests that this theory is correct, at least for concentrated solutions. The properties of metal–ammonia solutions have also been interpreted in terms of solvated metal ions and solvated electrons. The system resembles that of a solution of a strong electrolyte except that electrons instead of negative ions occupy the cavities formed by the orientation of solvent molecules. On this theory, the blue paramagnetic species present in dilute solutions is the electron in a solvent cavity, not in an expanded metal orbital.

Many compounds which cannot be made in the presence of water have been prepared using metal–ammonia solutions to effect reduction. A powerful reducing agent in aqueous solution may reduce water to hydrogen instead of reducing the compound under examination. Solutions of sodium in ammonia, for instance, can reduce many compounds to the free elements and sometimes to intermetallic compounds. Silver salts are reduced to the metal; bismuth tri-iodide, on the other hand, is reduced to the metal and various inter-metallic substances which have been assigned the formulae Na_3Bi, Na_3Bi_3

and Na_3Bi_5. These formulae were deduced from potentiometric and conductometric titrations on ammoniacal solutions of BiI_3 with sodium in ammonia. The species indicated by the formulae do not necessarily correspond with compounds which can be isolated as distinct entities. Again, lead iodide is first converted to the metal, then the normal plumbide, Na_4Pb, and finally a polyplumbide, Na_4Pb_9, is formed. Sulphur reacts with alkali metal solutions in ammonia to give a variety of sulphides of general formulae M_2S_4, M_2S_2 and M_2S.

Reactions of special interest are those which lead to the preparation of compounds of metals showing an unusual oxidation state. One of the best known examples is the reduction of the complex cyanide $K_2M(CN)_4$ (where M is Ni, Co or Pd) using potassium in liquid ammonia. In the case of nickel, reduction of $[Ni^{II}(CN)_4]^{2-}$ proceeds via the complex $[Ni^I(CN)_3]_2^{4-}$ to $[Ni(CN)_4]^{4-}$, in which the nickel is zerovalent:

$$2[Ni(CN)_4]^{2-} + 2e^- = [Ni(CN)_3]_2^{4-} + 2CN^-$$
$$[Ni(CN)_3]_2^{4-} + 2CN^- + 2e^- = 2[Ni(CN)_4]^{4-}$$

The corresponding tetracyano-complexes of Co^0 and Pd^0, $[Co(CN)_4]^{4-}$ and $[Pd(CN)_4]^{4-}$ respectively, are also known.

Another example of the use of alkali metal solutions in ammonia is in the preparation of the metal carbonyl hydrides.

Hydrogen fluoride

This solvent has a high dielectric constant; it is an excellent ionising solvent and dissolves many inorganic and organic compounds to give highly conducting solutions. HF is a strongly acid solvent and its proton-donor tendency is so powerful that even a molecule like nitric acid becomes a proton-acceptor or base when dissolved in this solvent:

$$HF + HNO_3 \rightleftharpoons H_2NO_3^+ + F^-$$

The self-ionisation of hydrogen fluoride may be written as:

$$2HF \rightleftharpoons H_2F^+ + F^-$$

Ionic fluorides like potassium fluoride dissolve readily and are bases because they increase the concentration of fluoride ion in solution. Usually the existence of basic solutes only need be considered because, with the possible exception of very strong acids like perchloric, no substances capable of increasing the H_2F^+ concentration in liquid hydrogen fluoride have been reported. Certain fluorides, notably SbF_5, AsF_5 and BF_3 dissolve in HF to give solutions which are capable of dissolving an electropositive metal like magnesium. This has been suggested as evidence in support of the acid nature of these solutes.

Many metallic salts of other acids (chlorides, bromides, iodides, cyanides, etc.) react with HF and are converted to fluorides with the liberation of the appropriate hydrogen halide:

$$\text{e.g.} \quad MCl + HF = M^+ + F^- + HCl$$

The reaction proceeds further with salts of oxo-acids such as nitrates:

$$KNO_3 + 2HF = K^+ + H_2NO_3^+ + 2F^-$$

Solvolysis occurs with sulphuric acid with the formation of fluorosulphonic acid, $HO.SO_2F$.

One of the many preparations which have been carried out in anhydrous hydrogen fluoride as solvent is that of anhydrous silver tetrafluoroborate, $AgBF_4$. This is strongly hygroscopic and so can only be made anhydrous in the complete absence of water. $AgBF_4$ is precipitated when silver nitrate and boron trifluoride solutions in HF are mixed. On the assumption that BF_3 acts as an acid, the reactions may be written:

$$AgNO_3 + 2HF = Ag^+ + H_2NO_3^+ + 2F^-$$
$$BF_3 \quad + 2HF = H_2F^+ + BF_4^-$$
$$Ag^+ \quad + BF_4^- = AgBF_4 \downarrow$$

Dinitrogen tetroxide

This is the diamagnetic compound formed by the dimerisation of two molecules of paramagnetic nitrogen dioxide, NO_2.

$$2NO_2 \rightleftharpoons N_2O_4$$

The extent of dimerisation is temperature-dependent and is greater the lower the temperature. At high temperatures in the gas phase, NO_2 is the more important but in the liquid state N_2O_4 predominates. Dissociation to.NO_2 is only 0·13 per cent at the boiling point.

N_2O_4 is not an ionising solvent and no simple salts are soluble in it. As a solvent it has been compared with diethyl ether. Many organic compounds are soluble in dinitrogen tetroxide and it has been widely used as a medium for carrying out organic reactions.

The pure liquid is a poor electrical conductor (*Table 5.4*) but ionisation to NO^+ and NO_3^- is promoted by the presence of a second solvent of higher dielectric constant. Thus in nitromethane, dielectric constant = 37, the electrical conductivity rises to $\approx 10^{-3}\,S\,m^{-1}$ and in pure sulphuric acid the ionisation to NO^+ is complete.

The salt nitrosyl bisulphate, $NO.HSO_4$, can be crystallised from solutions of ·N_2O_4 in excess sulphuric acid. Other nitrosyl salts have been made by solution of the tetroxide in the appropriate acid. For example, nitrosyl perchlorate, $NOClO_4$, is prepared by the reaction

$$N_2O_4 + HClO_4 = NOClO_4 + HNO_3$$

Dinitrogen tetroxide is of current interest because of its use in the preparation of anhydrous metal nitrates, the properties of which are strikingly different from those of the more familiar hydrated salts. A number of nitrates have been prepared by C. C. Addison and co-workers using mixed solvents consisting of N_2O_4 diluted with an organic solvent of higher dielectric constant to promote ionisation. This type of mixed solvent attacks metals which are unattacked by pure N_2O_4. For example, copper dissolves in an ethyl acetate/N_2O_4 mixture (1:1 by volume): from this solution the solvated compound $Cu(NO_3)_2 \cdot N_2O_4$ can be precipitated by the addition of excess N_2O_4. This loses N_2O_4 at 358 K to give pale-blue anhydrous copper nitrate, a solid which is appreciably soluble in many organic solvents containing oxygen or nitrogen. It sublimes readily when heated *in vacuo* above 473 K. Electron diffraction measurements on the compound in the vapour phase have shown that it has the structure

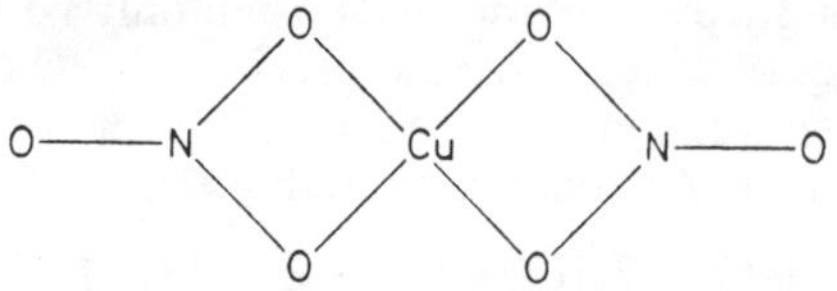

The copper atom has a square-planar co-ordination by four oxygen atoms each at a distance of 200 pm.

Other anhydrous nitrates such as those of lithium and sodium have been made by the action of liquid N_2O_4 on the metal carbonates. One of the most recently prepared nitrates is basic beryllium nitrate, $Be_4O(NO_3)_6$. Beryllium chloride undergoes solvolysis in ethyl acetate/dinitrogen tetroxide mixtures to give the crystalline addition compound, $Be(NO_3)_2 \cdot 2N_2O_4$. On heating this, anhydrous beryllium nitrate is first formed, but this decomposes at 398 K to N_2O_4 and the volatile basic beryllium nitrate of the above formula. The composition is analogous to that of basic beryllium acetate and it is probable that the two structures are similar with the nitrate group acting as a bridging group in exactly the same way as the acetate group. The properties of the nitrate group are becoming more fully understood as the result of these investigations.

Sulphur dioxide

This solvent is used industrially as a refrigerant and as an extractive solvent in petroleum refining. It is thus readily available and extensive chemical investigations have been carried out using it. The relatively low value of its dielectric constant (*Table 5.4*) accords with its property of dissolving covalent rather than ionic compounds. For instance, the halogen compounds Br_2, $AsCl_3$, BCl_3, CCl_4, ICl, PCl_3, $POCl_3$ and SO_2Cl_2 are all miscible with SO_2. Iodides and thiocyanates are the most soluble inorganic salts.

Solvolysis of solutes can occur, for example:

$$4KBr + 4SO_2 = 2K_2SO_4{\downarrow} + S_2Br_2 + Br_2$$

Preparations carried out in this solvent include thionyl bromide:

$$2KBr + SOCl_2 = SOBr_2 + 2KCl$$

and potassium hexachloroantimonate (III)

$$6KI + 3SbCl_5 = 2K_3SbCl_6{\downarrow} + 3I_2 + SbCl_3$$

Boron trifluoride, BF_3, reacts with acetyl fluoride, CH_3COF, to form acetyl fluoroborate:

$$CH_3COF + BF_3 = CH_3COBF_4$$

The solvent has been used experimentally to demonstrate the existence of the oxonium ion. H_2O and HBr dissolve in sulphur dioxide in equimolar proportions; the addition of more water merely causes the excess to separate as a second phase. This suggests some kind of compound formation. On electrolysis of the sulphur dioxide phase, the cathode products are water and hydrogen, the anode product is bromine. These observations can be explained only on the assumption of the formation of the oxonium compound H_3OBr, ionised to H_3O^+ and Br^-.

The reaction between thionyl compounds and sulphites takes place according to

$$SOCl_2 + Cs_2SO_3 = 2CsCl + 2SO_2$$

It was believed that this reaction supported the idea of self-ionisation according to:

$$2SO_2 \rightleftharpoons SO^{2+} + SO_3^{2-}$$

and therefore that a thionyl compound would be defined as an acid and a sulphite as a base in the sulphur dioxide solvent system. However, the nature of the products of the above reaction does not establish the mechanism of the reaction by which they are formed and there appears at present to be no undisputed evidence that sulphur dioxide can ionise as shown.

Sulphuric acid

Sulphuric acid has a high dielectric constant and so is a good solvent for electrolytes. For example, alkali metal sulphates are fully ionised in solution in sulphuric acid. They give rise to the hydrogen sulphate ion, HSO_4^-, and are classed as strong bases, cf. hydroxides in water.

$$KHSO_4 = K^+ + HSO_4^-$$

Normal sulphates are converted to hydrogen sulphates and hence are the analogues of metal oxides in water.

$$K_2SO_4 + H_2SO_4 = 2K^+ + 2HSO_4^-$$

Many compounds which behave as acid solutes in water are readily protonated in sulphuric acid. Thus, phosphoric acid acts as a strong base:

$$H_3PO_4 + H_2SO_4 = H_4PO_4^+ + HSO_4^-$$

Many organic compounds are similarly protonated, e.g.,

$$R.COOH + H_2SO_4 = R.COOH_2^+ + HSO_4^-$$

Acids in the sulphuric acid solvent system are limited in number. Some weak acids are disulphuric acid, $H_2S_2O_7$, and the higher polysulphuric acids present in oleum. Boric acid is very soluble in sulphuric acid and conductivity measurements have shown the presence of six ions for each molecule of boric acid originally added.

$$H_3BO_3 + 6H_2SO_4 = B(HSO_4)_4^- + 3H_3O^+ + 2HSO_4^-$$

The $B(HSO_4)_4^-$ ion is derived from the acid $HB(HSO_4)_4$ and this appears to be relatively strong.

Solutions of nitric acid, metal nitrates and dinitrogen pentaoxide in sulphuric acid are well-known reagents for aromatic nitration. In each case, the active nitrating species is the nitryl ion, NO_2^+.

$$HNO_3 + 2H_2SO_4 = NO_2^+ + H_3O^+ + 2HSO_4^-$$

This solvent has also been used to prepare stable solutions of ions which cannot exist in more basic solvents like water. For example, when I_2 and HIO_3 in the mole ratio of 7·0 are dissolved in sulphuric acid, the I_3^+ cation is formed:

$$HIO_3 + 7I_2 + 8H_2SO_4 = 5I_3^+ + 3H_3O^+ + 8HSO_4^-$$

Iodic acid alone dissolves in sulphuric acid to produce iodyl hydrogen sulphate, $IO_2 . HSO_4$,

$$HIO_3 + H_2SO_4 = IO_2 . HSO_4 + H_3O^+HSO_4^-$$

although this compound is polymeric and there is no evidence for the existence of the simple ion, IO_2^+.

HIGH-TEMPERATURE REACTIONS IN LIQUID MEDIA

Many important metallurgical reactions are carried out at high temperatures where at least some of the reactants or products are liquid. Reactions are possible under these conditions which cannot be effected in the presence of water. For example, the reactive metals sodium, magnesium, calcium and aluminium are prepared industrially from their fused compounds by electrolytic reduction, and fusion electrolysis is also used in the preparation of tantalum metal on a small scale and in the refining of titanium.

Most pyrometallurgical processes involve the chemical reduction of a metallic oxide to the metal. Only in a few cases are the compounds to be reduced available in the pure state and usually the metal produced must be separated from the impurities present. This separation is more easily effected when the metal is formed as a vapour (e.g. Zn, Mg) or as a liquid (Fe, Cu) than when it is produced as a solid (Ti, Zr, Mo). However, it is common practice to remove impurities by slag formation during the final reduction process.

Where the impurities are refractory, that is they are of high melting point, it is necessary to add materials (known as fluxes) which will combine with them to form a mixture, or slag, of compounds. Maximum removal of impurity is achieved when the slag is a liquid immiscible with the molten metal at the operating temperature. For example, lime is added as a flux to the blast furnace charge used in the reduction of iron oxide; the flux forms a slag with SiO_2 and P_2O_5 consisting of calcium silicate and phosphate.

Most industrial slags are formed by the mutual solution or combination of oxides which are classified according to their chemical behaviour as acidic, amphoteric or basic. SiO_2, P_2O_5, B_2O_3, Sb_2O_3 and TeO_2 are among the acidic oxides. Al_2O_3, Fe_2O_3, SnO_2 and ZnO are amphoteric because they behave as acids in the presence of a basic oxide and as bases in the presence of an acidic oxide. The oxides of Mg, the alkaline earths and the alkali metals are basic.

Slag formation can be regarded as the neutralisation of a Lewis base by a Lewis acid. For example, a basic oxide (or a carbonate which thermally decomposes to the oxide) supplies oxide ions. These co-ordinate with acidic oxides such as SiO_2,

$$O^{2-} + SiO_2 = SiO_3^{2-}$$

and P_2O_5,

$$3O^{2-} + P_2O_5 = 2PO_4^{3-}$$

Similar types of neutralisation are known in qualitative analysis in the borax bead and metaphosphate bead tests. For example, the reaction of a metaphosphate, $(PO_3)_x^{x-}$, a Lewis acid, with the oxide ion, a Lewis base, produces an orthophosphate:

$$(PO_3)_x^{x-} + xO^{2-} = xPO_4^{3-}$$

One further example is the conversion of refractory, insoluble oxides such as titanium dioxide to a soluble derivative by acid fusion with disulphate or hydrogen sulphate:

$$O^{2-} + S_2O_7^{2-} = 2SO_4^{2-}$$

SUGGESTED REFERENCES FOR FURTHER READING

BELL, R. P. *Acids and Bases*, Methuen, London, 1952.

FIELD, B. O. and HARDY, C. J. 'Inorganic Nitrates and Nitrato-Compounds', *Quart. Rev. chem. Soc., Lond.*, 18 (1964) 361.

FOWLES, G. W. A. and NICHOLLS, D. 'Inorganic Reactions in Liquid Ammonia', *Quart. Rev. chem. Soc., Lond.*, 16 (1962) 19.

SISLER, H. H. *Chemistry in Non-aqueous Solvents*, Reinhold, New York, 1961.

WADDINGTON, T. C. *Non-aqueous Solvents*, Nelson, London, 1969.

6　Co-ordination chemistry

INTRODUCTION

The term *co-ordination compound* is used to describe a great range of chemical substances. Some of these exist only in the solid state; others dissolve to give molecules or ions and can be recovered from solution unchanged; others again can be studied in solution only.

A co-ordination compound contains a central metal atom or ion surrounded by a number of oppositely charged ions or neutral molecules known as *ligands*. The number of atoms directly attached to the metal is its *co-ordination number*. When the group comprising the metal and its ligands carries a positive or negative charge it constitutes a *complex ion*.

Each ligand molecule or ion has at least one pair of unshared electrons which is donated to the metal to form a co-ordinate bond. The metal must have vacant orbitals of appropriate energy which can accept these electrons. Complexes are most commonly formed by transition metals and then the d orbitals of the penultimate quantum shell of the metal are generally involved in the bonding. Hexaamminecobalt (III) chloride, $Co(NH_3)_6Cl_3$, is a well-known co-ordination compound readily prepared by the oxidation of an ammoniacal solution of cobalt (II) chloride. The complex ion $[Co(NH_3)_6]^{3+}$ is made up of a tripositive cobalt ion and six ammonia ligands, each nitrogen atom donating its lone pair of electrons to the metal. Water acts as a ligand in exactly the same way as ammonia because the oxygen atom can donate one of its lone pairs to a metal. Many salts in aqueous solution and in the crystalline state contain metal ions complexed (hydrated) with water molecules. There is a useful analogy here with acid–base reactions for the metal ion can be regarded as a Lewis acid with the ligand as base. The basic behaviour of ammonia

$$H^+ + NH_3 \rightleftharpoons (H \leftarrow NH_3)^+$$

is then directly comparable with co-ordination to a metal:

$$M^{n+} + NH_3 \rightleftharpoons (M \leftarrow NH_3)^{n+}$$

When the ligand molecule or ion contains two atoms which each have a lone pair of electrons, it may be stereochemically possible for the molecule to form two co-ordinate bonds with the same metal ion. For instance, tris-(ethylenediamine) cobalt (III), formula (I), contains 6 co-ordinate cobalt with each ethylenediamine molecule occupying two co-ordinate positions. In this complex, three rings which each contain five atoms have been established. The process of ring formation is called *chelation*. The adjective *chelate* is derived from the great claw or 'chela' of the lobster and other crustaceans. It was originally applied to a compound whose molecule is capable of attaching itself directly or after loss of a proton or protons to a metal atom at more than one point but it is now often used to describe the metal complex so formed as well. Any group which contains more than one potential co-ordinating atom is known as a multidentate group, the number of such atoms being indicated by the terms unidentate, bidentate, etc. As a chelating ligand is one attached to a single central atom through two or more co-ordinating atoms, we may describe ethylenediamine in tris(ethylenediamine)cobalt (III) as a bidentate chelating ligand. Multidentate ligands can sometimes act as a bridging group linking two or more centres of co-ordination. For example, the bridging action of ethylenediamine is illustrated in the complex ion, $[Ag_2(en)]^{2+}$.

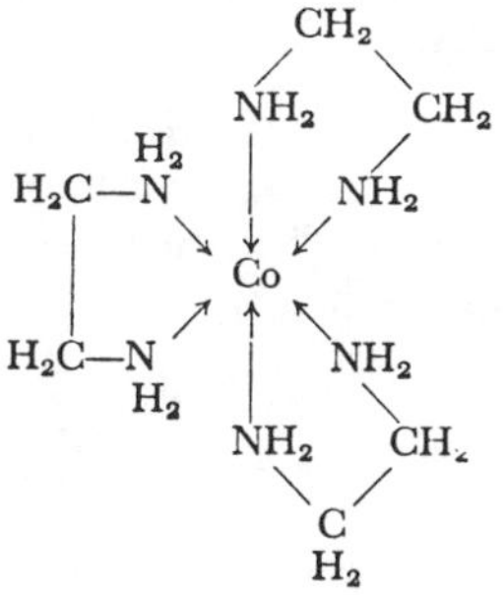

(I) $[Co(en)_3]^{3+}$

STEREOCHEMISTRY

Present ideas on the shapes of co-ordination compounds owe their origin to the work of A. Werner who in 1893 suggested what proved to be the correct explanation for the structures of the co-ordination compounds of tripositive cobalt and dipositive platinum. Using the ammines of these metals as examples, the evidence on which Werner's theory was based will now be reviewed.

The cobaltammines and other 6 co-ordinated complexes

In addition to hexaamminecobalt (III) chloride, a number of other complex salts have been prepared in which ammonia is strongly bound to tripositive cobalt. The distinctive properties of these are summarised in *Table 6.1*, together with Werner's formulations. Chemical evidence to support Werner's formulations came, for example, from conductance measurements on aqueous solutions which indicated what ions were present; also from the proportion of chlorine present as the chloride ion, that is, the fraction which was precipitated by silver nitrate.

Werner proposed that each metal had both a primary and a secondary valency. In the ammines of cobalt, the primary valency (or electrovalency) is 3 and the secondary valency is 6. He also suggested that the secondary valencies were directed in space about the central ion and that for 6 co-ordinate cobalt the ligands were arranged octahedrally about the metal ion. Then the two compounds of empirical formula $CoCl_3.4NH_3$ were regarded as

Table 6.1. THE AMMINES OF COBALT

Compound	Salt	Colour	No. of ions	Werner's formulation
$CoCl_3.6NH_3$	Luteo	Orange	4	$[Co(NH_3)_6]Cl_3$
$CoCl_3.5NH_3.H_2O$	Roseo	Pink	4	$[Co(NH_3)_5H_2O]Cl_3$
$CoCl_3.5NH_3$	Purpureo	Purple	3	$[Co(NH_3)_5Cl]Cl_2$
$CoCl_3.4NH_3$	Praseo	Green	2	$[Co(NH_3)_4Cl_2]Cl$
$CoCl_3.4NH_3$	Violeo	Violet	2	$[Co(NH_3)_4Cl_2]Cl$
$CoCl_3.3NH_3$		Blue-green	0	$[Co(NH_3)_3Cl_3]$

geometrical isomers (formulae (II) and (III)). Two isomers only are possible for an octahedral arrangement and in practice no more than two are known. If the arrangement were either of the other possibilities—namely trigonal prismatic or planar hexagonal—then three isomers would be expected. For

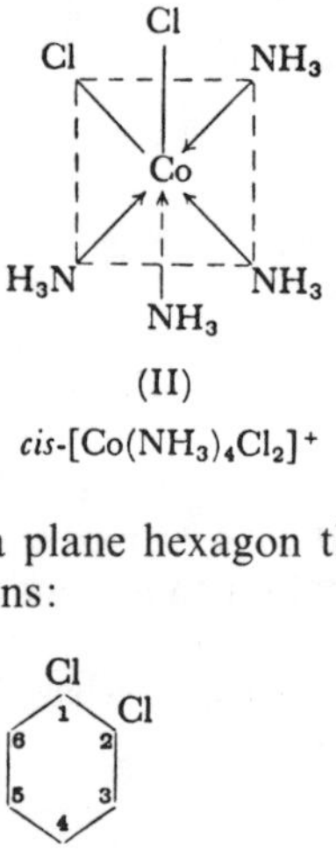

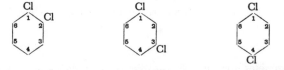

(II)

cis-$[Co(NH_3)_4Cl_2]^+$

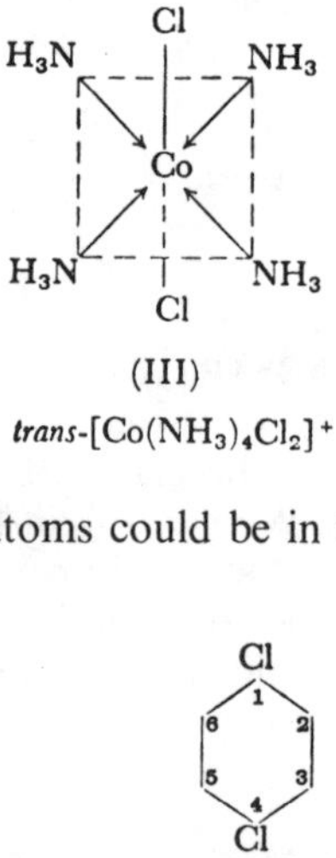

(III)

$trans$-$[Co(NH_3)_4Cl_2]^+$

example, in a plane hexagon the chlorine atoms could be in the 1,2 or 1,3 or 1,4 positions:

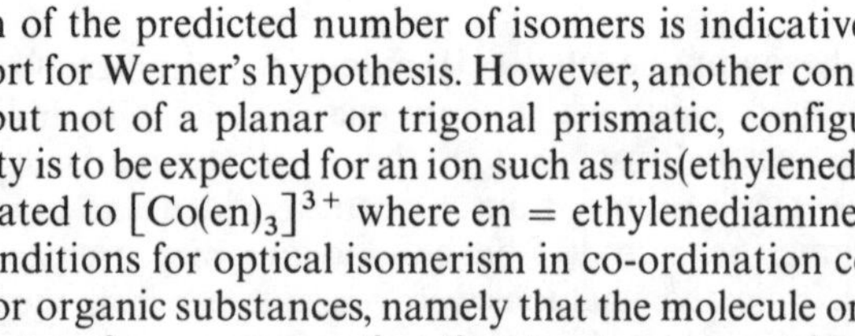

The isolation of the predicted number of isomers is indicative but not con-clusive support for Werner's hypothesis. However, another consequence of an octahedral, but not of a planar or trigonal prismatic, configuration is that optical activity is to be expected for an ion such as tris(ethylenediamine) cobalt (III) (abbreviated to $[Co(en)_3]^{3+}$ where en = ethylenediamine). The limiting symmetry conditions for optical isomerism in co-ordination compounds are the same as for organic substances, namely that the molecule or ion must lack a plane or centre of symmetry so that the structure cannot be superimposed on its mirror image. The ion has some symmetry elements, namely a threefold rotation axis of symmetry (C_3) and three twofold rotation axes (C_2) perpendicular to this, and is correctly described as dissymmetric. In this example two configurations, related as object and mirror image, formulae (IV) and (V), are possible. Again the dichlorobis(ethylenediamine) cobalt (III) ion, $[Co(en)_2Cl_2]^+$, exists in *trans*- and *cis*-forms (formulae (VI) and (VII) respectively) for an octahedral configuration. Moreover, whilst the *trans* isomer has a plane of symmetry, the *cis* does not and should therefore show two optically active forms (formulae (VII) *a* and *b*). The resolution of complex

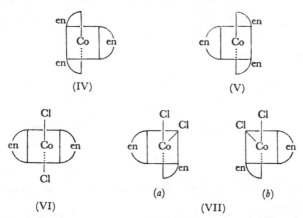

(IV)   (V)

(VI)   (a)   (b)
         (VII)

cations and anions is most commonly achieved by reaction with an optically active ion of opposite charge, so producing two diastereoisomers which usually show significant differences in their physical properties. For example the aerial oxidation of an aqueous solution of cobalt (II) sulphate in the presence of excess ethylenediamine (en) gives a solution containing* $(\pm)Co(en)_3^{3+}$. If barium $(+)$tartrate is added to the hot solution, the two diastereoisomers are produced.

$$4CoSO_4 + 12\ en + 4HCl + O_2 = 4[Co(en)_3]ClSO_4 + 2H_2O$$
$$2Co(en)_3ClSO_4 + 2Ba(+)tartrate =$$
$$(+)Co(en)_3Cl(+)tartrate + (-)Co(en)_3Cl(+)tartrate + 2BaSO_4$$

On allowing to cool $(+)$tris(ethylenediamine) cobalt (III) chloride($+$) tartrate. $5H_2O$ crystallises out whilst the other diastereoisomer stays in solution. Resolution is completed by removal of the optically active resolving agent. To do this in the above example, sodium iodide is added to a solution containing one of the diastereoisomers to precipitate out the sparingly soluble iodide. In this way, the two optically active compounds $(+)Co(en)_3I_3.$ H_2O and $(-)Co(en)_3I_3.H_2O$ are separately prepared. Both are deep orange crystalline solids stable at ordinary temperature though racemising when heated to 373 K in the solid state or in hot aqueous solution in the presence of charcoal or finely divided metals as catalyst.

In addition to $(+)$tartaric acid, $(+)$bromocamphorsulphonic acid has proved very useful for resolving cations. Alkaloids such as $(-)$ strychnine and $(-)$ brucine have been used to resolve anions. More recently, resolved metal

---

* $(\pm)$ denotes a racemate: $(+)$, dextro, and $(-)$, laevo, refer to the sign of the rotation of the optical isomer at the sodium D line wavelength.

complexes like $(+)[Co(en)_3]^{3+}$ have themselves been used as resolving agents.

X-ray crystal analysis has confirmed the octahedral arrangement around tripositive cobalt and the same orientation of groups has been shown for many other 6 co-ordinated metal ions.

The platinum ammines and other 4 co-ordinated complexes

The shape of a 4 co-ordinated complex may be either square planar or tetrahedral. Platinum (II) complexes are some of the best-known examples of the first of these.

Two isomers of diamminedichloroplatinum(II) can be made. The α-form is prepared by the treatment of K_2PtCl_4 with aqueous ammonia

$$[PtCl_4]^{2-} \xrightarrow{NH_3} [PtCl_3(NH_3)] \xrightarrow{NH_3} [PtCl_2(NH_3)_2]$$

and the β-form is made by the reaction between $[PtCl_2(NH_3)_4]$ and HCl

$$[Pt(NH_3)_4]^{2+} \xrightarrow{Cl^-} [PtCl(NH_3)_3]^+ \xrightarrow{Cl^-} [PtCl_2(NH_3)_2]$$

Werner proposed that these compounds were geometrical isomers, formulae (VIII) and (IX) below, and these are only possible for four bonds coplanar with the metal.

Chemical support for these configurations came from the preparation of the corresponding diamminedinitratoplatinum (II) isomers. When these were treated with oxalic acid, the nitrato-groups were replaced and two complexes were formed. The one from the α-isomer had the composition $Pt(NH_3)_2C_2O_4$; the one from the β-isomer had the formula $Pt(NH_3)_2(C_2O_4H)_2$. Assuming that the oxalato-group acts as a bidentate ligand only if it replaces two nitrato-groups in the *cis*-position, then the reactions may be represented as follows:

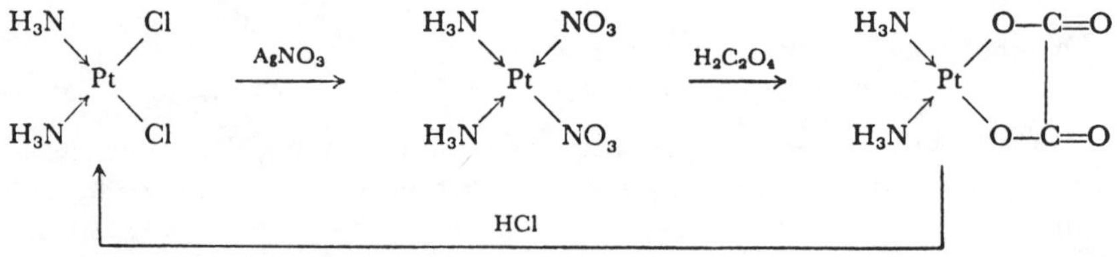

α-isomer

(VIII)

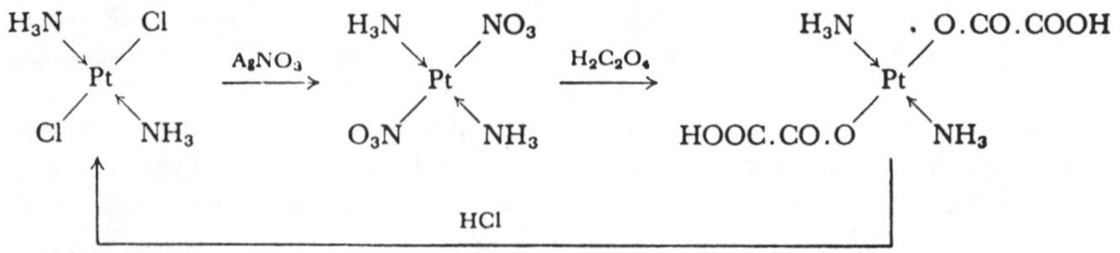

β-isomer

(IX)

They are consistent with Werner's formulation provided that no change in configuration occurred during the sequence of reactions.

When two unsymmetrical chelate groups such as isobutylenediamine (formula (X)) are tetrahedrally co-ordinated to a central atom, the resulting complex exists in mirror-image but no *cis-trans* forms. When the chelate

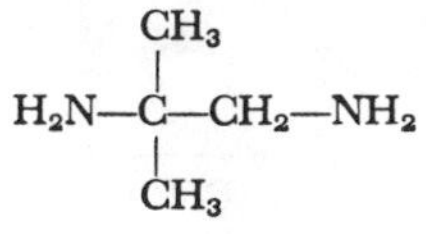

$$CH_3$$
$$|$$
$$H_2N-C-CH_2-NH_2$$
$$|$$
$$CH_3$$

(X)

groups are coplanar, *cis-trans* isomerism is possible but not optical activity. Although various claims of optical isomerism in Pt^{II} complexes have been made, in every case the activity appears to have been associated with the presence of the resolving acid or base as impurity and there is consequently no evidence for tetrahedral co-ordination.

However, one Pt^{II} complex has been synthesised which possesses asymmetry if the 4 co-ordinating groups are arranged at the corners of a square plane but not if the grouping is tetrahedral. This is the complex salt, isobutyl-enediamine-*meso*stilbenediamine platinum (II) chloride (formula (XI)). The

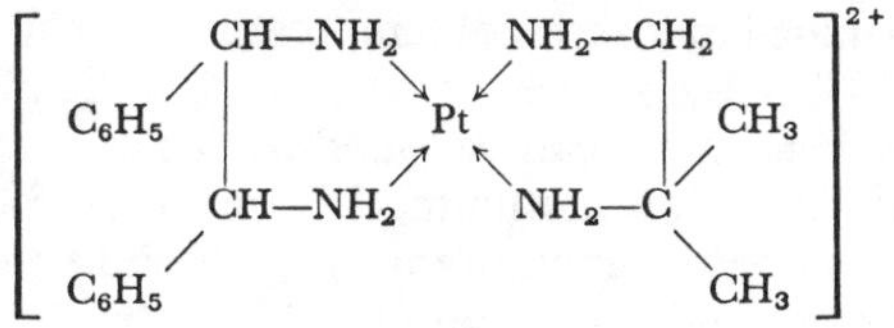

(XI)

asymmetry and hence planarity of the ligand atoms around the metal was demonstrated by the resolution of this compound in 1935 by Mills and Quibell.

Finally, it has proved possible to prepare three geometrical isomers of compounds such as $[PtBrCl(NH_3)py]$ (where py = pyridine). These, on a planar configuration, are represented by formulae (XII) *a*, *b* and *c*. In a tetrahedral model, the metal atom would be asymmetric in exactly the same way as carbon is when attached to four different groups. However, the two optically active isomers expected for this model cannot be produced.

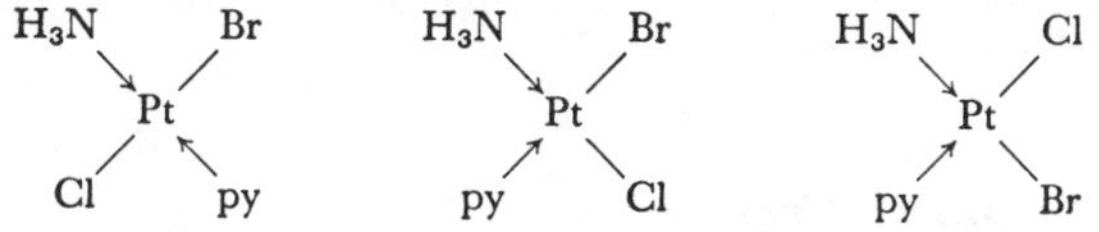

(XII)

Dipole moment measurements have often been useful in distinguishing between *cis* and *trans* geometrical isomers. The dipole moment is a vector quantity and the overall moment of a molecule containing several dipoles is the resultant of the component vectors. When the molecule has a centre of symmetry, the vector sum of the bond dipole moments will be zero.

The dipole moment of a compound may be calculated from the effect which it has on the capacitance of a condenser. The ratio of the capacitance of the condenser when the compound is placed between the plates to the capacitance in a vacuum is the dielectric constant, ε_r. This is related to the molar polarisation, P, of a substance of molar mass, M, and density ρ, by the equation

$$P = \frac{\varepsilon_r - 1}{\varepsilon_r + 2} \cdot \frac{M}{\rho}$$

When a polar substance is placed in a uniform electric field, this tends to orient the permanent dipoles. There is the opposing thermal effect increasing with temperature, which tends to give random orientation of the molecules. The field also induces a dipole in the molecule which lasts only as long as the field is present.

Debye showed that P is of the form

$$P = A + \frac{B}{T}$$

where A is the contribution due to the induced dipole and B is that due to the permanent dipole.

When ε_r is determined at a series of temperatures, P can be calculated and plotted against $1/T$. If B, the slope, is zero, $p_e = 0$. If the plot has a measureable slope, the dipole moment can be found from B.

α and β diamminedichloroplatinum (II) are not sufficiently soluble in non-polar solvents to permit measurement of their dipole moments. However, Jensen's work on related compounds of formula PtA_2X_2, where A is a substituted phosphine, arsine or stibine such as $(C_2H_5)_3P$, $(C_3H_7)_3As$ or $(C_2H_5)_3Sb$, and X is a halogen, showed a clear division into two kinds of isomer, those with zero moment and those with moments between 8 and 12 Debye units. In the compounds with zero moments, the individual moments have cancelled one another and so these must be the *trans* isomers. The compounds with large moments must be *cis*.

X-ray crystal analyses and dipole moment measurements of several Pt^{II} complexes have confirmed the square-planarity of the bonds around the metal atom. This arrangement has also been established for 4 co-ordinated Pd^{II}, Ag^{II}, Cu^{II} and Au^{III}. Tetrahedral configurations have been assigned to Cu^I, Ag^I, Au^I, Be^{II}, Al^{III}, Zn^{II}, Cd^{II}, Hg^{II} and Co^{II} in their co-ordination compounds. Some metals, for example, Ni^{II}, appear to show either configuration for co-ordination.

5, 7, 8 and 9 co-ordinated complexes

Relatively few of these are known compared with the numerous 4 and 6 co-ordinated complexes.

Co-ordination number 5 – Two stereochemical arrangements are found. The first, a trigonal bipyramid, is exemplified by pentacarbonyl iron, $Fe(CO)_5$, and the pentahalides $NbCl_5$, $NbBr_5$ and $MoCl_5$ in the vapour state. (In the solid state, these halides exist as dimers so that the metal atoms are octahedrally co-ordinated with two of the halogens acting as bridging atoms.) The second, a square pyramid, is found in a few compounds such as bis-(acetylacetonato)oxo-vanadium (IV) (XIII) and tribromobis(triethylphosphine)nickel (III), (XIV). In some cases, the stoichiometry of a compound

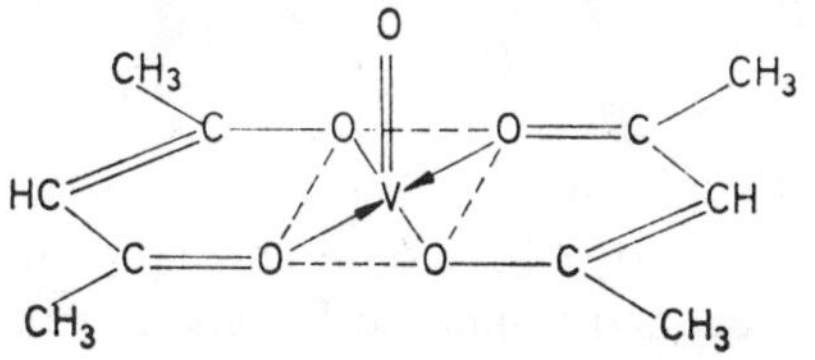

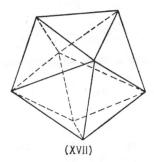

(XIII) 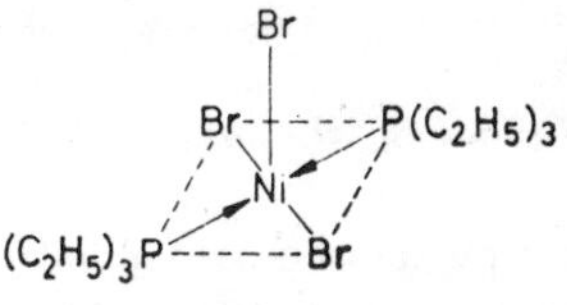 (XIV)

suggests 5 co-ordination but this is, in fact, absent. For example, the crystalline Cs_3CoCl_5 contains tetrahedral $CoCl_4^{2-}$ ions and separate Cl^- ions in its lattice.

Co-ordination number 7 — Various fluoro- or fluoro-oxo-complex ions are known with this co-ordination number. ZrF_7^{4-} and $UO_2F_5^{3-}$ (XV) have the shape of a pentagonal bipyramid. TaF_7^{2-} and NbF_7^{2-} (XVI) on the other hand have a trigonal prismatic arrangement of six fluorines around the metal with the seventh situated on the normal to one face.

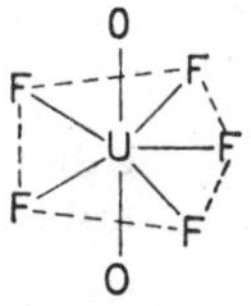

(XV) 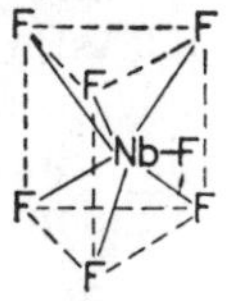 (XVI)

Co-ordination number 8 — Three arrangements are possible, the dodecahedron (XVII), the square anti-prism and the hexagonal bipyramid.

(XVII)

Dodecahedral complexes include $W(CN)_8^{4-}$, $Mo(CN)_8^{4-}$, $Zr(C_2O_4)_4^{4-}$ and $TiCl_4 \cdot (diarsine)_2$, where diarsine is $C_6H_4[As(CH_3)_2]_2$ (a bidentate ligand with the two dimethylarsine groups in the *ortho* position to one another).

The square anti-prism is found for TaF_8^{3-} and in various tetrakis(acetylacetonato) complexes such as those of Zr(IV), Hf(IV), Th(IV) and U(IV).

The hexagonal bipyramid is found whenever there are two particularly short metal–ligand bonds to determine the stereochemistry.

Co-ordination number 9 – Only one arrangement is known. Six ligands are at the corners of a trigonal prism and a further three ligands are above the rectangular faces. This is found in $Nd(H_2O)_9^{3+}$ ions and similar hydrated ions of other rare earths.

Other types of isomerism

The complexes of 6 co-ordinated cobalt (III) and 4 co-ordinated platinum (II) are the classical illustrations of optical and geometrical isomerism respectively. Other types of isomerism which occur are summarised below.

Ionisation isomerism – Different ions are produced when the complex is dissolved in water; e.g. $[Pt^{IV}Cl_2(NH_3)_4]Br_2$ and $[Pt^{IV}Br_2(NH_3)_4]Cl_2$.

Hydrate isomerism – This is a special example of ionisation isomerism e.g. the three hydrates of chromium (III) chloride: $[Cr(H_2O)_6]Cl_3$ (violet), $[CrCl(H_2O)_5]Cl_2H_2O$ (green) and $[CrCl_2(H_2O_4)]Cl.2H_2O$ (green). These are differentiated, in aqueous solution, by the proportion of total chlorine precipitated by silver nitrate. Only the ionic chlorine, that outside the complex ion, is immediately precipitated.

Linkage isomerism – Here the ligand is such that it can co-ordinate by one of two different atoms; e.g. nitro $[Co(NH_3)_5NO_2]^{2+}$ and nitrito $[Co(NH_3)_5 ONO]^{2+}$; thiocyanato $[Pd(AsPh_3)_2(SCN)_2]$ and isothiocyanato $[Pd(AsPh_3)_2 (NCS)_2]$ ($AsPh_3$ = triphenylarsine).

Co-ordination isomerism – Both anion and cation are complex and any given ligand may be co-ordinated in one or the other; e.g.

$$[Co(NH_3)_6][Cr(C_2O_4)_3] \quad \text{and} \quad [Cr(NH_3)_6][Co(C_2O_4)_3];$$
$$[Cu(NH_3)_4][PtCl_4] \quad \text{and} \quad [Pt(NH_3)_4][CuCl_4]$$

Polymerisation isomerism – The same empirical formula but the units are of different complexity; e.g. $[PtCl_2(NH_3)_2]$ and $[Pt(NH_3)_4][PtCl_4]$.

Co-ordination position isomerism – e.g. $(R_3P)_2Pt\begin{smallmatrix}Cl\\ \\Cl\end{smallmatrix}PtCl_2$ and

$Cl(R_3P)Pt\begin{smallmatrix}Cl\\ \\Cl\end{smallmatrix}Pt(PR_3)Cl$ (R_3P) is tri-substituted phosphine.

THE STABILITY OF COMPLEX COMPOUNDS

For many years the stereochemistry of co-ordination compounds was of the greatest interest. Increasing attention has been paid more recently to the way in which complexes are formed and to the quantitative study of their stabilities. Like the cobaltammines, some complexes are evidently very stable because they can be kept indefinitely without decomposition and in solution they do not give the usual analytical reactions of their constituent groups. Other complexes, such as the metal carbonyls, are unstable, being destroyed by the action of heat or water.

Apart from the work of a few chemists, notably Abegg, Bodlander and N. Bjerrum in the early part of this century, the chief impetus behind the current intensive study of the stability of metal complexes was the work, first published in 1941, of J. Bjerrum on the formation of metal ammines in aqueous solution. According to Bjerrum, the formation of a complex in solution proceeds by the step-wise addition of ligands to the metal. Thus, when a neutral ligand A complexes with the metal ion M^{n+}, a number of successive equilibria can be formulated. These are represented by equations (1), (2), ... (X) and for each of these an equilibrium constant $K_1, K_2 ... K_X$ can be defined. X is the maximum number of ligands attached to one metal ion for a particular set of experimental conditions and therefore represents its co-ordination number.

$$M + A \rightleftharpoons MA \quad K_1 = \frac{[MA]}{[M][A]} \tag{1}$$

$$MA + A \rightleftharpoons MA_2 \quad K_2 = \frac{[MA_2]}{[MA][A]} \tag{2}$$

$$MA_2 + A \rightleftharpoons MA_3 \quad K_3 = \frac{[MA_3]}{[MA_2][A]} \tag{3}$$

$$MA_{(X-1)} + A \rightleftharpoons MA_X \quad K_X = \frac{[MA_X]}{[MA_{(X-1)}][A]} \tag{X}$$

$K_1, K_2, ... K_X$ are the step-wise formation or stability constants and are related to the total formation constant, β_X, by the equation:

$$K_1 . K_2 . K_3 ... K_X = \beta_X = \frac{[MA_X]}{[M][A]^X}$$

It should be noted that, for the sake of clarity, charges have been omitted. Also it must be remembered that when the reaction between a ligand and a metal ion takes place in an aqueous medium, the metal ion is hydrated and complex-formation therefore involves the replacement of water by ligand molecules. It is again customary to omit the co-ordinated water molecules from the equilibrium equations.

As defined above, the K values are constant only at a specific ionic strength because they are related to the concentrations of the participating species. In general, such concentration equilibrium constants will have the dimensions of (concentration)$^{-X}$ where X will have different values depending on whether reference is made to a total or to a step-wise formation constant. Hence $\log_{10}$ K values in the tabulated data should be strictly regarded as $\log_{10}[K \times$ (concentration)$^X]$. If we refer to relative concentrations, $c/c°$ where $c°$ is some standard concentration (e.g. 1 mol dm^{-3} if other concentrations are expressed in the same manner) the equilibrium constant, now $K_{c/c°}$, is dimensionless. True thermodynamic stability constants, defined in terms of the activities of the reacting species, do not vary with the ionic strength and these can be calculated from the experimentally measured K values provided that the activity coefficients of the various ionic species are known at the ionic strength concerned. Alternatively, it may prove more accurate to measure the concentration constants, K_1, K_2, ... K_X at different ionic strengths and then to extrapolate the data to zero ionic strength to obtain values for the thermodynamic constants.

The calculation of a series of stability constants is usually a difficult procedure and involves considerable mathematical manipulation. This is because of the inherent complexity of the equilibrium systems being studied. When a metal salt and a ligand are present together in aqueous solution, equilibrium is established between the metal ion M^{n+} and its various complexes MA_1^{n+}, MA_2^{n+}, ... MA_X^{n+}. ('A' represents an uncharged ligand.) Experimentally, it may be possible to measure one of the following:

(a) the concentration of free uncomplexed metal ion $[M]$;

or (b) the concentration of one or more of the complexes $[MA_1]$, etc.;

or (c) the concentration of free ligand $[A]$.

The choice between these is largely determined by the nature of the system under investigation.

The experimental side of the determination of stability constants is essentially a matter of the measurement of (a), (b) or (c) in a range of equilibrium mixtures which contain the metal ion and ligand in different proportions. The ionic strength is generally maintained constant by the addition of some electrolyte (sodium perchlorate is frequently used) which does not form competitive complexes with the metal to any significant extent in the conditions of the experiment.

A wide variety of methods has been used in the measurement of the required concentrations. The more important of these methods are now considered very briefly.

(a) *The concentration of metal ion*—Several electrochemical procedures are used, particularly polarography and potentiometry with suitable electrodes. Ion-exchange resins have also been utilised. Thus a cation-exchange resin, when brought to equilibrium with an aqueous solution containing metal ion, will sorb a certain proportion of this ion from solution. If a complexing anion is now added to the aqueous solution, complexes of lower positive and even of zero or negative charge may be formed. The amount of metal taken up by the resin at equilibrium will therefore decrease. The magnitude of the decrease for a given concentration of complexing anion will be related to the concentration of free metal ion remaining in solution. For example, Zn^{2+} carries a double positive charge and is therefore appreciably sorbed from aqueous solutions of zinc salts by a cation-exchange resin. In the presence of chloride ion, the complexes $[ZnCl]^+$, $ZnCl_2$, $[ZnCl_3]^-$ and $[ZnCl_4]^{2-}$ will be formed and there will be a decrease in the amount of metal taken up by the resin.

Solubility measurements have also been used. These are of value where salts which are sparingly soluble in water are rendered more soluble by the presence of complex-forming ions. For example, silver chloride is more soluble in sodium chloride solutions than in pure water because of the formation of complex chloro-ions:

$$AgCl + Cl^- \rightleftharpoons [AgCl_2]^- \text{ etc.}$$

For the same reason sodium thiosulphate solution will dissolve AgCl and AgBr:

$$AgBr + S_2O_3^{2-} \rightleftharpoons [Ag(S_2O_3)]^- + Br^- \text{ etc.}$$

(b) *The concentration of metal complex* — Spectrophotometric methods are used where a colour change is associated with complex-formation. The metal ion, ligand or one of the complexes may be the coloured species. The measurement of the absorption of light by this species at a particular wavelength, suitably chosen so that the absorption of other coloured species is minimal, serves as a measure of its concentration. For example, when aqueous copper (II) sulphate is treated with HCl, the initial blue colour of the hydrated Cu^{2+} ion changes to green because of the formation of chloro-complexes:

$$[Cu(H_2O)_4]^{2+} + Cl^- \rightleftharpoons [CuCl(H_2O)_3]^+ + H_2O$$

and so on up to $[CuCl_4]^{2-}$.

Spectrophotometry may be combined with solvent extraction in the special case of the formation of an uncharged coloured metal complex. This can be formed in aqueous solution and then extracted into an immiscible organic solvent. Important applications are in the study of inner complexes formed between metals and various organic reagents.

In many systems, for example, those in which the reacting species are colourless, a radioisotope of the metal can be used for measuring the total concentration of metal in aqueous or organic phases or the quantity of metal sorbed by an ion-exchange resin. Stability constants can be calculated from such measurements.

(c) *The concentration of free ligand* — When A is a weak base, its concentration may be determined by a pH measurement on the solution containing the metal and ligand. A and its conjugate acid HA^+, are related as follows:

$$A + H^+ \rightleftharpoons HA^+$$

From a knowledge of the pH value and of the dissociation constant of HA^+, it is possible to calculate the concentration of A. This is measured in Bjerrum's method which involves the potentiometric titration of a standard metal perchlorate solution, also containing perchloric acid, with a solution of the ligand of known concentration. The titration is carried out under nitrogen atmosphere with mechanical stirring. After the addition of a small aliquot of ligand solution, the mixture is allowed to come to equilibrium (i.e. a constant pH) and the pH recorded. Successive additions of ligand are followed each time by pH measurements. The experimental results are then plotted as a titration curve of volume of ligand solution added against pH.

If they are not known, then the proton association constants of the ligand must also be determined. This can be done by performing another titration at the same ionic strength as before but with no complexing metal ion present.

Typical curves for the co-ordination of Cu^{2+} by ethylenediamine and for the protonation of this ligand are shown in *Figure 6.1*.

The calculation of stability constants from this experimental data is carried out using the formation function $\bar{n}$ defined by Bjerrum as the average number of ligands, A, attached to the metal, M.

For the system

$$M + XA \rightleftharpoons MA_X$$

(charges are omitted for clarity),

$$\bar{n} = \frac{[\text{Ligand bound in metal complexes}]}{[\text{Total metal ion}]}$$

$$= \frac{[MA] + 2[MA_2] + 3[MA_3] + \cdots X[MA_X]}{[M] + [MA] + [MA_2] + \cdots [MA_X]}$$

Inserting the step-wise stability constants,

$$\bar{n} = \frac{K_1[M][A] + 2K_1K_2[M][A]^2 + \cdots XK_1K_2\ldots K_X[M][A]^X}{[M] + K_1[M][A] + \cdots K_1K_2\ldots K_X[M][A]^X}$$

or

$$\bar{n} + (\bar{n}-1)K_1[A] + (\bar{n}-2)K_1K_2[A]^2 + \cdots (\bar{n}-X)K_1K_2\ldots K_X[A]^X = 0$$

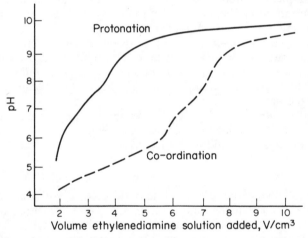

Figure 6.1. Titration curves for the protonation of ethylenediamine and its co-ordination with Cu^{2+}. *[Adapted from* D. E. GOLDBERG *J. chem. Educ.*, 39 (1962) 329]

Another way of expressing $\bar{n}$ is by

$$\bar{n} = \frac{C_A - [A] - [AH^+] - [AH_2^{2+}]}{C_M}$$

where C_A = total ligand concentration,
$\quad C_M$ = total metal concentration,
$\quad [A]$ = concentration of free ligand, assumed in this case to be a diacid base such as ethylenediamine which forms the two protonated species, AH^+ and AH_2^{2+}.

When the proton association constants are known or are experimentally determined, $[A]$, $[AH^+]$ and $[AH_2^{2+}]$ can all be calculated from the pH of the solution at equilibrium. As C_A and C_M are known from the composition of the mixture, $\bar{n}$ can be calculated. If the value of $\bar{n}$ is determined for at least X values of $[A]$, then x linear equations containing the unknowns $K_1, K_2, K_3 \dots K_X$, are obtained. These may be solved by determinants or by the method of successive approximations using approximate values of the constants to calculate more exact values until constancy is reached.

Bjerrum's method is most useful and accurate when the concentration of free ligand is significantly different from the total concentration of ligand, i.e. for moderately strong complexes.

For systems where weak complexes are formed, a large excess of ligand must be used and the concentration of free metal ion is then the more appropriate quantity to measure.

In the subsequent pages, the stability constants quoted will, unless otherwise stated, be concentration constants and are therefore only valid for a particular ionic strength.

CHELATION

Many compounds and ions, most of them organic in nature, form chelate rings on reaction with metal ions. Ring closure may take place with the establishment of covalent or co-ordinate bonds or both. The ability to form a covalent bond requires the presence of an acidic group in the ligand such as —COOH (carboxyl), —SO₃H (sulphonyl), —OH (enol), —SH (thioenol) or

$\text{\Large>}$N—OH (oxime). The most common functional groups containing donor

atoms and therefore able to co-ordinate with a metal are:

—NH₂, $\text{\Large>}$NH and $\text{=\!\!\!\!=}$N (amines); $\text{\Large>}$N—OH (oxime),

—OH (alcohol), $\text{\Large>}$CO (carbonyl), —O— (ether) and —S— (thioether).

Chelation is favoured when any two of these groups occur in the 1,4 or 1,5 positions of an organic molecule, with the formation of 5- and 6-membered rings respectively.

Multidentate ligands can thus be attached to a metal atom by two kinds of functional group and chelate compounds are classified according to the number and kind of attachments involved. For example, the bidentate chelating agents, which have been the most extensively studied, have been divided into three classes, each of which will be dealt with in turn.

1. Two acidic groups

In this class are included inorganic acids like carbonic and sulphuric; organic dicarboxylic acids like malonic, oxalic and phthalic; α-hydroxycarboxylic acids like glycollic and salicylic; and dihydroxy-compounds like the glycols and pyrocatechol. Formulae (XVIII), (XIX) and (XX) are examples of metal complexes formed by this type.

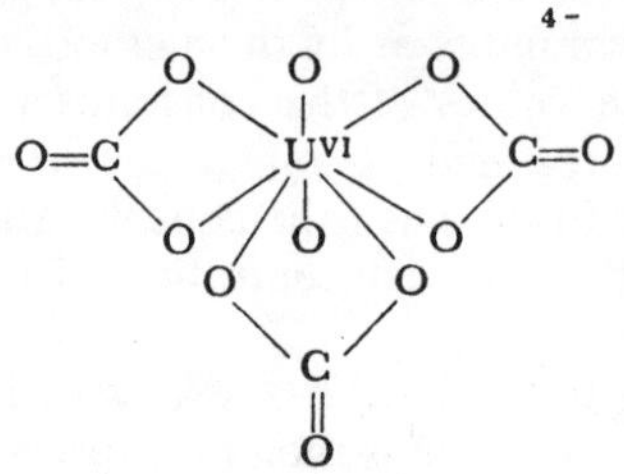

tris(carbonato)uranyl ion

(XVIII)

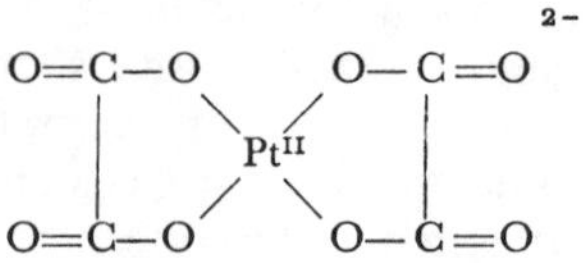

bis(oxalato)platinate (II) ion

(XIX)

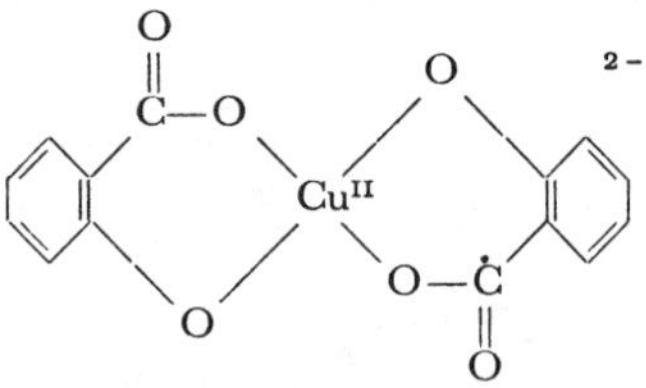

bis(salicylato)cuprate (II) ion

(XX)

2. One acidic and one co-ordinating group

This is an especially useful class of organic reagents which can, through chelation, often satisfy simultaneously the oxidation number and the co-ordination number of a metal ion. When this occurs, an unchanged species or *inner complex* is produced. 8-Hydroxyquinoline (formula (XXI)) is one of many examples. In its complex with Fe^{III} (formula (XXII)) the tripositive iron has a co-ordination number of 6. The reaction between the reagent HOx and Fe^{3+} can be represented as a series of steps:

$$Fe^{3+} + HOx \rightleftharpoons [FeOx]^{2+} + H^+$$
$$[FeOx]^{2+} + HOx \rightleftharpoons [Fe(Ox)_2]^+ + H^+$$
$$[Fe(Ox)_2]^+ + HOx \rightleftharpoons FeOx_3 + H^+$$

When the reaction is carried out in aqueous solution the uncharged complex, $FeOx_3$, is precipitated because of its low solubility. When excess reagent is present, complex-formation proceeds with the virtually complete precipitation of the metal from solution. Since the reactions are reversible, metal complexes of this type can often be redissolved by the addition of acid. The extent to which the dissociation of a metal complex proceeds at a given pH depends on its stability and the more stable a complex the lower the pH at which it can be precipitated from aqueous solution. When the stabilities of the complexes formed between 8-hydroxyquinoline and two different metals are very different, it may be possible to select a pH at which one is precipitated

and the other is not and so to effect an analytically useful separation. However, this reagent like many others is not very selective and it usually complexes
with several different metals under a given set of conditions. Its practical
value will depend on whether it can be rendered more specific in its reactions.
There are a number of ways in which this can be done, such as pH control by
buffering, competitive complexing with a second reagent to remove interfering metal ions or extraction with a variety of immiscible organic solvents.

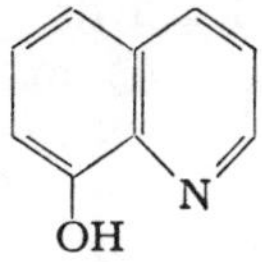

(XXI)

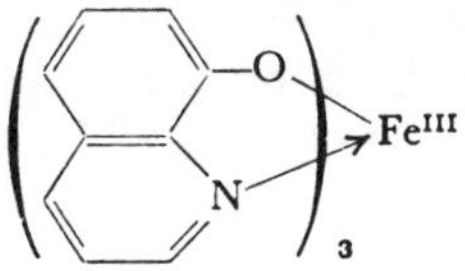

(XXII)

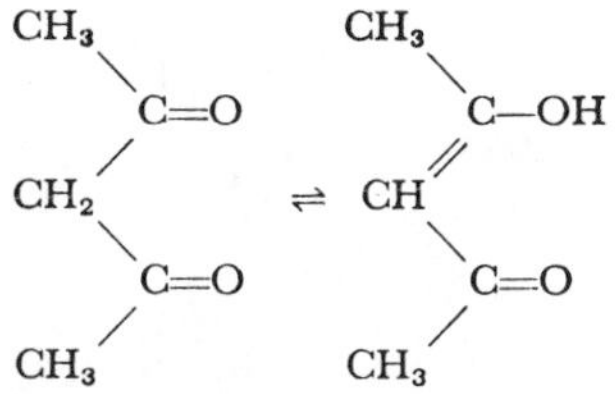

(XXIII)   (XXIV)

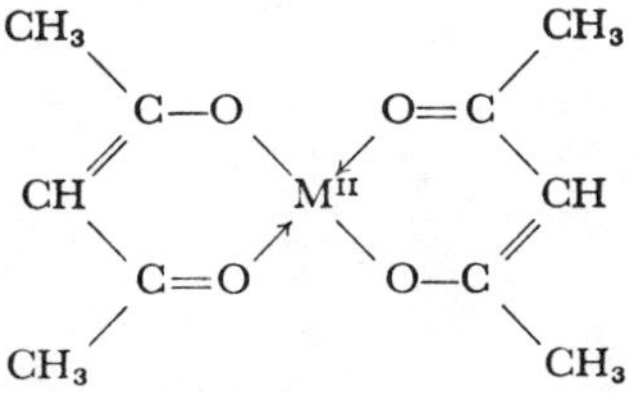

(XXV)

β-Diketones, such as acetylacetone (formula (XVIII)), are also valuable
chelating agents. Here, inner complex-formation involves the participation
of the enol form (formula XXIV) of the ketone. The complex formed with a
dipositive 4 co-ordinate metal is illustrated in formula (XXV). Another
diketone, trifluoroacetylacetone (which differs from formula (XXIV) only in
the replacement of one $-CH_3$ by $-CF_3$) has been used to realise the difficult
separation of zirconium and hafnium by selective extraction of the neutral
zirconium complex from aqueous hydrochloric acid into benzene. The
hafnium complex is not so readily extracted and the zirconium is thereby
freed from some of its hafnium content. The extraction process must be
repeated many times to effect an appreciable increase in the purity of the
zirconium.

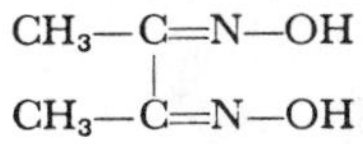

$$CH_3-C=N-OH$$
$$CH_3-C=N-OH$$

(XXVI)

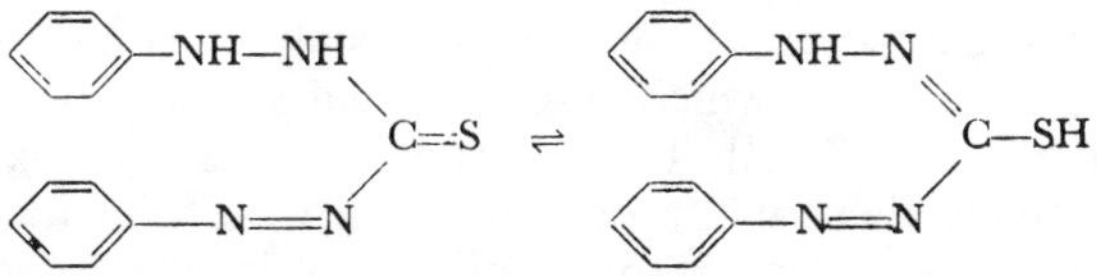

(XXVII)

Many other inner-complexing reagents have been developed and are widely used in analysis. These include dimethylglyoxime (formula XXVI), dithizone (formula XXVII) and salicylaldehyde (formula XXVIII). For instance dimethylglyoxime forms a water-insoluble red complex with nickel ions (XXIX) which is used in the gravimetric determination of this metal. Dithizone, itself soluble in organic solvents to give deep-green solutions, reacts with many metals to give highly coloured complexes which are variously pink, yellow, orange, etc. The striking colour change when the complex is formed is the basis of many colorimetric analytical methods. The hydrogen atom of the thioenol group in dithizone is replaced by a metal and ring closure takes place by co-ordination of one of the nitrogen atoms to the metal.

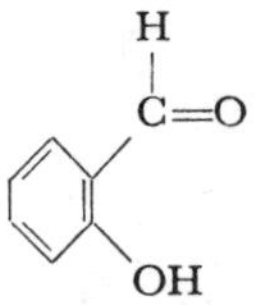

(XXVIII)

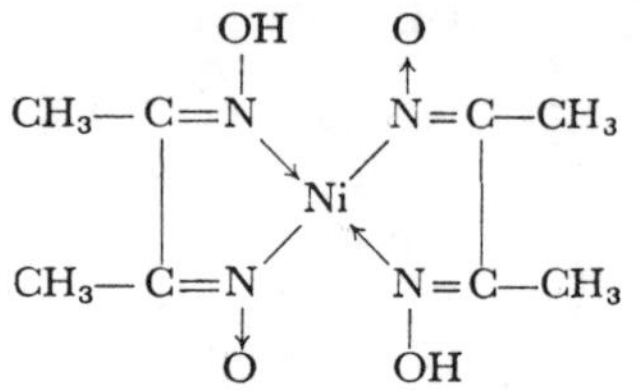

(XXIX)

3. Two co-ordinating groups

These include diamines like ethylenediamine, 2,2′-bipyridyl (XXX), and 1,10-phenanthroline (XXXI). 1,10-Phenanthroline is a heterocyclic base which has

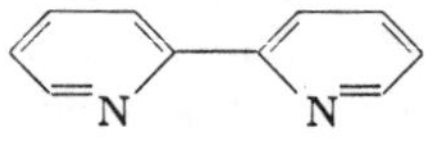

(XXX)

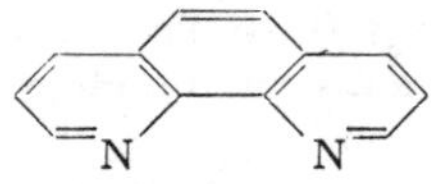

(XXXI)

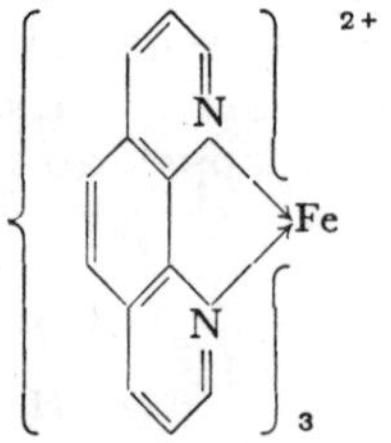

(XXXII)

two co-ordinating nitrogen atoms in its molecule. It forms an intensely red complex with iron (II) in which three molecules of the base are attached to one metal atom (XXXII). This chelate is a useful redox indicator for titrations involving dichromate or ceric ions because of the sharp colour difference between the tris(1,10-phenanthroline) iron (II) complex, intense red, and that of tris(1,10-phenanthroline) iron (III), pale blue.

Other chelating agents in this class are dihydroxy-compounds like 1,2-glycols and glycerol and α-hydroxy oximes like α-benzoin oxime (XXXIII).

Bidentate chelating agents of the above three classes have been extensively investigated because of their particular utility in analytical chemistry.

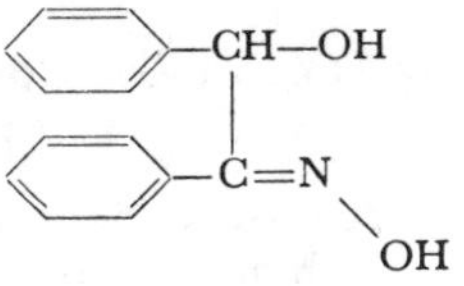

(XXXIII)

COMPLEXONES

These are a class of aminopolycarboxylic acids which have been much studied during the past 25 years by Schwarzenbach and many other workers. They have proved to be especially valuable in the rapid volumetric analysis of many elements.

The best known is ethylenediaminetetra-acetic acid (XXXIV), usually abbreviated to EDTA. The molecule contains four carboxyl groups and two

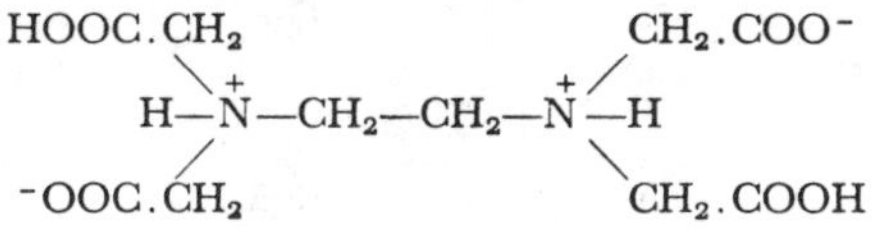

(XXXIV)

basic nitrogen atoms and hence there are up to six points of attachment available for the co-ordination of a metal ion. Although a single carboxyl group, for example in the acetate ion, forms only weak complexes with metals, the incorporation of four such groups into one molecule means that chelation occurs with the formation of several rings when a metal–EDTA complex is formed. Thus, if EDTA acts as a sexidentate ligand, five 5-membered rings are established. This complexone appears to be sexidentate towards Co^{III} but is only quinquedentate towards Ni^{II} and many other metal ions. This suggests that the co-ordination of all six groups in EDTA by one metal ion is critically dependent on stereochemical requirements. In other words, the size of the metal ion must be within quite narrow limits for the EDTA molecule to 'wrap around' it completely.

EDTA is a tetrabasic acid (H_4Y). In aqueous solution dissociation takes place to an extent which is determined by the pH value:

$$H_4Y \rightleftharpoons H_3Y^- + H^+$$
$$H_3Y^- \rightleftharpoons H_2Y^{2-} + H^+ \text{ etc.}$$

The acid dissociation constants for the 4 protons, K_1, K_2, K_3 and K_4 respectively have been determined experimentally to be:

$\log_{10}K_1 = -2.0$, $\log_{10}K_2 = -2.67$, $\log_{10}K_3 = -6.16$ and $\log_{10}K_4 = -10.26$

From these figures it can be deduced that the anion predominating in aqueous solutions of EDTA of pH 4 to 5 is H_2Y^{2-} and that in solutions of pH 7 to 9 is HY^{3-}.

Complex-formation between EDTA and a dipositive metal occurs as follows:

$$M^{2+} + H_2Y^{2-} \rightleftharpoons MY^{2-} + 2H^+ \quad (\text{pH} = 4 \text{ to } 5)$$
$$M^{2+} + HY^{3-} \rightleftharpoons MY^{2-} + H^+ \quad (\text{pH} = 7 \text{ to } 9)$$

In most cases, 1:1 complexes are formed. Some tetravalent metal ions complex with two molecules of EDTA but these complexes will not be considered here. *Table 6.2* indicates the relative stabilities of a number of 1:1 EDTA

Table 6.2. STABILITY CONSTANTS, $\log_{10} K_{MY}$, FOR A NUMBER OF METAL–EDTA COMPLEXES

Cation	$\log_{10} K_{MY}$
Li^+	2·79
Na^+	1·66
Mg^{2+}	8·69
Ca^{2+}	10·70
Sr^{2+}	8·63
Ba^{2+}	7·76
Mn^{2+}	13·79
Fe^{2+}	14·33
Co^{2+}	16·31
Ni^{2+}	18·62
Cu^{2+}	18·80
Zn^{2+}	16·5

complexes. From this it will be seen that those of the transition metals are the most stable and that even in the case of the alkaline-earth metals, complexes of appreciable stability are formed.

The chief practical application of EDTA is the use of its solutions in volumetric analysis and it has proved especially valuable for the rapid determination of magnesium and calcium. The disodium salt of EDTA, Na_2H_2Y, is commonly used in aqueous solution because the parent acid is not very soluble in water. This EDTA reagent is added to the solution of the metal salt to be determined and it converts the hydrated metal ion to its EDTA complex. At the equivalence point, usually when the ratio of metal ions to EDTA ions is 1:1, there is a sharp decrease in the concentration of free metal ion. This change can be detected potentiometrically, amperometrically or by a suitable metal indicator.

A metal indicator is an organic dyestuff which, like the complexones themselves, acts as a chelating agent. It possesses several ligand atoms suitably arranged for co-ordination and it can also take up or lose protons, depending on the pH of the solution in which it is dissolved. An essential property of a metal indicator is that its complex with a metal must be weaker than the metal–EDTA complex. It is also necessary that the colour of the indicator ion which predominates at the pH of the titration be different from that of the metal–indicator complex. Then, as a solution containing metal ion and its indicator is titrated with the aqueous solution of Na_2H_2Y, the metal is progressively complexed by the EDTA and at the equivalence point the colour changes from that of the metal–indicator complex to that of the indicator ion.

Eriochrome Black T (XXXV), a derivative of o,o'-dihydroxyazonaphthalene, is one of many metal indicators which have been developed for EDTA titrations. In this dyestuff, there are three acidic hydrogens, one sulphonic and two phenolic. The sulphonic group loses its proton at a low pH and the phenolic groups are those involved in colour changes above pH 7. The anion

H_2D^-

(**XXXV**)

shown as (XXXV) is, for these reasons, often abbreviated to H_2D^-. Below pH 6, Eriochrome Black T exists largely in the form of this ion, which is red in aqueous solution. Between pH 7 and 11, the indicator is blue due to the predominance of HD^{2-} Over this range of pH, metal ions such as magnesium, calcium and the lanthanoids give red-coloured complexes. Above pH 12, the solution becomes yellow–orange owing to the formation of D^{3-}, the fully ionised species. When the titration of a metal salt is carried out over the pH range 7 to 11 using EDTA in the presence of Eriochrome Black T, the solution is red until the equivalence point is reached, when it assumes the blue colour of the HD^{2-} ion. The pH of the titrated solution must be controlled by buffering for if it should fall below 6 there would be no significant colour change at the equivalence point because of the similar colours of the metal–indicator complex and the H_2D^- ions.

For details of other complexones and metal indicators and of their analytical uses, reference must be made to texts on analytical chemistry.

FACTORS INFLUENCING THE STABILITY OF COMPLEXES

Several factors determine the stability of a complex formed between a particular metal ion and ligand. The values of stability constants vary over an extremely wide range and it is not always easy to estimate the relative importance of the various factors which contribute towards stability. However, some trends are apparent in the experimental data and these may be discussed in relation to the properties of the metal ion and of the ligand.

Properties of the metal

1. Metals which form ions with a noble-gas structure

These ions are formed by Li, Na and the metals of Group IA; Be, Mg and the metals of Group IIA; Al and the metals of Groups IIIA, including the lanthanoids and actinoids. Complexes of these ions are not nearly as numerous nor as stable as those containing transition metal ions. The most stable appear to be those formed with small ionic ligands like fluoride, for example AlF_6^{3-}, and with ligands which contain oxygen as a donor atom, such as acetylacetone (XXIII), its various derivatives and aminopolycarboxylic acids like EDTA. Co-ordination by nitrogen also occurs in EDTA complexes, of course, and in some metal complexes of biological importance; for example, the magnesium complex with a porphyrin ring system in the chlorophyll pigments (XXXVI).

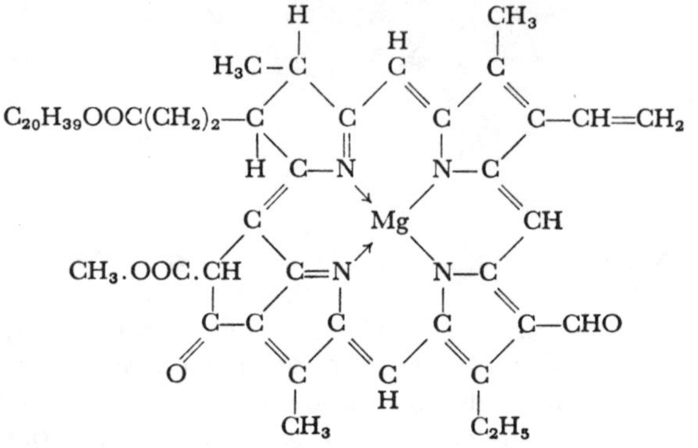

Chlorophyll b
(XXXVI)

The metals of this class are generally present as ions in their simple compounds and it is logical to suppose that in their complexes as well it is the characteristics of the ion which are of paramount importance. This is borne out by the observation that the stability of the complexes formed by one group of metals and a certain ligand usually increases as the size of the metal ion decreases. Thus the complexes of the alkali metals decrease in stability in the sequence

$$Li^+ > Na^+ > K^+ > Rb^+ > Cs^+$$

and those of Group IIA similarly

$$Mg^{2+} > Ca^{2+} > Sr^{2+} > Ba^{2+}, \text{ etc.}$$

Table 6.3. STABILITY CONSTANTS OF THE LANTHANOID–EDTA COMPLEXES

Atomic number	Ion	Stability constant $(\log_{10} K_{MY})$	Atomic number	Ion	Stability constant $(\log_{10} K_{MY})$
57	La^{3+}	15·5	65	Tb^{3+}	17·93
58	Ce^{3+}	15·98	66	Dy^{3+}	18·3
59	Pr^{3+}	16·4	67	Ho^{3+}	18·74
60	Nd^{3+}	16·61	68	Er^{3+}	18·85
62	Sm^{3+}	17·14	69	Tm^{3+}	19·32
63	Eu^{3+}	17·35	70	Yb^{3+}	19·51
64	Gd^{3+}	17·37	71	Lu^{3+}	19·83

Data from SCHWARZENBACH, G. et al., Helv. chim. acta, 37 (1954) 937.

There are exceptions to these general rules. For instance, the EDTA complex with Mg is less stable than that with Ca. The reason for this may well be the difficulty of co-ordinating a small ion like Mg^{2+} with such a large ligand. However, the lanthanoid–EDTA complexes illustrate clearly (*Table 6.3*) the increase in stability with decrease in size of the trivalent metal ion.

2. Metals which form ions with an outer shell of 18 electrons (d^{10} ions)

This class includes Cu^+, Ag^+, Au^+, Zn^{2+}, Cd^{2+}, Hg^{2+}, Ga^{3+}, In^{3+}, Tl^{3+} and Sn^{4+}.

Their general behaviour is quite different from ions of the preceding class. For example, stable complexes are formed with halide ions, ammonia and cyanide. Some representative constants are given in *Table 6.4*.

Table 6.4

Metal ion	Ligand	$\log_{10} K_1$	Medium	T/K
Zn^{2+}	Cl^-	-0.19	$NaClO_4$, 3·0 mol dm^{-3}	298·15
Cd^{2+}	Cl^-	1·54	$NaClO_4$, 3·0 mol dm^{-3}	298·15
Hg^{2+}	Cl^-	6·74	$NaClO_4$, 0·5 mol dm^{-3}	298·15
Ag^+	F^-	0·36	corr. to $I = 0$	298·15
Ag^+	Cl^-	3·3	corr. to $I = 0$	298·15
Ag^+	Br^-	4·4	corr. to $I = 0$	298·15

Often there is an increase in stability in a given sub-group with increase in size of metal ion: also the more stable complexes are usually those with ligand atoms of lower electronegativity. Both these trends support the conclusion that electrostatic forces cannot be decisive in determining stability so that account must also be taken of covalent bonding between d^{10} ions and ligands.

3. Transition metal ions

The most extensively studied complexes are those of the cations of the first transition series. The complexes formed by the stable divalent ions (that is, from manganese onward) with upwards of 80 ligands, co-ordinating generally through either oxygen or nitrogen, follow the sequence of stabilities

$$Mn^{2+} < Fe^{2+} < Co^{2+} < Ni^{2+} < Cu^{2+} > Zn^{2+}$$

This, after the original proposers, is usually referred to as the Irving–Williams order. An approximately linear correlation has been established of stability with the sum of the first and second ionisation potentials of the metal. The ionisation potential is a measure of the electron affinity (or power of attraction for electrons) of the metal ion and hence of its tendency to accept electrons from a ligand.

When a transition metal shows different valencies with the same ligand, the complexes of higher valency are nearly always the more stable. This again is related to the charge on the metal ion; the greater the charge the stronger the power of attraction for electrons.

Most transition metals form their strongest complexes with ligands carrying oxygen or nitrogen as donor atoms. There is a small group of ions, however, including Pd^{2+}, Pt^{2+}, Au^+, Ag^+, Cu^+ and Hg^{2+}, which form their most stable complexes with elements of the second or succeeding periods (for example P, S and As). These are the same metals which form stable olefin complexes and are characterised by d^8 or d^{10} electronic configurations. It has been proposed that in the complexes formed by this group of ions there is transfer (back-donation) of electrons from the metal to the ligand as well as from the ligand to the metal. For this to be possible, the ligand must possess vacant orbitals capable of receiving electrons and so back-donation is usually found only with ligand atoms of the second and later periods.

Properties of the ligand

1. Nature of the ligand atom

The atoms which are bound directly to metal ions in complexes are those of the more electronegative elements on the right-hand side of the Periodic Table, namely C; N, P, As, Sb; O, S, Se, Te; F, Cl, Br, I; H. Any of these atoms present in a ligand molecule or ion may co-ordinate with a metal ion. In the case of Group VII non-metals, the complexes formed by the monatomic anions have been widely studied. For most metals the sequence of stabilities is

$$F^- > Cl^- > Br^- > I^-$$

but this order is reversed for a few metals including Pt^{II}, Cu^I, Ag^I, Hg^{II} and Tl^I.

2. Basicity

When the ligand shows basic properties with respect to water as a solvent, a correlation is often noted between the base strength (i.e. the proton affinity) of a ligand and the stability of its metal complexes (i.e. its cation affinity). Bjerrum first pointed out this relationship from the results of his studies on the amine complexes of Ag^+ and Hg^{2+}. In general, unless steric effects interfered, the ratio of the metal/amine stability constant to the base strength of the ligand was approximately constant when considering a series of ligands of the same type, for example, primary amines.

3. Chelation

The establishment of a chelate ring increases the stability of a complex over that of comparable complexes where chelation is not possible. For example, two ethylenediamine molecules, like four ammonias, occupy four co-ordinate positions around the Cu^{II} ion. In the case of ethylenediamine (en), the overall stability constant for $[Cu(en)_2]^{2+}$ was found to be: $\log_{10}[\beta_2 \times (\text{concentration})^2] = 20 \cdot 07$ (measured in KNO_3(aq) $1 \cdot 0$ mol dm^{-3}). For the complex $[Cu(NH_3)_4]^{2+}$, $\log_{10}[\beta_4 \times (\text{concentration})^4] = 12 \cdot 63$ (in NH_4NO_3 (aq) $1 \cdot 0$ mol dm^{-3}). The enhanced stability of $[Cu(en)_2]^{2+}$ is associated with the presence of two 5-membered chelate rings (cf. $[Co(en)_3]^{3+}$, formula (I), (p. 161).

EDTA complexes are particularly stable because of the presence of several chelate rings. *Table 6.5* shows the variation with ring size of the stability constants of the calcium complexes of a series of acids of general formula

$$(HOOC.CH_2)_2.N.(CH_2)_n.N.(CH_2.COOH)_2$$

When $n = 2$, a 5-membered ring is formed between the two nitrogens and the metal; when $n = 3$, a 6-membered ring is set up, and so on. The greatest stability is found when $n = 2$.

Table 6.5. THE STABILITY CONSTANTS FOR THE CALCIUM COMPLEXES OF THE
AMINOCARBOXYLIC ACIDS
$(HOOC.CH_2)_2.N.(CH_2)_n.N.(CH_2.COOH)_2$

n	Ring size	Stability constant $(\log_{10} K_{MY})$
2	5	$10 \cdot 7$
3	6	$7 \cdot 1$
4	7	$5 \cdot 1$
5	8	$4 \cdot 6$

For conjugated ligands, a 6-membered ring is usually more stable than one containing only 5 atoms. The reason for this extra stability is the effect of the aromatic type of resonance which is possible with an even but not an odd number of atoms.

The chelate effect can be regarded thermodynamically as due to the entropy change of the reaction. When a solvated metal ion in solution reacts with a chelating ligand, solvent molecules in the co-ordination sphere of the metal ion are displaced. For example,

in $\qquad [Ni(H_2O)_6]^{2+} + 3 \text{ en} = [Ni(en)_3]^{2+} + 6H_2O$

the chelation process results in an increase of three in the number of molecules present. The replacement of unidentate by chelating ligands must always have this effect; hence chelation is associated with a positive entropy change and chelate complexes have a greater probability of formation than the corresponding complexes with unidentate ligands.

4. Steric effects

These are associated with the presence of a bulky group either attached to or near enough to a donor atom to cause mutual repulsion between the ligands and therefore a weakening of the metal–ligand bonds. This, for instance, accounts for the observation that the metal complexes of 2-methyl-8-hydroxy-quinoline are less stable than those of either 8-hydroxyquinoline itself or its 4-methyl derivative. The values for $\log_{10}\beta_2$ for a number of metal complexes are given in *Table 6.6*. Here the pK_{HOx} values in the second column represent the acid strengths of the ligands and are approximately constant. One would therefore expect similar stabilities of their metal complexes. The lower stability which is in fact shown by the complexes in which a methyl group is present in the 2-position is attributable to the greater difficulty of chelation caused by

Table 6.6. THE STABILITY CONSTANTS OF 8-HYDROXYQUINOLINE COMPLEXES

Substituent	pK_{HOx}	$\log_{10}\beta_2$				
		Mn	Co	Ni	Cu	Zn
None	11·5	15·5	19·7	21·4	26·2	18·9
2-CH$_3$	11·7	14·0	18·5	17·8	23·8	18·7
4-CH$_3$	11·6	15·5	20·0	22·3	—	20·2

the steric hindrance of this group. 4-Methyl-8-hydroxy quinoline behaves like the unsubstituted reagent because the methyl group is too far away from the nitrogen to affect sterically the process of complex-formation.

Hard and soft acids and bases

All metal ions act as Lewis acids and they are generally found in co-ordination with ligands acting as Lewis bases. This concept of acid–base interaction has been widely applied in inorganic chemistry and has been further extended by a differentiation into various kinds of acids and bases.

An acid, and similarly a base, can be described as being 'hard' or 'soft'. Hardness is defined as the property of retaining valency electrons very strongly. Thus a hard acid may be a cation which has a small size, a high positive charge and no electrons which are easily polarised or removed. A soft acid is one in which the acceptor atom is large, it carries a small or zero charge or it has a number of valency electrons which are easily polarised or removed. A hard base is one which retains its electrons strongly and a soft base is one which possesses valency electrons which are easily polarised or removed.

If acids and bases are classified in this manner it becomes clear that, in general, hard acids tend to react with hard bases and soft acids with soft bases. The preceding discussion on stability constants provides several illustrations of this general rule. Typically, hard acids include metal ions with a noble-gas electronic structure, transition metal ions such as Mn^{2+}, Cr^{3+}, Co^{3+}, Fe^{3+}, La^{3+}, Gd^{3+} and Ce^{3+} and molecules such as BF_3, $AlCl_3$ and SO_3. Soft acids include transition metal ions like Pd^{2+}, Pt^{2+}, Au^+, Ag^+, Cu^+, Hg^{2+}, Cd^{2+} and Tl^{3+}, metals in a zero oxidation state, and molecules like I_2, Br_2 and BH_3. Hard bases include NH_3, H_2O, R_2O, OH^-, F^-, Cl^-, PO_4^{3-}, SO_4^{2-}, CO_3^{2-}, ClO_4^- and NO_3^-. Soft bases include I^-, SCN^-, $S_2O_3^{2-}$, R_3P, R_3As, R_2S, CN^-, CO, RSH and RS^-. Some acids and bases are of borderline character between these two classes; for example, bases of this kind include C_5H_5N, N_3^-, Br^-, NO_2^- and SO_3^{2-}. Thus Br^- is intermediate between Cl^- (hard) and I^- (soft).

It must be remembered that the terms 'hardness' and 'softness' have qualitative significance only for there is no physical property of a Lewis acid or base which represents exactly its degree of hardness or softness. Nevertheless, they serve to correlate a great number of chemical reactions. The hardness of an element depends on a number of factors including its oxidation state and the nature of the groups attached to the acceptor atom. In low oxidation states, an element tends to be 'soft' and to co-ordinate with soft bases like CO, CN^- and phosphines. In high oxidation states an element is most stable when combined with hard bases like O^{2-}, OH^- and F^-. As an illustration of the second factor, boron in BF_3 is a hard acid and boron in BH_3 is soft. Thus experimentally it is found that $BF_3 . OR_2$ is more stable than $BF_3 . SR_2$ whereas the reverse is the case for BH_3, which also shows its softness in forming a complex with the soft base CO.

The bonding between hard acids and hard bases is, of course, chiefly ionic and that between soft acids and soft bases is mainly covalent. In many complexes, bonding is intermediate between the two extremes and so there is a whole range of softness or hardness.

MAGNETISM AND CO-ORDINATION COMPOUNDS

Two types of chemical compound can be distinguished according to their behaviour when placed in a magnetic field. *Diamagnetic* substances are less permeable to magnetic lines of force than is a vacuum and tend to move from the strong to the weak part of the magnetic field: *paramagnetic* substances show the opposite behaviour, that is they move from the weak to the strong part of the field.

Magnetism arises because of the interaction between the external magnetic field and the orbital electrons of the atoms. A special case of paramagnetism, known as ferromagnetism, occurs in a few metals, notably iron. This is the acquisition of a permanent magnetic moment by the metal because of the parallel orientation of magnetic particles in the crystalline lattice.

Diamagnetism is found for all chemical substances. It is caused by the interaction of an external field with the magnetic field produced by the movement of electrons in filled atomic orbitals. This results in an induced field which is directionally opposed to the applied field. Diamagnetism is independent of temperature but increases with atomic or molecular size and the diamagnetism of covalent compounds is approximately additive. Generally, however, its effect is much smaller than that of paramagnetism.

Paramagnetism arises whenever an atom, ion or molecule possesses one or more unpaired electrons. The presence of these causes the element to behave like a small permanent magnet. An applied magnetic field tends to align all these magnets in parallel. This tendency is opposed by thermal effects which favour a random arrangement of the magnets. Paramagnetism is found in the compounds of the transition metals, due to the presence of unpaired d electrons, in the lanthanoids and actinoids, due to the presence of unpaired f electrons, and in a few molecules such as O_2, NO, NO_2 and ClO_2, which contain an odd number of valency electrons.

Experimentally, a quantity known as the paramagnetic susceptibility is measured. This can be done using Gouy's method which is described below.

The Gouy method

The application of a magnetic field to a paramagnetic substance causes the magnetic dipoles to be lined up in the field direction; with diamagnetic materials the effect is one of polarisation of the electron charge cloud. In both cases a magnetic flux density (magnetic induction) B is set up, given by $B = \mu H$, where μ is the permeability of the medium and H is the field strength applied.

For chemists, the interesting quantity is the magnetic susceptibility (χ) defined as:

$$\chi = \mu_r - 1$$

where μ_r is the relative permeability (μ/μ_0) of the material. The magnetic susceptibility χ has been formerly known as the volume susceptibility κ though it differs from this latter cgs quantity by the factor 4π, i.e. $4\pi\kappa = \chi$. Two further quantities previously used were the mass susceptibility and the molar susceptibility (κ/ρ and $\kappa M/\rho$ respectively); these are now replaced by the products χv and $\chi v M$ respectively, where v is the specific volume and M the molar mass of the material. The relationships between the quantities in the two systems are thus:

$$\chi = 4\pi\kappa \text{ (SI units 1)}$$
$$\chi v = 4\pi\kappa/\rho \times 10^{-3} \text{ (SI units m}^3\text{ kg}^{-1})$$
$$\chi v M = 4\pi\kappa M/\rho \times 10^{-6} \text{ (SI units m}^3\text{ mol}^{-1})$$

The Gouy method measures the effect of an inhomogeneous magnetic field upon the specimen suspended from the arm of a sensitive balance by a fine silver chain (*Figure 6.2*). The specimen is in the form of a rod of uniform cross-sectional area (A) where each end of the rod is in a region of uniform magnetic field, in this case that outside the magnet (H_0) and that at the centre of the magnetic field (H).

The magnetic susceptibility of the specimen is obtained from the relationship:

$$F = \tfrac{1}{2}\mu_0 A\chi(H^2 - H_0^2)$$

The specimen, which may be crystalline or liquid, is contained in a flat bottomed Gouy tube constructed of glass and about 10–15 cm in length and of uniform cross-sectional diameter (3–10 mm). The tube itself, being a hollow specimen, experiences a force that is always present and which must be subtracted from the measured force when the sample is weighed in the field. Also, since measurements are made in air, which has an appreciable susceptibility, account must be taken of the volume of air displaced by the sample. Taking these factors into account leads to the expression:

$$F = \tfrac{1}{2}\mu_0 A(\chi - \chi_a)(H^2 - H_0^2) + \delta$$

where χ_a is the magnetic susceptibility of air and δ is the tube correction factor.

Measurements are made in and out of the magnetic field, which may be either that from a permanent magnet or an electromagnet, generally in the range 400 mT to 1 T (T = tesla) and with a pole gap of 2–3 cm. To avoid measurements of the field strength, area of the specimen and other constant factors, a calibration is performed using a substance of known susceptibility, e.g. a standard nickel chloride solution or solid $HgCo(SCN)_4$. Powdered materials must be well packed in the tube to avoid errors from air blocks. The volume of the sample may be determined conveniently from the mass of a volume of water equal to that of the specimen volume.

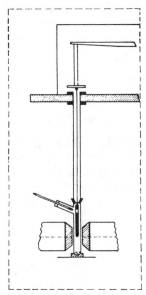

Figure 6.2. Gouy apparatus for measurement of magnetic susceptibility

For a constant length specimen the term:

$$\tfrac{1}{2}\mu_0 A(H^2 - H_0^2)$$

is a constant, i.e.

$$F' = (F - \delta) = C(\chi - \chi_a)$$

or

$$g\Delta m = C(\chi - \chi_a)$$

where Δm is the apparent change in mass on applying the magnetic field.

Making measurements at 293·15 K, $\chi_a = 4\pi \times 0\cdot029 \times 10^{-6}$

Hence, $= 0\cdot364 \times 10^{-6}$

$$g\Delta m = C(\chi - \chi_a)$$
$$= C\left(\frac{\chi v \times m_s}{V} - \chi_a\right)$$

where m_s = sample mass and V = sample volume. Therefore,

$$C' = \frac{gV}{C} = \frac{\chi v \times m_s - \chi_a V}{\Delta m}$$

$$\text{or } \chi v = \frac{C'\Delta m + \chi_a V}{m_s}$$

This value of χv then has to be converted to $\chi v M$ and corrections applied for diamagnetic ligand contributions thereby yielding a corrected value $\chi' v M$. This latter quantity is related to the magnetic moment (m) of the specimen by the Langevin equation:

$$\chi' v M = \frac{N_A \mu_0 m^2}{3kT}$$

where $m = X \times m_B$ and X is the magneton number of the specimen. The Langevin equation may be rewritten

$$X = \left(\frac{3k}{N_A \mu_0 m_B^2}\right)^{\frac{1}{2}} (\chi' v M T)^{\frac{1}{2}}$$

$$= 797 \cdot 9 \, (\chi' v M T)^{\frac{1}{2}} \, m^{-\frac{3}{2}} K^{-\frac{1}{2}} mol^{\frac{1}{2}}$$

and since the units of $(\chi' v M T)^{\frac{1}{2}}$ are $m^{\frac{3}{2}} K^{\frac{1}{2}} mol^{-\frac{1}{2}}$, X is dimensionless.

As an example of this method, the following results for tris(acetylacetonato)-manganese (III) at 293·15 K may be cited.

(i) Calibration using $HgCo(SCN)_4$.

	Field applied 10^3 m/kg	Field removed 10^3 m/kg	$10^3 \Delta m$/kg
Gouy tube	14·987 6	14·988 9	−0·001 3
Gouy tube + $HgCo(SCN)_4$	24·206 8	24·134 8	+0·072 0
Gouy tube + water		21·594 3	

Hence the mass of the standard material = $9·145 \times 10^{-3}$ kg and the volume of the specimen = $6·62 \times 10^{-6}$ m³, taking the density of water at 293·15 K to be $0·998 \times 10^3$ kg m⁻³.

$$\Delta m = 0·073 \, 3 \times 10^{-3} \text{ kg} \quad \chi v = 4\pi \times 10^{-3} \times 16·44 \times 10^{-6} \text{ m}^3 \text{ kg}^{-1}$$
$$= 20·65 \times 10^{-8} \text{ m}^3 \text{ kg}^{-1}$$

Thus,

$$C' = \frac{(20·65 \times 10^{-8} \text{ m}^3 \text{ kg}^{-1} \times 9·145 \times 10^{-3} \text{ kg} - 6·62 \times 10^{-6} \text{ m}^3 \times 0·364 \times 10^{-6})}{0·0733 \times 10^{-3} \text{ kg}}$$

$$= 25·73 \times 10^{-6} \text{ m}^3 \text{ kg}^{-1}$$

(ii) Results for tris(acetylacetonato)manganese (III).

	Field applied 10^3 m/kg	Field removed 10^3 m/kg	$10^3 \Delta m$/kg
Gouy tube + $Mn(acac)_3$	18·302 6	18·258 3	+0·044 3

$\Delta m = 0.045\ 6 \times 10^{-3}$ kg and the mass of the specimen $= 3.26^{\circ}\ 4 \times 10^{-3}$ kg. Thus,

$$\chi v = \frac{(25.73 \times 10^{-6}\,\text{m}^3\,\text{kg}^{-1} \times 0.0456 \times 10^{-3}\,\text{kg} + 6.62 \times 10^{-6}\,\text{m}^3 \times 0.364 \times 10^{-6})}{3.2694 \times 10^{-3}\,\text{kg}}$$

$$= 35.96 \times 10^{-8}\,\text{m}^3\,\text{kg}^{-1}$$

The molar mass of the compound is 352×10^{-3} kg mol^{-1} and therefore,

$$\chi v M = 126\ 579 \times 10^{-12}\,\text{m}^3\,\text{mol}^{-1}$$

Table 6.7. DIAMAGNETIC CORRECTION FACTORS $(-10^{12}\chi v M/\text{m}^3\ \text{mol}^{-1})$

Li$^+$	12·56	F$^-$	114·30	H	36·80
Na$^+$	85·41	Cl$^-$	293·90	C	75·36
K$^+$	187·14	Br$^-$	434·58	N (ring)	57·90
Rb$^+$	282·60	I$^-$	635·54	N (open chain)	69·96
Cs$^+$	439·60	NO$_3^-$	237·38	N (mono-amide)	19·34
Tl$^+$	448·40	ClO$_3^-$	379·31	N (di-amide, imide)	26·50
NH$_4^+$	167·05	ClO$_4^-$	401·92	O (ether, alcohol)	57·90
Hg$_2^{2+}$	502·40	BrO$_3^-$	487·33	O (ketone, aldehyde)	−21·73
Mg^{2+}	62·80	IO$_3^-$	651·86	O (carboxylic)	42·20
Zn^{2+}	188·84	IO$_4^-$	645·58	Constitutive corrections	
Pb^{2+}	401·92	CN$^-$	163·28		
				C=C	−69·08
Ca^{2+}	130·62	SCN$^-$	389·36		
				C≡C	−10·05
Fe^{2+}	160·77	OH$^-$	150·72		
				C=C—C=C	−133·14
Cu^{2+}	160·77	SO$_4^{2-}$	503·66		
				C (benzene)	−3·01
Co^{2+}	160·77	CO$_3^{2-}$	370·52		
				C—Cl	−38·94
Ni^{2+}	160·77				
				C—Br	−51·50
Mn^{3+}	125·60				
				C—I	−51·50
				N=N	−22·61
				C=N—R	−102·99

This value now has to be corrected for the diamagnetic contributions from the ligands. The value of this correction factor, calculated from tables is:

$$\chi_d = -1\ 478 \times 10^{-12}\,\text{m}^3\,\text{mol}^{-1}$$

Hence, $\chi' v M = 128\ 057 \times 10^{-12}\,\text{m}^3\,\text{mol}^{-1}$

The magneton number for the compound is therefore given by:

$$X = 797.9(128\ 057 \times 10^{-12}\,\text{m}^3\,\text{mol}^{-1} \times 293.15\ \text{K})^{\frac{1}{2}}\,\text{m}^{-\frac{3}{2}}\,\text{K}^{-\frac{1}{2}}\,\text{mol}^{\frac{1}{2}}$$
$$= 4.89$$

This may be compared with the literature figure in *Table 6.8* of 4.95.

Both the spin and orbital motion of an electron contribute to the paramagnetic moment of an atom or ion. The spins of individual electrons are regarded as coupled together to give a resultant spin momentum (S), where n, the number of unpaired electrons, equals $2S$. Coupling of the orbital angular momenta of the electrons similarly gives a resultant orbital angular momentum (L). It has been shown theoretically that S and L combine together in such a way that the resultant magneton number is $\sqrt{[4S(S+1)+L(L+1)]}$.

For an ion of the first transition series the unpaired 3d electrons are unshielded from the effect of the immediate environment of the ion in the solid lattice or in solution and their orbital momenta are not affected significantly by an external magnetic field. Under such circumstances, the expression for the magneton number reduces to $\sqrt{[4S(S+1)]}$ or $\sqrt{[n(n+2)]}$. The evidence

Table 6.8. MAGNETON NUMBERS OF SOME COMPLEXES OF THE FIRST ROW TRANSITION METALS

	Compound	*Magneton number*
d^1	Bis(acetylacetonato)oxovanadium (IV)	1·74
	Barium manganate (VI)	1·80
d^2	**Ammonium hexafluorovanadate (III)**	2·79
	Potassium hexafluorochromate (IV)	2·80
d^3	Pentaamminechlorochromium (III) chloride	3·88
	Potassium hexafluoromanganate (IV)	3·90
d^4	Tris(acetylacetonato)manganese (III)	4·95
	Potassium hexacyanomanganate (III)	3·18
d^5	Potassium hexathiocyanatomanganate (II) . $3H_2O$	6·06
	Potassium hexacyanomanganate (II) . $3H_2O$	1·80
	Tris(acetylacetonato)iron (III)	5·95
	Potassium tris(oxalato)ferrate (III)	5·75
	Potassium hexacyanoferrate (III)	2·40
d^6	Hexaammineiron (II) chloride	5·45
	Potassium hexafluorocobaltate (III)	4·26
d^7	Hexaamminecobalt (II) perchlorate	5·04
	Tris(o-phenylenebisdimethylarsine)cobalt (II) perchlorate	1·92
	Bis(triethylphosphine)tribromonickel (III)	1·82
d^8	Hexaamminenickel (II) chloride	3·11
	Bis(acetylacetonato)nickel (II)	3·20
	Potassium hexafluorocuprate (III)	2·83
d^9	Bis(acetylacetonato)copper (II)	1·95
	Potassium tetrachlorocuprate (II) . $2H_2O$	1·88

on which this conclusion is based is the close correlation observed between experimental magneton number and the theoretical values calculated for spin interaction only (*Table 6.9*). We should note especially in *Table 6.9* that complexes of ions with configurations between d^4 and d^7 inclusive fall into two classes, called 'spin-free' or 'spin-paired' according to whether their magnetic moments are respectively high or low. From such experimental data, it is possible to draw conclusions about the distribution of electrons within the metal ion.

For the metals of the second and third transition series and the lanthanoids and actiniods orbital contributions to the moment must be important, for only when these are taken into account as well as the spin component can reasonable agreement between predicted and experimental values be obtained. The observed paramagnetic moment is accordingly not simply related to the number of unpaired electrons present.

Table 6.9. MAGNETIC MOMENTS OF FIRST ROW TRANSITIONAL METAL IONS

Ion	No. of d electrons	No. of un-paired electrons	Spin-free complexes		Spin-paired complexes	
			Spin-only magneton number	Experimental magneton number	No. of un-paired electrons	Experimental magneton number
Ti^{3+}	1	1	1·73	1·73	—	—
V^{4+}				1·68–1·78	—	—
V^{3+}	2	2	2·83	2·75–2·85	—	—
V^{2+}	3	3	3·88	3·80–3·90	—	—
Cr^{3+}				3·70–3·90	—	—
Mn^{4+}				3·80–4·00	—	—
Cr^{2+}	4	4	4·90	4·75–4·90	2	3·20–3·30
Mn^{3+}				4·90–5·00	2	3·18
Mn^{2+}	5	5	5·92	5·65–6·10	1	1·80–2·10
Fe^{3+}				5·70–6·0	1	2·0–2·5
Fe^{2+}	6	4	4·90	5·10–5·70	0	—
Co^{3+}				—	0	—
Co^{2+}	7	3	3·88	4·30–5·20	1	1·8
Ni^{3+}				—	1	1·8–2·0
Ni^{2+}	8	2	2·83	2·80–3·50	0	—
Cu^{2+}	9	1	1·73	1·70–2·20	—	—

The above table is taken from Lewis and Wilkins 'Modern Co-ordination Chemistry' p. 406.   Interscience, New York, 1960.

THEORETICAL ASPECTS OF CO-ORDINATION CHEMISTRY

Co-ordination compounds have so far been described from the viewpoint of experimental data relating to some of their most distinctive properties. These include stereochemistry, the thermodynamics of formation and magnetic and spectroscopic properties. It is useful now to see how far these properties can be explained theoretically.

Early theories of co-ordination regarded this as the donation of an electron pair from the ligand to the metal atom or ion. Based on this approach, Sidgwick suggested that the maximum co-ordination number of a metal is reached when it accepts sufficient electrons to raise its total electronic content to that of the next highest noble gas. This total is called the *Effective Atomic Number* (E.A.N.) of the metal. *Table 6.10* gives some examples of complexes

Table 6.10. SOME COMPLEXES WHICH OBEY THE E.A.N. RULE

Complex	Atomic number of metal	Oxidation state	Number of donated electrons	E.A.N.
$[Fe(CN)_6]^{4-}$	26	2	12	$24+12=36$
$[Co(NH_3)_6]^{3+}$	27	3	12	$24+12=36$
$Fe(CO)_5$	26	0	10	$26+10=36$
$Ni(CO)_4$	28	0	8	$28+ 8=36$

which obey this rule. The rule has some usefulness in that it permits the correlation of a large number of co-ordination compounds. It has, however, only qualitative significance because there are numerous exceptions. For example, although hexacyanoferrate (II), $[Fe(CN)_6]^{4-}$, follows the rule, hexacyanoferrate (III), $[Fe(CN)_6]^{3-}$, does not, having an E.A.N. of 35.

THE ELECTRONEUTRALITY PRINCIPLE

The simple theory implies that when co-ordination occurs and the metal accepts several pairs of electrons there must be an accumulation of negative charge on the metal atom. For instance, Co^{III} in $[Co(NH_3)_6]^{3+}$ gains a half-share in 12 electrons on co-ordination and so the charge on the cobalt changes from $+3$ to -3. Such a build-up of negative charge on the metal is most unlikely and Pauling expressed this in his *Electroneutrality Principle*. This states that complexes are stable when each atom carries only a small electric charge in the range -1 to $+1$. The principle requires that the bonds in $[Co(NH_3)_6]^{3+}$ have some ionic character: in other words the electron pair is not shared equally between the cobalt and nitrogen atoms but is attracted more strongly by the non-metal. This prevents the accumulation of negative charge on the cobalt and is in keeping with the greater electronegativity of nitrogen compared with cobalt.

VALENCE-BOND APPROACH

The stereochemistry of complex compounds is here treated in terms of hybridised orbitals. In the transition metals, with which we are largely concerned here, the d orbitals of the penultimate quantum shell are near in energy to the s and p orbitals of the outermost shell and various hybridisations are possible. The same types of hybridisation can occur in the non-transition metals but here the d, s and p orbitals are all of the same principal quantum number.

Octahedral co-ordination (d^2sp^3 hybrids)

This approach may be illustrated with reference to the complexes of Fe^{II}.
In the free Fe^{2+} ion, the electronic arrangement in the 3d orbitals is:

$$3d$$

$$Fe^{2+} \quad \boxed{1\downarrow \mid 1 \mid 1 \mid 1 \mid 1}$$

and the presence of four unpaired electrons is shown by the strong paramagnetism of this ion. In the presence of excess CN^-, the complex hexacyanoferrate (II) is formed. This is octahedral and contains no unpaired electrons.

To account for these observations, Pauling proposed a redistribution of electrons in the metal orbitals. The six electrons, originally in the five 3d orbitals, are paired off so that they occupy only three. The remaining two d orbitals are now available for hybridisation with the vacant 4s and 4p orbitals. Each hybrid orbital accepts a pair of electrons donated by a cyanide ligand and the distribution of electrons in the complex becomes accordingly:

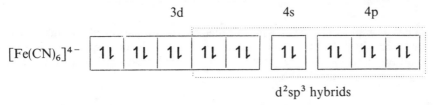

d^2sp^3 hybrids

The hexacyanoferrate (III) ion has one electron less. It therefore has one unpaired electron in the 3d sub-shell and this agrees with its observed paramagnetism.

The situation is different for the hydrated ion $[Fe(H_2O)_6]^{2+}$. This has a paramagnetic moment corresponding to four unpaired electrons and evidently no rearrangement of electrons within the 3d sub-shell occurs when hydration of Fe^{2+} takes place. Here Pauling regards the d orbitals required for hybridisation as of the same principal quantum number as the s and p orbitals involved. The arrangement of electrons is accordingly:

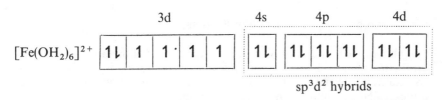

sp^3d^2 hybrids

$[Fe(OH_2)_6]^{2+}$ is an example of a complex described variously as 'ionic', 'outer-orbital' or 'spin-free'. The number of unpaired electrons is the same as in the isolated ion. $[Fe(CN)_6]^{4-}$ is classed as a 'covalent', 'inner-orbital' or 'spin-paired' complex. In this example, the number of unpaired electrons is four less than in the isolated ion.

The differentiation into 'ionic' and 'covalent' complexes is misleading because covalent-bond formation is an essential feature of Pauling's approach to all co-ordination compounds. It is not generally used now for this reason and complexes are described in one of the other ways.

Pauling made great use of his 'magnetic criterion of bond type'. The preceding examples of iron (II) complexes illustrate clearly how the distribution of electrons in the orbitals of the metal is deduced from the measured magnetic moment of the complex.

In some cases, Pauling's approach makes possible a logical explanation of the change in stability of the valency state of a metal which is associated with complex-formation. For example, although simple Co^{III} salts have strong oxidising properties, Co^{III} complexes have remarkable stability. In contrast, simple Co^{II} salts are quite stable but Co^{II} complexes are easily oxidised to

Co^{III}. The arrangement of electrons in the 3d level for Co^{II} is:

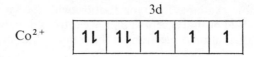

When complexing occurs with octahedral co-ordination of the metal, for instance with CN^-, the use of two of the 3d orbitals for d^2sp^3 hybridisation requires that one 3d electron is promoted to the nearest vacant orbital of higher energy, namely the 5s.

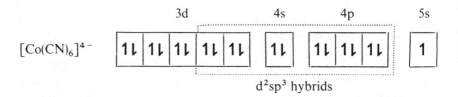

The electron is easily lost from this high level and so oxidation to $[Co(CN)_6]^{3-}$ readily occurs.

Tetrahedral and square-planar co-ordination

Two shapes are found for 4 co-ordinated complexes of the transition metals; tetrahedral with sp^3 hybridisation and square planar with dsp^2 hybridisation. For example, the simple Pt^{II} ion has eight 5d electrons which are distributed in the available orbitals in the following manner:

When complex-formation occurs these electrons are rearranged to leave one d orbital vacant. When hybridised with the 6s and two of the 6p orbitals, this provides four equivalent orbitals, each of which contains a pair of electrons in the square-planar complexes of Pt^{II}.

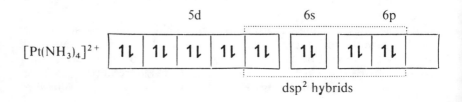

Pauling was able to predict correctly that certain diamagnetic nickel (II) complexes must be square planar, even before their structures were experimentally determined. The distribution of electrons in the outer orbitals of Ni^{II} before and after co-ordination can be represented as:

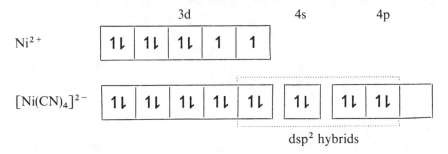

When the complete x-ray analysis of the compound $Na_2[Ni(CN)_4].3H_2O$ was carried out later, the square-planar arrangement of the four cyanides around the nickel was demonstrated. The complex of nickel with dimethylglyoxime is another example of the planar distribution around Ni^{II}. However, it has been observed that certain paramagnetic complexes of Ni^{II}, such as $[Ni(NH_3)_4]^{2+}$ and $[NiCl_4]^{2-}$, are tetrahedral and Pauling's theory cannot account for these.

When nickel is zerovalent, the tetrahedral shape of its complexes can be satisfactorily explained:

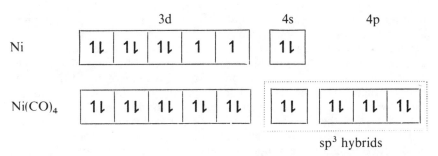

The formation of four single covalent bonds from the carbon atoms to the nickel would result in a large negative charge on the metal. Pauling suggested that, in such a situation, double bonding takes place with the back-donation of d electrons from the metal to the ligand to such an extent that the electroneutrality principle is obeyed. X-ray studies have shown that the nickel–carbon interatomic distance is distinctly shorter than the sum of the covalent radii of the two atoms. This is experimental support of Pauling's contention that the electronic structure of tetracarbonylnickel is best described in terms of a resonance hybrid of various possible forms with important contributions from double-bonded structures.

Cu^{II}, with one more electron than Ni^{II}, also forms planar complexes with paramagnetic moments corresponding with the presence of one unpaired electron. Pauling's magnetic criterion is of no help here in deciding between dsp^2 and sp^3 hybridisation because each of these would leave the complex ion with one odd electron either in a 4p or a 3d orbital. The x-ray analysis of Cu^{II} complexes has provided unambiguous evidence for their planarity and the electronic arrangement in terms of hybrid orbitals is therefore:

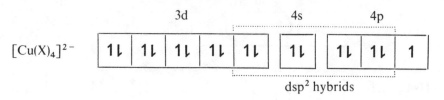

where X is an anionic ligand.

Cu^I forms tetrahedral 4 co-ordinated complexes. The electronic distribution therein is:

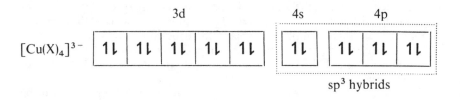

According to Pauling's theory, the stereochemistry of metal complexes is determined by the hybridised orbitals used in bond formation. In many cases, a comparison of the magnetic moment of the complex with that of the free metal ion indicates clearly the orbital distribution of the electrons which is consistent with the shape of the complex determined by x-ray analysis. The accumulation of negative charge on the metal is reduced by (i) the partial ionic character of the metal–ligand bonds and/or (ii) the formation of multiple bonds with back-donation of metal d electrons to the ligand.

Emphasis on the covalent nature of the bonds does impose certain limitations on the theory. Although it provides a satisfying pictorial representation of co-ordination compounds and their magnetic properties based on the concept of orbital hybridisation, quantitative interpretations are not possible. For example, the striking colour changes frequently associated with complex-formation cannot be explained.

Another unsatisfactory feature is that the theory sometimes requires the promotion of one electron to a vacant orbital of higher energy by the intake of an amount of energy which is unrealistically large. For instance, in accounting for the square planarity of Cu^{II} complexes the promotion of a 3d electron to a 4p energy level is postulated to allow for the dsp^2 hybridisation to occur. This process requires 1422·6 kJ mol^{-1} in the free Cu^{II} ion and probably a large fraction of this in the complex. It is not easy to see where this energy can come from and one would also expect, by analogy with the explanation for the stability of Co^{III} complexes, that Cu^{II} complexes would lose this unpaired 4p electron and be readily oxidised to Cu^{III}. Although a number of

Cu^{III} compounds are known, for example K_3CuF_6, their stabilities are certainly not comparable with those of Cu^{II} compounds.

CRYSTAL AND LIGAND FIELD THEORY

This is a theory which is concerned with electrostatic interactions in ionic lattices and particularly the effect of the surrounding ions on the electron distribution within a central ion. The concept of a crystal field, the electrostatic field existing within a crystal, was first proposed by Bethe in 1929 with reference to a sodium chloride type of lattice and the theory was subsequently developed by Van Vleck. Bethe considered charges on ions to be located at points coincident with their nuclei. Thus in sodium chloride, the environment of each sodium ion is regarded as being six negative point charges situated at the vertices of an octahedron. In crystal field theory, the total electrostatic potential from all surrounding ions is calculated for any point near the central ion and the effect of such a potential on its electrons is considered. The calculation is rather complex and details are not given

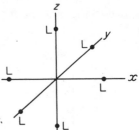

Figure 6.3. Representation of an octahedral complex with ligands, L, located at points on the Cartesian axes

here. Some of the conclusions of crystal field theory are of great usefulness in inorganic chemistry and are specially relevant to transition metal complexes.

One of the main conclusions which transpires is that there may be considerable degeneracy among the energy levels in the free ion but that this is generally removed when the ion is situated in an ionic lattice. Provided that covalent bonding can be ignored and the interactions between a metal ion and its ligands (which are either ions or dipolar molecules) can be regarded as entirely electrostatic, the effect on the energy levels of a metal ion in an octahedral complex is exactly the same as that of the six nearest halide ions on those of the sodium ion in sodium chloride. For many purposes, it is sufficient to consider the effect of the nearest neighbours only on the orbitals of the metal ion and to disregard the influence of other ions which are further away.

An octahedral complex may be represented as shown in *Figure 6.3*. The ligands are located as point charges on the axes of a Cartesian co-ordinate system which has its origin at the centre of the metal ion. The crystal field raises the energies of all electrons in the orbitals of the metal ion with the greatest effect on those in the valency shell.

The energy of an electron in an s orbital is increased but no splitting of this energy level is possible because there is, of course, only one such orbital in a given quantum shell. The three p orbitals have preferred directions in space. If we take these to be the x, y and z axes respectively in *Figure 6.3*, it is evident that electrons in these orbitals are affected equally by the crystal field. The p orbitals, degenerate in the free ion, remain so in the complex.

In the case of d orbitals, the situation is different. We have already seen that the five d orbitals are not all of the same shape. Another difference is apparent when we consider their spatial characteristics in relation to the Cartesian axes. The d_{z^2} and $d_{x^2-y^2}$ orbitals have their maximum amplitudes along the z and the x and y axes respectively whereas the d_{xy}, d_{xz} and d_{yz} orbitals have their maximum amplitudes along the directions between the respective pairs of axes denoted by their subscripts.

The d_{xy}, d_{yz} and d_{xz} orbitals constitute one set (labelled as d_ε or t_{2g}): d_{z^2} and $d_{x^2-y^2}$ constitute a second (d_y or e_g). Orbitals of the d_ε set are symmetrically placed with respect to ligands on the Cartesian axes and the energy of all three is raised by an equal amount. The d_y orbitals are pointing directly at the ligands and electrons in these orbitals will be repelled more strongly than those in the d_ε orbitals. This results in a greater increase in energy for the d_y compared with the d_ε orbitals. Although d_{z^2} and $d_{x^2-y^2}$ are not equivalent in shape, it may be shown theoretically that the energies of both are raised to the same extent in an octahedral environment.

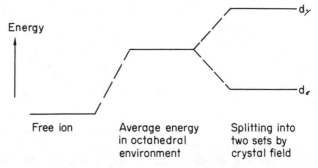

Energy

Free ion

Average energy
in octahedral
environment

Splitting into
two sets by
crystal field

Figure 6.4. Splitting of d orbital energies in an octahedral crystal field

The reason why splitting occurs can also be described rather differently. The average energy in the crystal field corresponds to a hypothetical situation in which the metal ion is surrounded by a spherical shell of negative charge at a radius equal to the metal–ligand distance. The charge is that due to the ligands and it raises the energy of all d electrons equally when distributed in this way. If we now alter the distribution and regard the charge as concentrated equally at six points on the sphere corresponding to the ligand positions in the actual complex then, although the energy of the d electrons is not changed as a whole, that of the d_y electrons is increased further and that of the d_ε electrons is lowered. Electrons in the d_ε orbitals are stabilised and those in the d_y orbitals are de-stabilised relative to the average energy.

The effect of the crystal field on the metal ion is therefore to raise the energy of all d orbitals and to split them into two sets of different energies. This process is illustrated in *Figure 6.4*. The magnitude of the splitting, that is, the difference in energy between d_ε and d_y, is usually given the symbol Δ.

The energy of the orbitals in each set can be conveniently referred to the average energy of all five orbitals. Each d_ε orbital is lower by $2/5\,\Delta$ and each d_γ orbital is higher by $3/5\,\Delta$ than the average.

We have been emphasising the increase in energy of d orbitals which accompanies the process of complex-formation but it is important to remember that the total energy of the system will fall because of the electrostatic attraction between the positive charges on the metal ion and the negative charges on the ligands. The energy released in this way may be much greater than that absorbed by the crystal field effect.

In a tetrahedral complex, d-orbital splitting again occurs. The spatial relationship between such a complex and the metal d-orbitals is illustrated in *Figure 6.5*. The ligands are located at opposite corners of the cube and so none of the d orbitals has its maximum amplitude in the direction of the

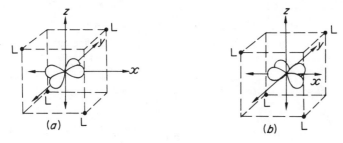

Figure 6.5. Spatial relationship between metal d orbitals and ligand positions in a tetrahedral complex

ligands. However, each lobe of the d_{xy} orbital is fairly close to one ligand and the d_{yz} and d_{xz} orbitals are similarly placed. The lobes of the $d_{x^2-y^2}$ orbital are directed along the axes towards the centres of the cube faces and are somewhat further away from the ligands: the same is true of d_{z^2}. As in the case of octahedral complexes, the crystal field splits the d orbitals into two sets (d_ε and d_γ or, in group theoretical notation, e and t_2; note that the subscript g is no longer used for this refers to orbitals which are symmetrical with respect to inversion through a centre of symmetry and a tetrahedron, unlike an octahedron, does not possess such a centre), but the order of these is inverted, the d_ε being of higher energy than the d_γ set (*Figure 6.6*). Calculations based on a simple electrostatic model show that the splitting in a tetrahedral complex would be $4/9$ of that in an octahedral complex of the same metal and the same ligand and with identical metal–ligand distance in both complexes.

The d-orbital splitting existing in a square-planar complex can be qualitatively deduced if we regard this as being derived from an octahedron by the removal of two ligands *trans* to one another. For example, when two ligands lying on the z axis are taken away, this stabilises the d_{z^2} relative to the $d_{x^2-y^2}$ orbital and the degeneracy between them is removed. Similarly, the d_{yz} and d_{xz} orbitals are stabilised relative to d_{xy}. The splitting in a square-planar complex is therefore as shown in *Figure 6.7*.

Before proceeding to consider how crystal field theory has been used to explain the properties of transition metal complexes, it is helpful to have some idea of the magnitude of Δ. This can be calculated from the electronic spectra of complexes. In general, these spectra consist of one or more absorption bands in the ultra-violet, visible or near infra-red region. Absorption bands arise from the absorption of energy which raises the complex from its initial energy state, E_1, to a higher state, E_2. The frequency, v, of energy absorbed is related to E_1 and E_2 by $E_2 - E_1 = hv$. Three main types of electronic transition give rise to absorption in this region: (a) within the metal ion between d orbitals of different energy resulting from crystal field splitting, (b) between

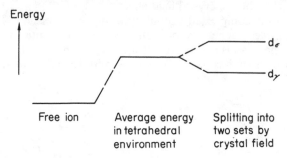

Figure 6.6. Splitting of d orbital energies in tetrahedral crystal field

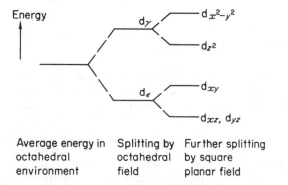

Figure 6.7. Splitting of d orbital energies in square planar crystal field

different energy levels on the metal and ligand such that a transfer of charge from metal ion to ligand or vice versa is involved, and (c) within the ligand itself, a type of transition which is particularly important for organic ligands which contain systems of delocalised π electrons. Absorption in the visible and near infra-red is commonly due to (a) and referred to as d-d-transitions. Absorption in the visible or ultra-violet arises from transitions of type (b) and (c). Type (b) are generally called charge-transfer bands. A well-known example is the intense red colouration of Fe(III) complexes with thiocyanate. It is the d-d-transitions with which we are concerned here because these are primarily responsible for the characteristic colours of transition metals in their complexes and the nature of such transitions is well understood theoretically.

From the point of view of electronic configuration, the simplest transition metal ion is d^1. One d^1 ion is Ti^{3+}. This is present in aqueous solution as the octahedral complex, $[Ti(H_2O)_6]^{3+}$. Its colour is purple due to a weak, broad absorption band centred at $2{\cdot}02 \times 10^6$ m^{-1} (at a wavelength of about 500 nm). In the ground state of the complex the single d electron will be located in the d_ε set; absorption of energy will transfer this to a d_γ orbital of higher energy. The energy absorbed is given by $E_2 - E_1 = h \times c \times \bar{v}$. For 1 mole this becomes $N_A(E_2 - E_1)$ or, substituting the appropriate values, $242{\cdot}2$ kJ mol^{-1}.

Values of Δ can also be calculated from the spectra of complexes of metal ions with more than one d electron. The number of energy levels in the free ion which arises for such configurations is fairly large because of splitting associated with inter-electronic repulsions and the interaction of electron spin and orbital angular momenta which are both significant whenever there is more than one electron in a quantum shell. A thorough understanding of the theory of the spectra of free ions is necessary before the effect of the crystal field on these levels can be understood. It must suffice here to say that it is possible theoretically to account (at least in a qualitative manner) for the spectra of numerous metal complexes using the concept of an electrostatic crystal field.

For tripositive ions of the first transition series, Δ is found to lie between 1×10^6 and 3×10^6 m^{-1}. For dipositive ions of the same series, Δ varies from $7{\cdot}8 \times 10^5$ m^{-1} for Mn^{2+} to $1{\cdot}1 \times 10^6$ m^{-1} for Cr^{2+}. Complexes of the second or third transition series metal have Δ values up to twice those of the complexes of the first series metal of the same sub-group.

A study of the spectra of many complexes has revealed that, for a given stereochemistry, ligands can be arranged in an order of increasing Δ which is generally independent of the metal ion concerned. This order is called the *spectrochemical series* and for a number of common ligands is as follows: $I^- < Br^- < Cl^- < F^- < H_2O < NH_3 \simeq$ pyridine $<$ ethylenediamine $< NO_2^- < CN^-$.

Magnetic properties

In an isolated transition metal ion, the five d orbitals are degenerate and electrons are distributed among them in accordance with Hund's Rule of maximum multiplicity. When the number of unpaired electrons (those which have parallel spins) is the greatest possible, the potential energy is lowest and this arrangement is the most stable. For d^1 to d^5 ions inclusive, all electrons are unpaired. If more than 5 d electrons are present, then some pairing-up necessarily takes place. An amount of energy, sometimes called the pairing energy, is required to overcome the inter-electronic repulsion and to bring about a reversal of the spin of one electron which must take place before two electrons can occupy the same orbital.

We now consider the electronic distribution in an octahedral environment for metal ions as the number of d electrons is increased.

For d^1, d^2 and d^3 ions, the electrons are distributed singly in the d_ε orbitals and complexes of these ions show the paramagnetism expected for one, two and three unpaired electrons respectively.

For a d^4 ion, the fourth electron may be located in one of the less stable d_γ orbitals, thereby preserving maximum multiplicity, or it may be paired up in a d_ε orbital with the electron already there. Which of these two distributions is actually found depends on the relative magnitude of Δ and the pairing energy. When the field due to the ligands is weak, Δ is small and less energy is required for the fourth electron to enter a d_γ orbital than for it to pair up in a d_ε orbital. The number of unpaired electrons is the same as in the isolated ion and such a complex is known as 'high-spin' or 'spin-free'. If the ligand field is strong, then all four electrons are located in the d_ε set and only two are unpaired. This is described as a 'low-spin' or 'spin-paired' complex. In *Table 6.8* the magneton numbers of two manganese (III) complexes are given. The magneton number of tris(acetylacetonato)manganese (III) is 4·95 and it is clearly a high-spin complex. That of potassium hexacyanomanganate (III) is 3·18, close to the value of 2·83 expected for two unpaired electrons and hence this complex is low-spin. The experimental value is somewhat larger than expected theoretically because of an orbital contribution to the magneton number. The difference between the magneton numbers of the two complexes must mean that acetylacetonate lies below cyanide in the spectrochemical series.

For transition metal ions with five, six or seven d electrons there are again

Table 6.11. ELECTRON DISTRIBUTION IN OCTAHEDRAL COMPLEXES

Number of d electrons	Arrangement in weak ligand field					Number of unpaired electrons
	d_ε			d_γ		
1	1					1
2	1	1				2
3	1	1	1			3
4	1	1	1	1		4
5	1	1	1	1	1	5
6	1↓	1	1	1	1	4
7	1↓	1↓	1	1	1	3
8	1↓	1↓	1↓	1	1	2
9	1↓	1↓	1↓	1↓	1	1

Number of d electrons	Arrangement in strong ligand field					Number of unpaired electrons
	d_ε			d_γ		
1	1					1
2	1	1				2
3	1	1	1			3
4	1↓	1	1			2
5	1↓	1↓	1			1
6	1↓	1↓	1↓			0
7	1↓	1↓	1↓	1		1
8	1↓	1↓	1↓	1	1	2
9	1↓	1↓	1↓	1↓	1	1

two possible electronic arrangements, corresponding in each case with high- and low-spin. In a d^5 ion like Mn^{2+}, its hexacyano complex is low-spin whereas its hexathiocyanato complex is high-spin. Similarly, $[Fe(CN)_6]^{4-}$ is a low-spin complex of the d^6 ion, Fe^{2+}, and the hydrated ion, $[Fe(H_2O)_6]^{2+}$, is high-spin. It is instructive to compare this description of iron complexes, which is a natural consequence of d-orbital splitting in a crystal field, with their more arbitrary classification, according to valence-bond theory, as inner- and outer-orbital respectively.

For d^8 and d^9 ions, the d_ε set is completely filled irrespective of the magnitude of Δ and there is only one possible arrangement of electrons. The electronic distributions found in octahedral complexes of d^1 to d^9 ions are summarised in *Table 6.11*.

A similar pattern of behaviour is found for tetrahedral complexes. The number of unpaired electrons in d^1, d^2, d^7, d^8 and d^9 ions is the same whether the ligand field is weak or strong. For d^3 to d^6 ions inclusive, there is again a possibility of either a spin-free or a spin-paired complex. The position is summarised in *Table 6.12*.

The number of tetrahedral complexes known for first series transition metals is relatively small. Some examples, with their magneton numbers measured at 300 K given in brackets, are $[(C_2H_5)_4N]_2[MnCl_4]$ (5·94) and

Table 6.12. ELECTRON DISTRIBUTION IN TETRAHEDRAL COMPLEXES

Number of d electrons	Arrangement in weak ligand field					Number of unpaired electrons
	d_γ		d_ε			
1	1					1
2	1	1				2
3	1	1	1			3
4	1	1	1	1		4
5	1	1	1	1	1	5
6	1↓	1	1	1	1	4
7	1↓	1↓	1	1	1	3
8	1↓	1↓	1↓	1	1	2
9	1↓	1↓	1↓	1↓	1	1

Number of d electrons	Arrangement in strong ligand field					Number of unpaired electrons
	d_γ		d_ε			
1	1					1
2	1	1				2
3	1↓	1				1
4	1↓	1↓				0
5	1↓	1↓	1			1
6	1↓	1↓	1	1		2
7	1↓	1↓	1	1	1	3
8	1↓	1↓	1↓	1	1	2
9	1↓	1↓	1↓	1↓	1	1

$[(C_2H_5)_4N] [FeCl_4]$ (5·88) — both of which contain d^5 ions — and Cs_2CoCl_4 (4·71), which contains a d^7 ion. On the basis that all three are spin-free, the theoretical spin-only value for the manganese and iron complexes is 5·92 and that for the cobalt complex is 3·88. There is evidently a significant orbital contribution to the magnetic moment of Cs_2CoCl_4.

All known tetrahedral complexes of first series transition metals are spin-free. This is not surprising in view of the much smaller crystal field in tetrahedral compared with octahedral complexes. It seems unlikely that a complex will be discovered in which the field is strong enough to force electrons to pair in the d_y set before d_ε is occupied at all.

Jahn–Teller effect

The effect of a crystal field has so far been discussed with reference to complexes of regular octahedral or tetrahedral symmetry and it has been implicit that we have been concerned with identical ligand atoms at all vertices. Co-ordination compounds rarely possess such regular shapes and there are very many complexes which contain two or more different ligands. If the six ligands in an octahedral complex are not equivalent (in other words they occupy different positions in the spectrochemical series) the metal ion must be in a field of lower than octahedral symmetry. The electronic energy levels then show further splitting.

Even when all ligands are the same, other factors may cause a distorted structure to be more stable than a regular one. In the solid state, crystal packing forces may influence the structure in this way: in solution, the solvent molecules may affect the local symmetry around the metal. One important cause of distortion is the *Jahn–Teller Effect*. The basis for this is a theorem propounded by Jahn and Teller which states that 'any non-linear ion or molecule which would appear, from its proposed symmetry in the ground state to be orbitally degenerate, will actually be distorted so as to remove the degeneracy'. The meaning of this theorem is best understood by considering an example.

The Cu^{2+} ion in a regular octahedral environment would have the configuration $d_\varepsilon^6 d_y^3$. This is orbitally degenerate because of the unequal occupation of the d_y orbitals. Thus the distribution $(d_{x^2-y^2})^1(d_{z^2})^2$ has the same energy as the alternative $(d_{x^2-y^2})^2(d_{z^2})^1$. We consider the first of these and the effect of a tetragonal distortion on the energies of the d electrons. This distortion involves an increase in the metal–ligand distances along the $\pm z$ axes and stabilises the d_{z^2} orbital and de-stabilises the $d_{x^2-y^2}$ orbital. These are now separated by an amount of energy, δ (*Figure 6.8*). Similarly, the d_ε orbitals are split into d_{xy} (of higher energy) and a degenerate pair, d_{yz} and d_{xz} (of lower energy). The splitting is such that the centres of gravity of the d_ε and d_y sets are not affected by the distortion. The energy of each electron in d_{z^2} is lowered by $\delta/2$ and that of the electron in $d_{x^2-y^2}$ is raised by $\delta/2$. The result is a net stabilisation of $\delta/2$ compared with the undistorted complex. As each d_ε orbital is filled to an equal extent there is no change in the overall energy of these electrons when distortion occurs. We can see, however, that the distortion removes the degeneracy of the d_y orbitals and produces an electronic distribution of lower energy than that of the regular octahedron.

We could equally well have represented $d\gamma^3$ as $(d_{x^2-y^2})^2(d_{z^2})^1$. In this case the distortion expected would be a lengthening of the metal–ligand distances in the xy plane. Although the Jahn–Teller theorem enables us to predict that an octahedron of ligands around Cu^{2+} will distort it does not indicate which of the two possible distortions actually occurs or what the magnitude of the distortion will be. Crystallographic data show that copper

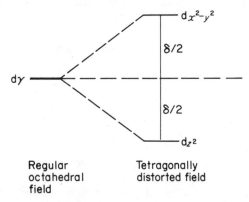

Figure 6.8. Splitting of $d\gamma$ orbitals by operation of the Jahn–Teller Effect

complexes are invariably distorted tetragonally and the d_y electrons must therefore be distributed according to the first scheme.

The same type of distortion is found in high-spin octahedral complexes of d^4 ions $(d_\varepsilon^3 d_y^1)$ such as Cr^{2+} and Mn^{3+}. It is expected for any configuration where there is an unequal occupation of orbitals in the d_ε or the d_y set but not all possibilities have been realised experimentally.

Crystal field stabilisation energy

The process of placing a free ion in a crystal field results in a considerable stabilisation of the system because of the electrostatic attraction between the ligands and the metal ion. At the same time, there is some de-stabilisation (much smaller in magnitude) due to the repulsion by the ligands of the d electrons of the metal which serves to raise their average energy. As we have seen, the d orbitals are split into two sets of different energy in an octahedral or in a tetrahedral field. In ions where all the d orbitals are equally populated (d^0, d^5 and d^{10}), the total energy of the electrons in these orbitals is the same as the average energy because the stabilisation resulting from electrons in the lower energy set is equal and opposite to the de-stabilisation associated with electrons in the higher energy set. When all the d orbitals are not equally populated, an additional amount of energy is released by the electrons preferentially entering the lower energy level. This is known as crystal field stabilisation energy and its magnitude varies from one metal ion to the next.

It is easy to see what the crystal field stabilisation energies are in the case of metal ions in a weak ligand field. For octahedral co-ordination, each electron in a d_ε orbital is stabilised by $-2/5\Delta$ and each one in the d_γ set is de-stabilised by $3/5\Delta$. The stabilisation energies for d^0 to d^{10} inclusive are given below.

Number of d electrons		Crystal Field Stabilisation Energy
0		0
1	6	$-2/5\ \Delta$
2	7	$-4/5\ \Delta$
3	8	$-6/5\ \Delta$
4	9	$-3/5\ \Delta$
5	10	0

The d^n and d^{n+5} configurations lead to the same stabilisation energy and in this sense they are equivalent. To derive stabilisation energies for ions in a strong field we should need to take another factor into account, namely a de-stabilisation due to the pairing-up of electrons in the d orbitals.

The splitting of d orbitals by crystal fields is of the order of 10^6 m^{-1} whereas the energy of formation of an octahedral complex from gaseous ions and ligands may be as much as 10^7 m^{-1}. Crystal field stabilisation energy is therefore quite small in comparison. Nevertheless, we should expect to find some evidence for the effect of the crystal field on the energies of formation or reaction of transition metal complexes.

Lattice enthalpy

The lattice enthalpies of transition metal difluorides are plotted against the number of d electrons in *Figure 6.9*. Data for calcium and zinc fluorides are also included. The lattice enthalpy does not show a steady increase with the number of electrons but the curve through the experimental values has a characteristic double-humped shape. If we draw a straight line between the points for d^0, d^5 and d^{10} ions, we can interpolate expected values for the energies of the remaining fluorides. The plot shows that the experimental values are all higher than the interpolated ones. These compounds are more stable than expected because of the crystal field stabilisation energy. When the experimental values are adjusted by applying a correction for the stabilisation energy, based on the spectroscopic measurements of Δ, the corrected values lie on or very close to the straight line. It is strictly correct to use the stabilisation energies given above only for those metal ions which are in a regular octahedral environment. They are inapplicable to ions such as Cr^{2+} and Cu^{2+}.

Similar effects of crystal field stabilisation energy are to be found in the lattice enthalpies of the transition metal dichlorides and in the enthalpies of hydration and of complex-formation for the metal ions.

In the absence of spectroscopic data, crystal field stabilisation energies and hence Δ can be estimated from curves like that shown in *Figure 6.9*. This method is useful, for example, where a compound is opaque and does not give an absorption spectrum from which Δ can be evaluated.

The value of Δ for nickel (II) fluoride is found in the following way. A value for the lattice enthalpy of nickel fluoride without crystal field stabilisation energy, $\Delta H^0_{LE(calc.)}$, is found by linear interpolation between the lattice enthalpies for two fluorides of metal ions which have no stabilisation

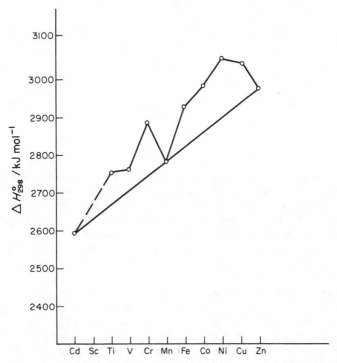

Figure 6.9. Lattice enthalpies of the first series transition metal difluorides

energy. These are MnF_2 and ZnF_2, for which the lattice enthalpies are $2781 \cdot 9$ and $2986 \cdot 1$ kJ mol^{-1} respectively.

$$\text{Hence } \Delta H^\circ_{LE(calc.)} = 2986 \cdot 1 - 2/5(2986 \cdot 1 - 2781 \cdot 9)$$
$$= 2904 \cdot 4.$$

For nickel fluoride, $\Delta H^\circ_{LE(expt.)} = 3064 \cdot 4$ kJ mol^{-1}. The crystal field stabilisation energy is $6/5\Delta$ and this is equal to $\Delta H^\circ_{LE(expt.)} - \Delta H^\circ_{LE(calc.)}$.

$$6/5\, \Delta = 3064 \cdot 4 - 2904 \cdot 4 = 160 \text{ kJ mol}^{-1}$$
$$\text{and} \qquad \Delta = 133 \cdot 3 \text{ kJ mol}^{-1} \, (\equiv 1 \cdot 12 \times 10^6 \text{ m}^{-1})$$

Ionic radii

In the absence of crystal field effects, the radii of successive transition metal ions would be expected to show a steady decrease. This is because an additional d electron has only a poor shielding effect of the increased nuclear charge on the other d electrons. The d orbitals and hence the ionic radius should show a contraction along the series.

The observed internuclear distances for first series transition metal oxides of formula MO are given in *Table 6.13*.

Table 6.13

MO	M—O distance (r/pm)	$r_{M^{2+}}$ (r/pm)
CaO	240	100
TiO	212	72
VO	205	65
MnO	222	82
FeO	217	77
CoO	212	72
NiO	208	68

These compounds all have the sodium chloride structure. If a constant radius of 140 pm for O^{2-} is assumed, then the radius of M^{2+} shows an irregular decrease as the atomic number increases. This irregularity is a consequence of the unsymmetrical distribution of electronic charge around the nucleus when the d orbitals are only partially filled.

In the three ions Ca^{2+}, Mn^{2+} and Zn^{2+} (d^0, d^5 and d^{10} respectively), the distribution of d electron density around the metal ion is spherical because all five orbitals are either unoccupied or equally occupied. The radii of these three ions show the steady decrease expected ($Zn^{2+} = 74$ pm).

In the case of Ti^{2+}, however, this is a d_ε^2 ion and the negative charge of the d electrons is concentrated in those regions of space away from the metal–ligand bond axes. These electrons thus provide unusually little shielding between the metal ion nucleus and the negative ligands; the ligands are attracted closer to the metal than they would be if the two d electrons were spherically distributed. The same effect, to a greater extent, is observed in V^{2+} (d_ε^3) because there are now three d electrons all providing little shielding.

Proceeding onwards from V^{2+}, the d_γ orbitals are filled up. d_γ electrons provide more screening than would be provided by a spherically-distributed electronic charge and an increase in radius is observed. The radius of Cr^{2+} in a regular octahedral environment is unknown because distortion of the environment is always found.

A similar pattern is followed as the five electrons after Mn^{2+} are added successively. Again Cu^{2+} cannot be accurately determined because of distortion from a regular octahedral arrangement.

Limitations of crystal field theory

One fundamental criticism of the crystal field theory is that the point charge model does not truly represent the situation of a metal ion in the electrostatic field of the surrounding ligands. The model ignores the attractive effect of the nuclear charge in the ligand atom on the metal ion d electrons. In an octahedral complex, this effect is much greater on the d_γ electrons than on the d_ε electrons (because the ligand nucleus is nearer to the former). When this, as well as the repulsion between electron clouds, is considered, the stability of the d_γ orbitals is very close to and may even be somewhat less than that of the d_ε orbitals. Thus when a physically realistic electrostatic model is used, there is no satisfactory explanation for the d orbital splittings observed experimentally.

Furthermore, a completely electrostatic interaction between metal ions and ligands does not occur in complexes. Even in fluoro complexes, where one would expect the maximum ionic character of the bond, modern experimental techniques such as electron spin resonance and nuclear magnetic resonance provide conclusive evidence for at least some overlap of metal d orbitals and ligand orbitals.

Ligand field theory

The effect of overlap between metal and ligand orbitals is taken into account in ligand field theory. Bonding in complexes is described in terms of molecular orbitals formed by the interaction between these orbitals. Just as in simpler systems such as the homonuclear diatomic molecules considered earlier, the nature of the molecular orbitals is determined by the directional and symmetry properties of the orbitals from which they are formed. Thus only orbitals with the same symmetry with respect to the bond axis will interact with each other.

In an octahedral complex of a metal of the first transition series, the metal atom or ion has 3d, 4s and 4p orbitals which can participate in bonding. Six of these, d_{z^2}, $d_{x^2-y^2}$, s, p_x, p_y and p_z, have components of their wavefunctions directed along the metal–ligand bond axes. These overlap with appropriate orbitals on the six ligands to form six σ-bonding orbitals and six σ^*-antibonding orbitals. To construct these molecular orbitals we no longer distinguish between orbitals on individual ligands and it is necessary to derive the wave-functions of various groups of ligand orbitals. According to group theory, there are six linear combinations of ligand orbitals, referred to as 'ligand group orbitals', each of σ symmetry, which interact with metal orbitals of the same symmetry to form molecular orbitals.

The bonding orbitals are more stable (of lower energy) than the orbitals in the uncombined metal and ligands. The antibonding orbitals are less stable than these. In the case where the metal orbitals are empty and each ligand orbital contains two electrons, a total of twelve electrons is distributed among the six bonding orbitals.

The energy level diagram for an octahedral complex, in terms of σ-bond formation only, is shown in *Figure 6.10*. The metal t_{2g} orbitals have π symmetry with respect to the bond axes and, unless the ligands also have orbitals of this symmetry, they will not be involved in bonding and will have substantially the same energy in the complex as in the free ion. In the complex, any electrons in the t_{2g} orbitals can be regarded as originating with the metal. In the figure the letters *a*, *e* and *t* refer respectively to energy levels which are singly, doubly and triply degenerate: the subscripts g and u refer respectively to orbitals which are symmetric and antisymmetric with respect to inversion through their centres of symmetry: the subscripts 1 and 2 refer to symmetry and asymmetry respectively with respect to a twofold axis through the centre of symmetry of the orbital.

The bonding σ orbitals are closer in energy to the ligand group orbitals than to the metal orbitals. They resemble the former more than the latter and electrons in them are regarded as being predominantly ligand electrons. Similarly, the antibonding e_g^* orbitals are mainly of metal d orbital character.

It is possible to regard the d electrons of the free metal as distributed between the t_{2g} and e_g^* molecular orbitals. The energy separation between

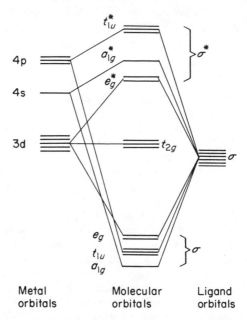

Metal orbitals   Molecular orbitals   Ligand orbitals

Figure 6.10. Molecular Orbital Energy Level diagram for an octahedral complex (σ-bond formation only)

these two sets is the counterpart in ligand field theory of the crystal field splitting energy and electrons will be distributed between them in exactly the same manner as described in crystal field theory.

For some complexes, there will be overlap also between metal t_{2g} orbitals and certain orbitals on the ligands leading to the formation of π molecular orbitals. This will necessarily modify the energy-level diagram of *Figure 6.10* although the changes in energy are unlikely to be large because π-bonds are in general weaker than σ-bonds.

Three possible situations for π-bonding can be recognised. Firstly, when the metal t_{2g} orbitals contain electrons and the ligand π orbitals are of higher energy and are not occupied. Examples are complexes containing ligand atoms like sulphur, selenium, phosphorus, arsenic or antimony. Secondly, when the ligands have filled π orbitals of lower energy than the metal t_{2g} orbitals. Ligands of this kind include fluorine and oxygen. Thirdly, when the the ligands, typified by chloride, bromide and cyanide ions, have both filled and unfilled π orbitals. The strengthening of metal–ligand bonding by contributions from π-bonds has special significance in complexes of ligands such as carbon monoxide and unsaturated hydrocarbons.

METAL CARBONYLS AND RELATED COMPOUNDS

The carbonyls are molecular co-ordination compounds between metals and carbon monoxide. This acts as a ligand by the formation of bonds between its carbon atom and a metal atom. Other neutral molecules such as nitric oxide, various substituted phosphines, stibines and arsines behave in the same way and are able to replace carbon monoxide partially or completely in complexes.

The known binary carbonyls of the transition metals are listed in *Table 6.14*. In simple, mononuclear carbonyls like $Cr(CO)_6$, $Fe(CO)_5$ and $Ni(CO)_4$,

Table 6.14. THE CARBONYLS OF THE TRANSITION METALS

Group	V	VI	VII	VIII		
	$V(CO)_6$	$Cr(CO)_6$	$Mn_2(CO)_{10}$	$Fe(CO)_5$	$Co_2(CO)_8$	$Ni(CO)_4$
				$Fe_2(CO)_9$	$Co_4(CO)_{12}$	
				$Fe_3(CO)_{12}$	$Co_6(CO)_{16}$	
		$Mo(CO)_6$	$Tc_2(CO)_{10}$	$Ru(CO)_5$	$Rh_2(CO)_8$	
				$Ru_2(CO)_9$	$Rh_4(CO)_{12}$	
				$Ru_3(CO)_{12}$	$Rh_6(CO)_{16}$	
		$W(CO)_6$	$Re_2(CO)_{10}$	$Os(CO)_5$	$Ir_2(CO)_8$	$Pt(CO)_4$
				$Os_2(CO)_9$	$[Ir(CO)_3]_4$	
				$[Os(CO)_4]_3$	$Ir_6(CO)_{16}$	

the metals have a formal oxidation state of zero and each compound may be regarded as formed from metal atoms and carbon monoxide molecules. In polynuclear complexes like $Mn_2(CO)_{10}$, $Fe_3(CO)_{12}$ and $Co_2(CO)_8$, pairs of metal atoms are sufficiently close to be able to form direct covalent bonds with each other. This feature of metal-to-metal bonding is sufficiently rare as to be worthy of special note. Indeed for many years the carbonyls were regarded as unique in this respect: more recently, the so-called 'cluster' compounds have been shown to contain similar metal-to-metal bonds. Another unusual aspect of the structures of polynuclear carbonyls is that the CO groups can function in different ways, not only as a unidentate ligand but also as a bridge between two and sometimes even three metal atoms.

Carbonyls and their derivatives have important industrial applications. The formation of nickel tetracarbonyl, $Ni(CO)_4$, first discovered by Mond in 1890, is the basis of the Mond Process for nickel refining. This carbonyl is formed as a vapour by interaction between impure nickel and CO at 333 K. Other metals do not form carbonyls under these conditions, nickel can therefore be separated very efficiently and then recovered in the pure state by thermal decomposition of $Ni(CO)_4$ vapour on nickel pellets at 453 K.

Carbonyl compounds are intermediates in many organic syntheses involving carbon monoxide and transition metal compounds as catalysts. For example, the reaction of an olefin with a mixture of carbon monoxide and hydrogen under high pressure to form an aldehyde (a process known as hydroformylation) is catalysed by cobalt, whether this is present in the form of the reduced metal, a cobalt salt or octacarbonyldicobalt, $Co_2(CO)_8$. The reactive intermediate is believed to be a carbonyl hydride of cobalt, $HCo(CO)_4$, which then reacts with the olefin to form the aldehyde:

$$HCo(CO)_4 + CH_2=CH_2 \longrightarrow C_2H_5Co(CO)_4 \text{ followed by}$$

$$C_2H_5Co(CO)_4 \xrightarrow{CO+H_2} C_2H_5CHO + Co_2(CO)_8$$

Only two carbonyls, $Ni(CO)_4$ and $Fe(CO)_5$, can be made by direct reaction between the finely-divided metal and CO. Unlike that of $Ni(CO)_4$, the formation of $Fe(CO)_5$ requires the use of elevated temperatures and a high pressure of CO.

Some carbonyls may be made by reaction between a metal compound and CO and H_2 at high temperatures and pressures. For example, octacarbonyldicobalt is made in this way from cobalt (II) carbonate:

$$2CoCO_3 + 2H_2 + 8CO \longrightarrow Co_2(CO)_8 + 2CO_2 + 2H_2O$$

Otherwise a general method of 'reductive carbonylation' must be used. In this a transition metal compound such as an anhydrous halide or a neutral complex is treated, in suspension or in solution in an organic solvent, with a strong reducing agent under high pressure of CO gas. For example, tris-(acetylacetonato)chromium (III) in pyridine is reduced by zinc or magnesium under high pressure of CO to give yields of $Cr(CO)_6$ of over 80 per cent. Vanadium (III) chloride or tris(acetylacetonato)vanadium (III) is reduced under similar conditions giving $V(CO)_6$ in 40–50 per cent yield.

Polynuclear carbonyls result from the action of ultra-violet radiation on some mononuclear carbonyls. For example, $Fe(CO)_5$ is converted to enneacarbonyldiiron, $Fe_2(CO)_9$, in this way and $Fe_2(CO)_9$ in turn is converted to dodecacarbonyltriiron, $Fe_3(CO)_{12}$, by heating it to 333 K. Alternatively, $Fe_3(CO)_{12}$ is prepared from $Fe(CO)_5$ by way of the hydride carbonyl anion, $[HFe(CO)_4]^-$. This anion is made by reaction between $Fe(CO)_5$ and sodium hydroxide in ethanol:

$$Fe(CO)_5 + 3NaOH \longrightarrow Na[HFe(CO)_4] + Na_2CO_3 + H_2O$$

$Na[HFe(CO)_4]$ is then oxidised by manganese dioxide to $Fe_3(CO)_{12}$:

$$3Na[HFe(CO)_4] + 3MnO_2 + 3H_2O \longrightarrow 4Fe_3(CO)_{12} + 3Mn(OH)_2 + 3NaOH$$

Tetracarbonylnickel and the pentacarbonyls of Fe, Ru and Os are liquids at ordinary temperature. Other carbonyls are generally low-melting solids which sublime at relatively low temperatures. All are typically covalent compounds and, with a few exceptions like $Fe_2(CO)_9$ which is remarkably insoluble and chemically inert, are readily soluble in non-polar solvents. Carbonyls burn easily and decompose when heated into the metal and CO. They are highly toxic substances and must be handled with great care.

With the exception of hexacarbonylvanadium, all carbonyls are diamagnetic. $V(CO)_6$ is monomeric in the solid state and is paramagnetic with one unpaired electron.

The mononuclear carbonyls of Cr, Fe and Ni obey the effective atomic number rule. The metal atom, in each compound, attains a krypton configuration of 36 electrons if it is assumed that each CO molecule donates two electrons to the zerovalent metal. $V(CO)_6$ is easily reduced to the anion, $[V(CO)_6]^-$, which is isoelectronic with $Cr(CO)_6$. The stereochemistry of the molecules has been established: $V(CO)_6$ and $Cr(CO)_6$, octahedral; $Fe(CO)_5$, trigonal bipyramid; $Ni(CO)_4$, tetrahedral.

Although the formulae of the mononuclear carbonyls may be rationalised by the Effective Atomic Number rule, the description of the bonding which this implies is an oversimplification in terms of modern theories of chemical bonding.

In carbon monoxide itself, as we have already seen, the electrons are distributed, according to molecular orbital theory, in the following way:

$$KK \ (z\sigma)^2 \ (y\sigma)^2 \ (x\sigma)^2 \ (w\pi)^4.$$

The triple bond between carbon and oxygen is composed of $(x\sigma)^2$ and $(w\pi)^4$. It is conventional to regard $(x\sigma)^2$ as formed by overlap between an sp hybrid (derived from a 2s and a 2p orbital) on carbon with a similar sp hybrid on oxygen. The other two σ pairs are essentially lone pairs situated in the other two sp hybrids, one in oxygen and the other in carbon. The lone pair on carbon is more strongly directed away from the C—O bond than that on oxygen (evidence for this comes from the very small dipole moment of carbon monoxide; on the basis of relative electronegativities of carbon and oxygen one would expect a polar molecule with a fairly large dipole but this 'natural' polarity is almost exactly offset by the directional property of the lone pair on carbon) and is the one involved in overlap with a vacant orbital on the metal to form a σ-bond (*Figure 6.11(a)*). The two π-bonding orbitals, represented by $(w\pi)^4$, are formed by overlap of p_y and p_z pairs of atomic orbitals on the two atoms. There are three unfilled antibonding molecular orbitals in carbon monoxide: the doubly-degenerate $(v\pi)$ and the $(u\sigma)$. Of these, $(v\pi)$ have the lower energy.

If the bond between the carbon and metal atoms were solely that due to the σ electron pair from carbon, we would expect carbon monoxide to behave in the same way towards protons as it does towards metal atoms, that is, we would expect it to behave as a Lewis base. There is no evidence for this and it appears difficult at first sight to account for the formation of metal carbonyls but not protonated carbon monoxide. The reason for the difference is that an additional kind of bonding exists between the metal and the ligand in carbonyls which cannot occur between the proton and carbon monoxide. In all the known metal carbonyls, the metal atom has at least one filled d orbital of the right symmetry and energy to overlap with an empty π-antibonding orbital of carbon monoxide, giving rise to a π-bond (*Figure 6.11(b)*). From the electroneutrality principle, the donation of an electron pair from the carbon to the zerovalent metal atom would be expected to proceed to a limited extent only so that the σ-bond set up would be weak. The additional π-bonding means that electronic charge is transferred in the opposite direction, from metal to carbon, and more extensive σ-bonding can take place. The net result is a strengthened σ-bond as well as a π-bond between the two atoms. Experimental evidence for the presence of electrons in the π-anti-bonding orbital of CO is obtained from infra-red spectra. The characteristic stretching frequency of the C—O bond in carbon monoxide itself is 215 500 m^{-1} but in metal carbonyls it is around 200 000 m^{-1}: the lower frequency shows that the bonding between carbon and oxygen is less strong.

In polynuclear carbonyls, there are often two kinds of carbon monoxide group. One is terminal and is bound to the metal in the same way as in a mononuclear compound; the other is bridging and is bound to two or more metal atoms. The presence of a bridging CO group is inferred from infra-red absorption due to C—O bond stretching at around 185 000 m^{-1}, an even lower frequency than that found for terminal groups. It is generally assumed

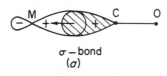

σ — bond
(a)

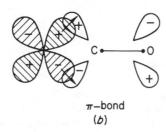

π—bond
(b)

Figure 6.11. The bonding in metal carbonyls

that these bridging groups contribute only one electron to each of the metal atoms, forming normal σ-bonds with them. X-ray crystallography on $Fe_2(CO)_9$ has shown that the two iron atoms are joined by three CO bridges. In addition, each metal atom is co-ordinated by three terminal CO groups, each contributing two electrons. If this were the complete picture of the bonding, each iron atom would reach an E.A.N. of $(26+6+3) = 35$ and the complex would be paramagnetic. In fact, $Fe_2(CO)_9$ is diamagnetic and the two iron atoms are separated by 246 pm. This is a sufficiently short distance for there to be direct bonding between them. The diamagnetism can be explained by postulating that an electron-pair bond does exist between the two iron atoms, both thereby reaching an E.A.N. of 36. The structure of $Fe_2(CO)_9$ is illustrated by (XXXVII). In $Mn_2(CO)_{10}$, a metal–metal bond is again found but in this molecule (XXXVIII), the bridging CO groups are absent. Similar intermetallic bonds and, in most cases, bridging CO groups are found in the more complex polycarbonyls such as $Fe_3(CO)_{12}$ and $Co_4(CO)_{12}$.

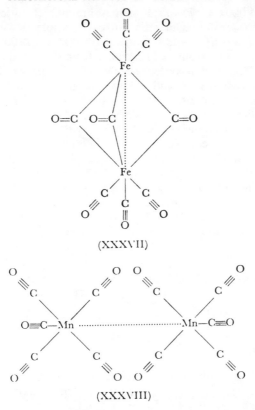

(XXXVII)

(XXXVIII)

One feature of carbonyl chemistry is the ease with which CO groups can be wholly or partly replaced by other ligands such as NO. A number of mixed carbonylnitrosylmetals are known. These are conveniently prepared by the action of nitric oxide gas on metal carbonyls.

For example, $Co_2(CO)_8 + 3NO \longrightarrow 2Co(NO)(CO)_3 + 2CO$;
$$Fe_3(CO)_{12} + 6NO \longrightarrow 3Fe(NO)_2(CO)_2 + 6CO$$

The manganese compound, $Mn(CO)(NO)_3$ is known. When $Fe(CO)_5$ is heated with NO under pressure, a complex of formula $Fe(NO)_4$, is obtained.

Bonding involving nitric oxide involves the transfer of one electron to the metal, decreasing its oxidation state by one, to give the nitrosyl cation, NO^+, which is isoelectronic with carbon monoxide and co-ordinates to a metal in the same way. Nitric oxide thus acts as a three-electron donor. There is a series of isoelectronic compounds comprising $Ni(CO)_4$, $Co(NO)(CO)_3$, $Fe(NO)_2(CO)_2$ and $Mn(NO)_3(CO)$, in which each metal atom has an E.A.N. of 36. The final member of this series would be $Cr(NO)_4$, but this has not yet been made.

As in metal carbonyls, back-donation of metal d electrons to the ligand is possible and this will serve to strengthen the metal–nitrogen bonding and stabilise low oxidation states of the metal. In the vibrational spectrum of $Fe(NO)_2(CO)_2$, absorption bands due to co-ordinated nitrosyl have been identified at 181 000 and 176 700 m^{-1}. These are characteristic of co-ordinated NO$^+$ since they are found in other nitrosyl complexes. A few complexes show additional absorption in the region 100 000 to 120 000 m^{-1}. This has been assigned to NO$^-$ groups which can be formed when the metal transfers an electron to the nitric oxide molecule. The compound Fe(NO)$_4$ shows absorption bands at 181 000 and 173 000 and at 114 000 m^{-1} and we therefore conclude that it contains both NO$^+$ and NO$^-$ groups. A formulation in accordance with the E.A.N. rule is Fe(NO$^+$)$_3$(NO$^-$).

The cyanide ion, CN$^-$, is also isoelectronic with carbon monoxide and it is not surprising that in complexes such as [Fe(CN)$_6$]$^{4-}$ the cyanide can be replaced by either CO or nitrosyl. For example, the reaction between CO and a solution containing [Fe(CN)$_6$]$^{4-}$ produces a salt of trihydrogen-carbonylpentacyanoferrate (II), [H$_3$Fe(CN)$_5$(CO)]. Replacement of one cyanide ion by nitrosyl produces the pentacyanonitrosylferrate (II) ion, [Fe(CN)$_5$(NO)]$^{2-}$. This ion forms deep-violet colours when it reacts with sulphides. The coloured species is believed to result from a two-stage reaction:

$$[Fe(CN)_5(NO)]^{2-} \xrightarrow{\ SH^-\ } \left[Fe(CN)_5N{\stackrel{\textstyle O}{\diagdown_{SH}}} \right]^{3-} \xrightarrow{\ OH^-\ }$$

$$\left[Fe(CN)_5N{\stackrel{\textstyle O}{\diagdown_{S}}} \right]^{4-}$$

Representatives of the class of substances known as carbonylhydrido compounds have already been mentioned. They can be prepared from solutions of carbonyls in alkali. For example, Fe(CO)$_5$ dissolves in ethanolic potassium hydroxide to give a solution containing [HFe(CO)$_4$]$^-$. This is characterised by powerful reducing properties. When acidified, the ethanolic solution reacts to give tetracarbonyldihydridoiron, Fe(CO)$_4$H$_2$. This is an unstable compound, gaseous at ordinary temperature. Alternatively, the carbonylhydride compound is prepared by the reduction with sodium of the parent carbonyl dissolved in liquid ammonia. The sodium salt is formed first and then the hydride by hydrolysis.

Fe(CO)$_4$H$_2$ is a yellow liquid at low temperatures which decomposes at 263 K and above. It behaves as a dibasic acid with dissociation constants of 3.6×10^{-4} and 1×10^{-14} for loss of the first and second protons respectively.

Other carbonylhydrido compounds for the first period transition metals include Mn(CO)$_5$H and Co(CO)$_4$H. Mn(CO)$_5$H is a stable liquid at ordinary temperature whilst Co(CO)$_4$H is unstable above its m.p., 247 K. Both are like the iron compound in having reducing and acidic properties. These compounds are of interest from a structural view-point because they contain direct metal–hydrogen bonds. The hydrogen atom, like the other ligands, occupies a definite co-ordination position. It has been established, for instance, that the five CO groups in Mn(CO)$_5$H occupy five of the six vertices of a nearly regular octahedron. The hydrogen atom cannot be located by x-ray diffraction but its position in the remaining vertex is inferred from other physical evidence such as infra-red absorption due to the Mn—H bond.

Metal carbonyl halides can be made by the direct reaction of polynuclear carbonyls with halogens or by reaction between a metal halide and carbon monoxide under high pressure. Most metals which form carbonyls also form carbonyl halides; in addition Pd, Au, Cu^I and Ag^I form halides although the simple carbonyls are unknown.

As a typical carbonyl halide, we may refer to $Mn(CO)_5Br$ which is formed by the direct reaction of $Mn_2(CO)_{10}$ with bromine at 313 K. At a higher temperature, 393 K, a dimeric compound, $[Mn(CO)_4Br]_2$, is produced. In $Mn(CO)_5Br$ octahedral co-ordination of the manganese atom is maintained; this is also found in the dimer because here the bromine atoms act as bridges between the two manganese atoms.

As an example of a metal carbonyl halide where the metal does not form a carbonyl, one of the best-known is $Cu(CO)Cl$, formed by the action of carbon monoxide on ammoniacal solutions of copper (I). For many years this reaction was exploited in gas analysis as a means of absorbing carbon monoxide from gas mixtures.

NITROGEN COMPLEXES

Molecular nitrogen is isoelectronic with carbon monoxide and a number of complexes are known in which it behaves in a similar manner as a ligand. Studies of these complexes have a special significance in relation to 'nitrogen-fixation', the processes whereby certain bacteria and other biological agents can react with nitrogen to form nitrogenous products like ammonia. Again, in reactions where nitrogen is catalytically reduced by hydrogen to ammonia, it appears likely that the nitrogen must be co-ordinated to a transition metal (catalyst) before it becomes susceptible to reduction.

The first stable complex between molecular nitrogen and a metal to be made and characterised was the cation, pentaammine-dinitrogen-ruthenium (II), $[Ru(NH_3)_5N_2]^{2+}$. This is produced when ruthenium (III) chloride is reduced by hydrazine and when an acidified solution containing $[Ru(NH_3)_5 Cl]^{2+}$ is reduced by zinc amalgam in the presence of nitrogen gas at 298 K.

The ruthenium (II) complex contains the linear grouping $Ru—N{\equiv}N$. Its infra-red spectrum shows intense absorption at $212\,900$ m^{-1}. This is assigned to the N—N stretching frequency of co-ordinated nitrogen because it is much lower than the stretching frequency in free nitrogen (observed only in the Raman spectrum and centred at $233\,000$ m^{-1}). The frequency shift is attributed to a decrease in the strength of the N—N bond due to back-donation of electrons from d orbitals on the metal to empty antibonding π^* orbitals in the nitrogen molecule.

The nitrogen co-ordinated in the ruthenium complex is inert towards reducing agents. In some systems, it has proved possible to convert nitrogen to ammonia and the reaction probably proceeds through the formation of metal–nitrogen complexes as intermediates. For example, when nitrogen gas is bubbled through a cell in which a solution of titanium tetraisopropoxide and aluminium chloride in 1,2-dimethoxyethane is being electrolysed, ammonia is produced. Reduction can also be effected chemically by an alkali metal or by the naphthalene radical anion (naphthalide). No nitrogenyl complex has, however, been isolated from this system.

ORGANOMETALLIC COMPOUNDS

Apart from the metal carbonyls, there are many other compounds known which contain metal–carbon bonds. These form a major branch of chemistry, known as organometallic chemistry. Developments in this branch have been particularly rapid during the last twenty years. It is usual to include within the classification of organometallic compounds those in which carbon is bonded to boron, silicon, the phosphorus sub-group elements, selenium and tellurium, although none of these elements has marked metallic character. The range of organometallic compounds includes the metal alkyls, organomagnesium halides (Grignard reagents), organosilicon compounds (including silicones) and compounds between metals and π-bonding organic ligands (like ferrocene and other cyclopentadienyl complexes).

Various types of organometallic compounds can be differentiated according to the nature of the bond present between the metal and carbon atoms. The bonding may be predominantly ionic, covalent (σ), multicentre or covalent (π).

Ionic bonding is found, as we would expect, only when carbon is combined with the most electropositive metals such as sodium and the heavier alkali metals. This type of bond is also favoured when an organic anion exists as a stable entity.

With the less electropositive metals, carbon can form a covalent (σ) bond. This is found in compounds like $B(CH_3)_3$, and $C_6H_5SiCl_3$. Metal–carbon bonds of this type occur throughout the Periodic Table and the polarity associated with them will depend on the relative electronegativities of the metal and carbon. The σ-bonds involving non-transition metals are usually more stable than those involving transition metals.

Some of the electropositive metals like lithium, beryllium and aluminium form organometallic compounds in which there are multicentre bonds of the same kind as in the boron hydrides. Examples include $[Li(CH_3)]_4$, $[(CH_3)_2Be]_n$ and $Al_2(CH_3)_6$. As the formulae indicate, none of these exists as simple, mononuclear molecules. In all three, some at least of the methyl groups act as bridges. This requires more bonds than there are electron pairs available for the formation of normal covalent bonds and so the compounds are electron-deficient. Dimethylberyllium has the structure:

and trimethylaluminium is:

Most transition metals can form complexes with unsaturated hydrocarbons like olefins and acetylenes and with hydrocarbon groups which have a system of de-localised electrons in π orbitals. These compounds have aroused great interest because of their unusual structures and because of the nature of the bonding between the metal and the ligands. The metal atom cannot be regarded as bound to a single atom on the ligand because it is symmetrically situated with respect to two or more atoms on the ligand. Bonding is between the de-localised π-electrons of the ligand and vacant orbitals on the metal and there is also the possibility of back-donation of electrons from metal to ligand. π-bonding ligands are listed in *Table 6.15*

Table 6.15. PI-BONDING LIGANDS AND THE NUMBER OF ELECTRONS WHICH THEY DONATE

Number of electrons	*Examples*
2	ethylene
3	allyl
4	cyclobutadiene
5	cyclopentadienyl
6	benzene, cycloheptatriene
7	cycloheptatrienyl
8	cyclo-octatetraene

according to the number of electrons which they formally donate to the metal.

The Effective Atomic Number rule applies to the formulae of many complexes which can be isolated. Their stability towards hydrolysis, oxidation and thermal decomposition can be related to the presence of completely full electron shells. Thus the E.A.N. of iron in bis(π-cyclopentadienyl)-iron, $Fe(C_5H_5)_2$, is $26+(2 \times 5) = 36$ and that of chromium in bis(benzene) chromium, $Cr(C_6H_6)_2$, is $24+(2 \times 6) = 36$. In both cases the metal atom is in a formal oxidation state of zero.

Preparative methods

Alkylmetals of the more electropositive metals can be made by reaction between metals and organic halides. The reaction is carried out in ether or in hydrocarbon solvents. For example, ethyllithium is made by the reaction:

$$2Li + C_2H_5Cl \longrightarrow C_2H_5Li + LiCl$$

Magnesium reacts with alkyl or aryl halides in ether solution to form Grignard reagents, RMgX:

$$RX + Mg \longrightarrow RMgX$$

The solution in ether very probably contains dialkyl (or diaryl) magnesium, R_2Mg, and magnesium chloride as well as RMgX but it is not generally necessary to isolate the Grignard reagent in the pure state if it is to be used in organic synthesis.

Another reaction used for the preparation of alkylmetals is that between the metal concerned and organomercury compounds. For example, dimethyl-mercury, itself prepared by the reaction:

$$2Hg + 2CH_3Br \longrightarrow (CH_3)_2Hg + HgBr_2$$

is used to prepare dimethylzinc and trimethylaluminium.

$$Zn + (CH_3)_2Hg \longrightarrow (CH_3)_2Zn + Hg$$
$$2Al + 3(CH_3)_2Hg \longrightarrow 2(CH_3)_3Al + 3Hg$$

Organo-lithium and -magnesium compounds are themselves used widely as reagents in the preparation of other organometallic compounds. They react vigorously with halides of the more electronegative elements, for example,

$$2C_2H_5MgCl + CdCl_2 \longrightarrow (C_2H_5)_2Cd + 2MgCl_2$$

Triethylaluminium is a useful alkylating agent for the preparation of organoboron compounds. Thus it reacts with ethyl orthoborate to form triethylborane:

$$B(OC_2H_5)_3 + (C_2H_5)_3Al \longrightarrow (C_2H_5)_3B + Al(OC_2H_5)_3$$

One of the most important metal alkyls, from an industrial aspect, is tetraethyllead, prepared by reaction between ethyl chloride and a sodium/lead alloy:

$$4C_2H_5Cl + 4Na/Pb \longrightarrow (C_2H_5)_4Pb + 3Pb + 4NaCl$$

This is used extensively as an anti-knock additive in petrol.

COMPLEXES BETWEEN METALS AND UNSATURATED HYDROCARBONS

The first compound of an unsaturated hydrocarbon with a metal, $PtCl_2 \cdot C_2H_4$, was prepared by the Danish chemist Zeise and was reported in the chemical literature as early as 1827. This compound may be prepared by reaction between a tetrachloroplatinate (II), $[PtCl_4]^{2-}$, and ethylene in aqueous solution, followed by extraction with ether. The reaction proceeds with a step-wise replacement of chloride by olefin:

$$[PtCl_4]^{2-} \longrightarrow \underset{\text{(the anion of Zeise's salt)}}{[PtCl_3(C_2H_4)]^-} \longrightarrow [PtCl_2(C_2H_4)]_2^0$$

The neutral compound is a dimer as shown by freezing point depression measurements in benzene and has a bridged structure with *trans* arrangement of the two ethylene molecules:

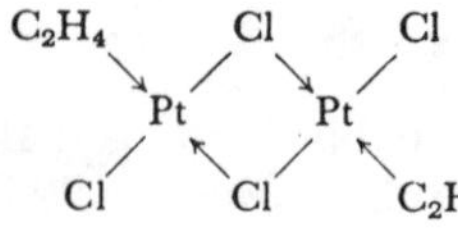

Other unsaturated hydrocarbons, for example acetylene, may be used instead of ethylene but the ability to form such complexes appears to be restricted to a limited group of transition metal ions: Rh^I, Ir^{II}, Pd^{II}, Pt^{II}, Cu^I, Ag^I and Hg^{II}. The last three ions have their full complement of ten d electrons so evidently empty d orbitals on the metal are not essential for binding the olefin molecules.

X-ray analysis has shown that in the ion $[PtCl_3(C_2H_4)]^-$, the two carbon atoms are equidistant from and on one side of the metal (XXXIX). In 1953, Chatt and Duncanson suggested that the bonding involves co-ordination of the two π electrons of the ethylene double bond. The π orbital of ethylene has, in fact, the same symmetry, with respect to the metal–ligand bond, as a σ orbital on a simple ligand. A co-ordinate bond of σ symmetry results. These workers also proposed that the strength of the bond is reinforced by back co-ordination of non-bonding d electrons of the platinum atom to the vacant (antibonding) π orbitals of the ethylene molecule. The orbitals used in the bonding are illustrated in *Figure 6.12*.

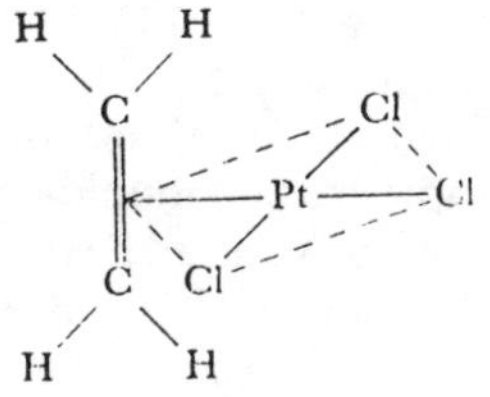

(XXXIX)

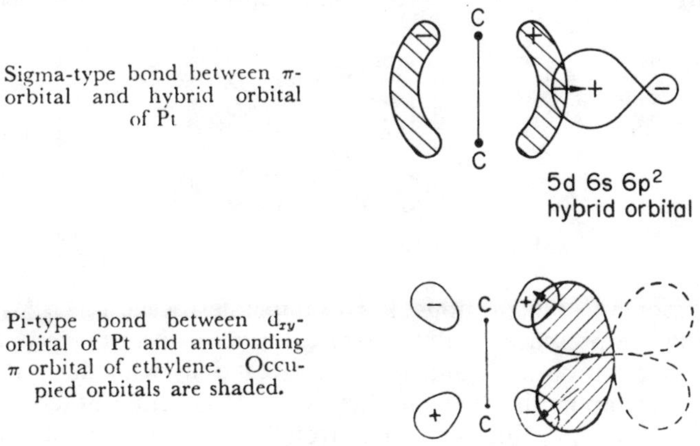

Sigma-type bond between π-orbital and hybrid orbital of Pt

5d 6s 6p² hybrid orbital

Pi-type bond between d_{xy}-orbital of Pt and antibonding π orbital of ethylene. Occupied orbitals are shaded.

Figure 6.12. The bonding in $[PtCl_3(C_2H_4)]^-$

This description of the bonding accounts satisfactorily for the observation that metal–olefin complexes are usually formed by transition metals which have several d electrons and which are in low oxidation states.

Of the other π-bonding ligands listed in *Table 6.15*, cyclopentadienyl has attracted most attention since the synthesis of its first transition metal complex, bis(π-cyclopentadienyl)iron, often known as ferrocene, in 1951. The compound resulted from an attempt by Kealy and Pauson to prepare dicyclopentadienyl by the reaction between a Grignard reagent, cyclopentadienylmagnesium bromide and iron (III) chloride in an organic solvent. Instead, ferrocene was obtained, $FeCl_3$ being first reduced to $FeCl_2$ which then reacted according to:

$$2C_5H_5MgBr + FeCl_2 \longrightarrow (C_5H_5)_2Fe + MgBr_2 + MgCl_2$$

Ferrocene has also been obtained by the direct action of cyclopentadiene on iron powder.

Ferrocene is an orange, crystalline solid, m.p. 446 K, which is insoluble in water but readily soluble in organic solvents. It is remarkably stable to heat and does not decompose below 743 K.

The cyclopentadienyl derivatives of many other transition metals are now known. For example, the dicyclopentadienyls of Co, Ni, Cr and V have all been synthesised and found to be isomorphous with ferrocene (crystallising in the monoclinic system).

Structural investigations have shown that these compounds have a 'sandwich' structure in which the metal atom lies between two planar C_5H_5 rings and is equidistant from all ten carbon atoms. In ferrocene itself, the iron atom lies at the centre of an antiprismatic 'double-cone' structure (XLa). Electron diffraction studies on ferrocene vapour have shown that in this phase the molecule has the 'eclipsed' structure, (XLb). Both ruthenocene and osmocene crystallise in the eclipsed form. In these molecules, the C—C bond lengths in the C_5H_5 rings are equal within the limits of experimental error.

It was soon remarked that the hydrogen atoms of the C_5H_5 rings in ferrocene can be replaced by other groups in reactions similar to those used for

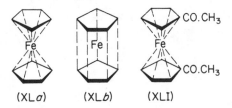

(XLa)   (XLb)   (XLI)

effecting aromatic substitutions. For example, ferrocene reacts with acetyl chloride in the presence of aluminium chloride to form diacetyl ferrocene (XLI). Condensation with aldehydes and reactions with diazonium compounds also support the view that the hydrocarbon rings in ferrocene have 'aromatic character'. This behaviour strongly suggests that each ring has a non-localised sextet of electrons. (The C_5H_5 group has 5 non-localised electrons and by gaining one electron from the iron atom is converted to $C_5H_5^-$.)

E. O. Fischer has proposed that the bonding in ferrocene involves the donation of three pairs of electrons from each $C_5H_5^-$ ring to the Fe^{2+} ion. The metal would thereby attain a krypton configuration and the electronic distribution would be exactly the same as the valence-bond description of $[Fe(CN)_6]^{4-}$. However, the aromatic substitution reactions of ferrocene imply a high electron availability in each ring and this is incompatible with the involvement of six electrons in bonding to the metal.

Alternatively the bonding can be described in terms of molecular orbitals. These are constructed from non-localised orbitals on the ligands and orbitals of appropriate symmetry on the metal. The various combinations possible lead to orbitals which extend over the entire molecule. The problem can be treated in a quantitative manner and the order and energies of the orbitals calculated. Several accounts have been published but we cannot consider the details here. One feature which does emerge from molecular orbital theory is that there appear to be six low-lying orbitals which are strongly bonding in character. The bonding between the metal atom and the rings is due chiefly to the complete filling of these with 12 electrons.

The direct action of cyclopentadiene on the metal is effective only for the preparation of ferrocene. A more general method is that based on the reaction (in benzene or ether) of the Grignard reagent cyclopentadienylmagnesium bromide with an anhydrous halide of the metal or one of its co-ordination compounds, for example the acetylacetonato complex. This method gives a satisfactory preparation of $Ni(C_5H_5)_2$ and $V(C_5H_5)_2$, but in the case of cobalt, $[Co(C_5H_5)_2]^+$ is obtained. This is isoelectronic with ferrocene and salts containing this ion strongly resist oxidation. It can, however, be reduced to bis(cyclopentadienyl)cobalt, $Co(C_5H_5)_2$, by $LiAlH_4$. Bis(cyclopentadienyl)-cobalt is a paramagnetic complex which is very sensitive to atmospheric oxidation to $Co(C_5H_5)_2^+$. In this sense, $Co(C_5H_5)_2$ and $Co(C_5H_5)_2^+$ resemble Co^{II} and Co^{III} complexes respectively.

Table 6.16. METAL COMPLEXES OF CYCLOPENTADIENE

Formula	Metals
$M(C_5H_5)$	Li, Na, K, Rb, Cs, Tl and In
$M(C_5H_5)_2$	Be, Mg, Ca, Zn, Hg, V, Cr, Mn, Fe, Co, Ni, Sn, Pb, Ru and Os
$M(C_5H_5)_3$	Sc, Ga, Y, In, Sb, Bi and the rare earths
Cations such as $Zr^{IV}(C_5H_5)_2^{2+}$, $Co^{III}(C_5H_5)_2^+$ and $Nb^V(C_5H_5)_2^{3+}$	Transition metals in high valency states

The metal cyclopentadienyls are summarised in *Table 6.16*. For complete-ness, the compounds with non-transition metals are included. Those which contain electropositive metals such as the alkalis and alkaline earths are typically salt-like. For instance, cyclopentadienylpotassium, KC_5H_5, first prepared by Thiele in 1901 by the action of the metal on a solution of cyclo-pentadiene in benzene, is a colourless salt which decomposes rapidly on exposure to air. It ionises in polar solvents and complete hydrolysis occurs with water:

$$KC_5H_5 + H_2O = KOH + C_5H_6$$

Its general properties indicate that this compound is ionic, $K^+C_5H_5^-$. Compounds of weakly electropositive metals such as Hg, Sn, Pb and Bi are intermediate in properties between the salt-like cyclopentadienyls of the alkali metals and the covalent complexes of the transition metals. Thus, $Sn(C_5H_5)_2$ is a colourless compound which is unstable in air, moderately soluble in benzene and ether, unaffected by cold water but decomposed by acids to cyclopentadiene. It has been suggested that this type of compound contains a relatively weak σ-bond between the metal and one carbon atom of the C_5H_5 ring.

(XLII)

Many more complex derivatives of the transition metal cyclopentadienyls have been prepared in which the cyclopentadiene is partly replaced by other ligands such as CO, NO and CN^-, but details of these are outside the scope of the present treatment.

It is very interesting to note that some complexes have also been made which contain 6-membered rings. For example in 1955 bis(benzene) chromium, $Cr(C_6H_6)_2$, was synthesised by heating anhydrous chromic chloride, aluminium chloride, aluminium and benzene in an autoclave to 453 K. This produced the $[Cr(C_6H_6)_2]^+$ ion:

$$3CrCl_3 + 2Al + AlCl_3 + 6C_6H_6 = 3[Cr(C_6H_6)_2]^+ AlCl_4^-$$

Reduction with sodium dithionite gave bis(benzene) chromium as a brown–black diamagnetic solid of m.p. 557 K. This compound has a sandwich structure (XLII), analogous to that of ferrocene.

SUGGESTED REFERENCES FOR FURTHER READING

ABEL, E. W. 'The metal carbonyls' *Quart. Rev. chem. Soc., Lond.*, 17 (1963) 133.

BELCHER, R. and NUTTEN, A. J. *Quantitative Inorganic Analysis—a Laboratory Manual*, 3rd edn, Butterworths, London,

COATES, G. E., GREEN, M. L. H., POWELL, P. and WADE, K. *Principles of Organometallic Chemistry*, Methuen, London, 1968.

DWYER, F. P. and MELLOR, D. P. *Chelating Agents and Metal Chelates*, Academic Press, New York and London, 1964.

FLASCHKA, H. *EDTA Titrations—an Introduction to Theory and Practice*, Pergamon Press, Oxford, 1959.

GRADDON, D. P. *Introduction to Co-ordination Chemistry*, Pergamon Press, Oxford, 1961.

ORGEL, L. *An Introduction to Transition Metal Chemistry—Ligand-Field Theory*, Methuen, London, 1960.

PAUSON, P. L. 'Ferrocene and related compounds', *Quart. Rev. chem. Soc., Lond.*, 3 (1949) 263.

7 The distribution and extraction of the chemical elements

DISTRIBUTION

Formation and structure of the Earth

Theories about the origin of the Earth fall into two categories, those which propose that the Earth was formed by the splitting off and gradual cooling down of incandescent matter from the Sun and those which take the primary stage to be the accretion of cold particles and gaseous material leading to a solid mass, constituting the nucleus of the Earth. It is now generally supposed that the Earth and other planets in the solar system were formed in the second of these ways. The accretion of matter was accompanied by the generation of heat from the kinetic energy of particles as they came together under the influence of gravity and from the decay of radioactive elements such as potassium, thorium and uranium. Although some loss of heat by radiation occurred, the rest was retained and resulted in an increase in the temperature of the Earth to such an extent that chemical changes and at least partial melting took place. Some segregation of matter by gravitational forces resulted and volatile materials moved towards the surface. A complexity of geological and chemical processes followed and produced the distribution of the elements throughout the Earth as we know it to-day.

Some support for the accretion theory comes from analytical data on the average compositions of cosmic and terrestrial matter. Estimates of the composition of the Universe, based mainly on spectrographic studies of stars, show that there is a greater proportion of volatile elements in the Universe than in the Earth. In the first stages of an accretion process, the gravitational pull of very small fragments would be insufficiently strong to retain the most volatile materials, the bulk of which would be easily lost.

From geophysical measurements, it has been established that the density of the Earth increases downwards from the surface. The distribution of materials inside the Earth has been inferred from data on the propagation of seismic waves. The generally accepted structural model for the Earth is essentially as follows: a high-density core, at about 2900 km depth, consisting probably of molten iron in slow turbulent motion; a solid mantle, probably composed mainly of iron and magnesium silicates and oxides; and a thin crust of rocks and minerals, of average thickness 35 km under the continents and 5–10 km under the oceans. The crust–mantle interface is defined by the Mohorovicic discontinuity (named after its discoverer). This is the depth at which compressional seismic waves change in speed from about 6 km s^{-1} in the crust to about 8 km s^{-1} in the uppermost part of the mantle. It is reasonable to suppose that the Earth's crust has originated from the top of the mantle which underwent fractional melting and the other geological changes in the course of the ageing of the Earth.

V. M. Goldschmidt proposed a⋅ classification of the elements on the presuppositions that the Earth was, at some time, largely in the molten state and that the cooling of this brought about a separation into three phases, respectively metallic, sulphidic and siliceous in type. Similar separations of three largely immiscible liquid phases are well-known in metallurgy. Goldschmidt classified elements according to their preferential concentration in one or more of these phases: *siderophile* (elements found with native iron and probably therefore concentrated in the core), *chalcophile* (concentrated in sulphides) and *lithophile* (concentrated in silicates). A fourth group can be added to complete the classification in relation to distributions in the Earth, the *atmophile* elements (important constituents in the atmosphere). The classification is summarised in *Table 7.1*.

Table 7.1. GOLDSCHMIDT'S CLASSIFICATION OF THE ELEMENTS

Siderophile	Chalcophile	Lithophile	Atmophile
Fe, Co, Ni,	Cu, Ag, (Au),*	Li, Na, K	H, N, (C),
Ru, Rh, Pd,	Zn, Cd, Hg,	Rb, Cs, Be,	(O), (F),
Re, Os, Ir,	Ga, In, Tl,	Mg, Ca, Sr,	(Cl), (Br)
Pt, Au, Mo,	(Ge), (Sn), Pb,	Ba, B, Al,	(I), noble
Ge, Sn, C	As, Sb, Bi,	Sc, Y, La,	gases
P, (Pb), (As),	S, Se, Te,	(C), Si, Ti,	
(W)	(Fe), (Mo),	Zr, Hf, Th,	
	(Re)	(P), V, Nb,	
		Ta, O, Cr,	
		W, U, (Fe),	
		Mn, F, Cl,	
		Br, I, (H),	
		(Tl), (Ga),	
		(Ge), (N)	

* Elements in parentheses belong chiefly to another class but are found to some extent in association with those of this class.

The different classes overlap and many elements are grouped in more than one. For example, iron, the principal element of the core, is also found associated with sulphides and in silicate minerals. In general, there are points of chemical similarity between elements in the same class. Thus siderophile elements have low electrode potentials, lithophile elements have high potentials and chalcophile elements have intermediate potentials.

The abundance of elements in the Earth's crust has been determined by analytical methods and the average composition of the whole of the Earth has been estimated on the assumption that it is similar to that of meteorites, regarded as being small fragments of the material from which the Earth was formed. The two sets of data are given in *Table 7.2*. This shows that the eight elements oxygen, silicon, aluminium, iron, magnesium, calcium, sodium and potassium constitute over 98·5 per cent by weight of the crust. Some lithophile elements, like potassium and titanium, are apparently much more abundant simply because they are concentrated in the Earth's crust.

The chemist and the metallurgist are primarily interested in the distribution of elements within the crust because this provides the source from which all elements can be extracted and isolated. Rocks on the Earth's surface have been formed in a number of different ways and each of these was associated with fractionation processes which affected distribution.

Table 7.2. COMPOSITION OF THE EARTH AND ITS CRUST
(MAJOR CONSTITUENTS)

| Element | Abundance (p.p.m.) | |
	In the Earth	In the crust
H	—	1 400
O	295 000	466 000
Na	5 700	28 300
Mg	127 000	20 900
Al	10 900	81 300
Si	152 000	177 200
P	1 000	1 050
S	19 300	260
K	700	25 900
Ca	11 300	36 300
Ti	500	4 400
Cr	2 600	100
Mn	2 200	950
Fe	346 300	50 000
Ni	23 900	75

The primary products of freezing molten material are *magmatic* (or *igneous*) rocks, either formed by solidification deep within the crust, when they are known as *intrusive*, or resulting from the extrusion of molten material on to the surface and its subsequent crystallisation to produce *volcanic* rocks. Composition of typical igneous rock is given in *Table 7.3*.

The action of heat and pressure beneath the surface of the Earth caused recrystallisations of the original igneous material leading to the formation

Table 7.3. COMPOSITION OF TYPICAL IGNEOUS ROCK

Oxide	Percentage
SiO_2	66·4
TiO_2	0·7
Al_2O_3	14·9
Fe_2O_3	1·5
FeO	3·0
MnO	0·08
MgO	2·2
CaO	3·8
Na_2O	3·6
K_2O	3·3
H_2O	0·6
P_2O_5	0·18

of *metamorphic* rocks. These differ from igneous rocks in that they were formed largely by reactions in the solid state. This is demonstrated by the frequently observed preservation of the structures of the original materials from which the metamorphic rocks were formed. Metamorphic rocks include quartzite, marble, slate, micas, and amphiboles. Some of these possess structures with distinctive orientations in planes and lines, pressumably caused by directional pressures during metamorphic processes.

A third important class of rocks comprises those formed by *sedimentary* processes which involve chemical changes such as hydrolysis, precipitation and oxidation and reduction. Sediments and their consolidated products cover igneous and metamorphic rocks over much more than half the Earth's surface. They are produced by the action of water, carbon dioxide, oxygen and organic acids and fractionation of many elements, between an aqueous and a solid phase, has occurred during their formation. The compositions of three kinds of sedimentary rocks — sandstone, limestone and carbonate — are given in *Table 7.4*.

Table 7.4. AVERAGE PERCENTAGE COMPOSITION OF SOME SEDIMENTARY ROCKS

Oxide	Sandstone	Limestone	Carbonate rock
SiO_2	70·0	6·9	8·2
TiO_2	0·58	0·05	—
Al_2O_3	8·2	1·7	2·2
Fe_2O_3	2·5	0·98	1·0
FeO	1·5	1·3	0·68
MnO	0·06	0·08	0·07
MgO	1·9	0·97	7·7
CaO	4·3	47·6	40·5
Na_2O	0·58	0·08	—
K_2O	2·1	0·57	—
H_2O	3·0	0·84	—
P_2O_5	0·10	0·16	0·07
CO_2	3·9	38·3	35·5
SO_3	0·7	0·02	3·1

The most abundant sedimentary rocks contain calcium carbonate. The conditions under which carbonate is formed or dissolved by natural waters containing dissolved carbon dioxide (carbonic acid) are determined by the equilibrium:

$$CaCO_3 + H_2CO_3 \rightleftharpoons Ca^{2+} + 2HCO_3^-$$

The forward reaction illustrates the solution and weathering of limestone whereas the backward reaction describes the precipitation of carbonate sedimentary rocks.

Sedimentary rocks have a greater content of water, carbon dioxide, chlorine and a higher ratio of ferric to ferrous ion than igneous rocks. The decrease in sodium content reflects the extent to which sodium ion has passed into solution in the oceans during weathering processes.

Ferrous compounds are oxidised to the ferric state by prolonged exposure to atmospheric oxygen. The product is generally Fe_2O_3 itself or one of its hydrates. Ferric oxide commonly remains unchanged because of its very low solubility. Under some conditions, it remains suspended as a sol and may be transported over long distances in this form. The colloid is positively charged and is able to absorb anions like phosphate. Flocculation and deposition results from the action of electrolytes when, for example, the stream carrying the sol enters the sea.

Manganese, present initially as manganese (II), dissolves to a slight extent in aqueous carbonic acid and is converted, if the conditions are oxidising, to the (III) and (IV) states. Mixed oxides such as Mn_3O_4 may be formed but the final product appears to be MnO_2, which, like Fe_2O_3, is unreactive because of its insolubility. The manganese oxides may also be formed first as sols. Unusually, these are negatively charged and preferentially take up cations such as K^+, Ni^{2+}, Co^{2+}, Pb^{2+}, Ba^{2+} and Cu^{2+}. During sedimentary processes, manganese often becomes very effectively separated from other elements, especially iron. This may be attributable to differences in solubility with increasing pH, ferric oxide precipitating out before oxides of manganese, but the precise mechanism remains obscure.

Sulphur is another common element which undergoes oxidation during weathering. In igneous rock, the element is present as sulphide (oxidation state $= -2$) and contacts with air and water convert it eventually to sulphate ($+6$). Metal sulphides only form as sedimentary minerals in very reducing environments, usually therefore in the presence of organic substances.

Weathering processes are responsible for the distribution of elements in sea-water (*Table 7.5*). This is a complex aqueous buffer system of pH between

Table 7.5. CONCENTRATION OF THE MAIN COMPONENTS OF SEA-WATER
(EXPRESSED IN p.p.m. FOR A SALINITY OF 35 PARTS PER THOUSAND)

Cl^-	18 980	K^+	380
Na^+	10 556	HCO_3^-	140
SO_4^{2-}	2 649	Br^-	65
Mg^{2+}	1 272	Sr^{2+}	8
Ca^{2+}	400	H_3BO_3	26

8.0 and 8.4. This range means that much of the calcium ion carried to the sea by rivers has been precipitated as calcium carbonate. Other elements, present as colloidal oxides, have been removed by coagulation by the electrolytes present. There is a selective concentration of the remaining elements in sea-water. These largely determine the composition of salt deposits formed by the evaporation of sea-water, the most abundant compounds in these deposits being calcium sulphate and sodium chloride.

Other elements are in steady circulation in the Earth's crust. Calcium and magnesium enter into biological processes, calcium to form bone and magnesium to form chlorophyll. Both potassium and sodium have very important biological roles. There are the three vitally important natural cycles involving carbon, phosphorus and nitrogen.

Goldschmidt's rules

As silicates are so predominant in the Earth's crust, the properties of these structures, such as their capacity for undergoing isomorphous replacement of one ion by another, are of great significance in relation to the distribution of the less common (trace) elements. This was recognised by V. M. Goldschmidt, who formulated a set of empirical rules to describe their distribution. He proposed that the factors principally controlling the distribution of a particular ion are its size and charge. His rules are as follows:

(1) if two ions have the same or similar radius and the same charge, they will be distributed in a given mineral in amounts proportional to their abundances. The isomorphous replacement of one ion by a different one occurs extensively as long as their radii do not differ by more than about 15 per cent;

(2) if two ions have similar radii and the same charge, the smaller ion is preferentially concentrated in the solids formed early in crystallisation;

(3) if two ions have similar radii but different charges, that with the higher charge will be preferentially concentrated early in crystallisation.

As illustrations of the first rule, it is found that Ba^{2+} (135 pm) and K^+ (133 pm) have extensively replaced each other in minerals, and that Fe^{3+} (64 pm) and Cr^{3+} (69 pm) constitute another pair of ions which show isomorphous substitution. The second rule accounts for the observation that Mg^{2+} (65 pm) enriches the early crystals in the isomorphous series of olivines compared with Fe^{2+} (82 pm). Again Li^+ (60 pm) and Mg^{2+} substitute extensively for each other in accordance with rule (1) but lithium is found to be concentrated in the minerals which are late to crystallise (rule 3).

Although Goldschmidt's Rules are a useful guide to the distribution of trace elements, there are enough exceptions to show that they should be regarded as strictly qualitative. For example, Zn^{2+} (74 pm) would be expected to occur with trace elements such as Ni^{2+} (76 pm) and Co^{2+} (78 pm) in ferromagnesium silicates. In fact, Zn^{2+} is not usually found in octahedral co-ordination in these silicates but occurs preferentially in minerals where it shows tetrahedral co-ordination. We recall here that, although the concept of ionic radius has proved of great utility in rationalising crystal structures, the value assigned to a given ion is not invariant and changes with the environment, for example, its co-ordination number.

Another limitation is that the rules take no account of the type of bond formed. Thus Cu^+ (96 pm) is not found with Na^+ (95 pm) despite their closeness in size because Cu^+ forms bonds of markedly less ionic character than Na^+ with common anions. Considerations similar to this caused Ringwood to extend the validity of Goldschmidt's Rules by proposing that another factor of importance is the electronegativity of an ion. His criterion is that 'for two ions with similar valencies and radii, the ion with the lower electronegativity will be preferentially taken up in a crystal structure because it forms the stronger (more 'ionic') bond than the other ion'.

The chief criticism of Goldschmidt's Rules has been on thermodynamic grounds. Implicit in the rules is the assumption that the crystalline lattice enthalpy is the determining factor governing distribution. It is certainly one important factor but account should also be taken of the thermodynamic properties of the ions in the phase(s) from which the lattice is formed. This is not often possible because we have insufficient information on the solvation energies or the changes in free energy associated with reactions taking place in molten media and so we cannot judge the relative stabilities of the ion in the melt and in the crystalline phase.

Composition and structure of the Moon

The Apollo 11 mission to the Moon made possible for the first time the sampling of lunar rocks and the return of this material to the Earth for intensive study and analysis. Since then, further Apollo flights have recovered samples from different parts of the Moon and scientists are now beginning to build up a comprehensive picture of the nature of the lunar surface. Geophysical and physical measurements are providing more and more data about the Moon as a whole and, together with chemical analysis, make it possible to formulate theories of the evolution and structure of the Moon on a much firmer basis than has ever been possible before.

The interior of the Moon appears to be unlayered and relatively cool compared with that of the Earth. Volatile elements, like lead, thallium and bismuth, are of lower abundance than in terrestrial rocks. Crystallisation of lunar rocks occurred in strongly reducing conditions unlike those on Earth. Thus in samples from the Sea of Tranquillity, Fe^{3+} is absent, metallic iron is present and Cr^{2+} occurs in olivine crystals. The age of lunar rock, that is the time which has elapsed since crystallisation, has been estimated at between 3.6 and 3.8×10^9 y. This has been done by measuring the relative amounts of a radioactive rubidium isotope, ^{87}Rb, and its daughter element, ^{87}Sr. The ratio of these two is related to the time when their chemical differentiation took place, presumably by crystallisation.

All these differences are indicative of very different evolutionary processes for the Earth and Moon and suggest the Moon may have been formed by accretion, either well outside the gravitational field of the Earth or while in orbit around the Earth.

The chemical composition of samples collected by Apollo 11 in the Sea of Tranquillity is summarised in *Table 7.6*. Although this cannot represent

Table 7.6. COMPOSITION OF LUNAR SAMPLES COLLECTED FROM THE SEA OF TRANQUILLITY

Composition range	Elements
10–100%	O, Si, Ca, Fe
1–10%	Mg, Al, Ti
0·1–1%	S, Na, K, Cr, Mn
100–1000 p.p.m.	C, N, P, Cl, Sr, Y, Zr, Ba
10–100 p.p.m.	F, Sc, V, Co, Ni, Zn, Nb, La, Ce, Pr, Nd, Sm, Gd, Dy, Er, Yb, Hf
less than 10 p.p.m.	Others

the average composition of the Moon, certain features are significant. Thus the high level of titanium, zirconium and rare-earth elements is noteworthy. The zirconium:hafnium ratio is less than 25 (compared with 40–50 for terrestrial and meteoritic samples). The composition shows the material sampled has almost certainly been through melting or crystallisation stages but whether this was caused by internal heating or impact has not yet been resolved.

EXTRACTION

In the extraction of the pure elements from their ores, chemical, economic and metallurgical factors are all of fundamental importance. These may, however, often conflict. For instance, the most elegant chemical method for the preparation of an element is rarely the cheapest. Again, the particular use envisaged for a metal may determine the choice of extraction procedure. In the ensuing pages we shall deal mainly with the chemical aspects of the extraction of metals with emphasis on the exploitation of chemical differences to facilitate their separation.

The extraction of a metal from its ore generally involves three major operations: concentration of the ore, extraction of the crude metal and refining. These are discussed in turn with some examples.

Concentration of the ore

Except for high-grade ores, this is a necessary first stage. Unwanted material can be removed by either physical or chemical methods or a combination of both. Physical methods include gravity separation and flotation processes; in some cases where the ore contains a magnetic constituent this property can be used in the concentration. Chemical methods are those of *hydrometallurgy*. This is the leaching of ores by aqueous solution to extract the required metal in the form of one of its soluble salts.

Important examples of hydrometallurgical operations are:
  (i) the treatment of cuprous sulphide ores with dilute sulphuric acid in the presence of atmospheric oxygen to produce copper sulphate;
  (ii) the leaching of silver ores with sodium cyanide solution, whereupon the silver is extracted as its complex, $[Ag(CN)_2]^-$;
  (iii) the sulphuric acid or sodium carbonate leach of uranium ores followed by further purification using an ion-exchange process;
  (iv) the ammoniacal leach of sulphide ore containing Ni, Co and Cu under oxidising conditions to form the metal ammine complexes;
  (v) the precipitation of magnesium from sea-water as its hydroxide;
  (vi) the digestion of bauxite with caustic soda under pressure to extract aluminium as soluble aluminate leaving insoluble materials as residue.

As the supplies of high-grade ores become exhausted, and greater use is made of low-grade, more complex ores, hydrometallurgical separations are being increasingly employed for preliminary concentration. Kinetic aspects are of prime importance in leaching processes and the chief disadvantage of older methods is the time-consuming nature of the operation. However, the application of the more versatile and rapid techniques of ion exchange and solvent extraction is resulting in the development of many new hydrometallurgical processes.

Element	Occurrence	Extraction method	Notes
HYDROGEN	Widely distributed as a constituent of water and other compounds	Small (lab.) scale: Zinc + acid Large scale: (a) $2H_2O + C$ $= CO_2 + 2H_2$ (at 1273 K) (b) $CH_4 + H_2O$ $= CO + 3H_2$ (at 1373 K) (c) Electrolysis of water (d) $4H_2O + 3Fe$ (steam) $\rightleftharpoons Fe_3O_4 + 4H_2$	
LITHIUM	*Spodumene*, $LiAl(SiO_3)_2$ *Lepidolite* (lithia mica)	Electrolysis of fused $LiCl/KCl$	The electropositive metals of Group I occur as water-soluble salts of strong acids and as cations in alumino-silicate rocks. Their high reactivity necessitates extraction of the metal under anhydrous conditions.
SODIUM	*Rock-salt*, $NaCl$ *Feldspar*, $NaAlSi_3O_8$ *Chile Saltpetre*, $NaNO_3$ *Borax*, $Na_2B_4O_7 . 10H_2O$	Electrolysis of fused $NaOH$ or $NaCl/CaCl_2$	
POTASSIUM	*Carnallite*, $KCl . MgCl_2 . 6H_2O$ Various alumino-silicates *Saltpetre*, KNO_3	Electrolysis of fused $KCl/CaCl_2$	
RUBIDIUM	Associated with K and Li	Both rubidium and caesium by displacement from their chlorides by calcium: $2RbCl + Ca$ $= 2Rb + CaCl_2$	
CAESIUM	*Pollucite*, caesium aluminium silicate		
BERYLLIUM	*Beryl*, $3BeO . Al_2O_3 . 6SiO_2$ *Chrysoberyl*, $BeO . Al_2O_3$	Electrolysis of fused BeF_2/NaF or magnesium reduction of BeF_2	Be is unique in this group in occurring as a mixed oxide.
MAGNESIUM	*Carnallite*, *Magnesite*, $MgCO_3$ *Spinel*, $MgAl_2O_4$ *Olivine*, Mg_2SiO_4	Electrolysis of fused $KCl/MgCl_2$ Carbon reduction of MgO	The more electropositive elements of Group II occur as silicates and as their sparingly soluble salts. Carbon reduction can be carried out with Mg, not with the alkaline earths because a carbide is formed.
CALCIUM	*Dolomite*, $MgCO_3 . CaCO_3$ *Limestone*, $CaCO_3$ *Gypsum*, $CaSO_4$ *Fluorspar*, CaF_2 *Apatite*, $CaF_2 . 3Ca_3(PO_4)_2$	Electrolysis of fused $CaCl_2/CaF_2$	
STRONTIUM	*Strontianite*, $SrCO_3$ *Celestine*, $SrSO_4$	Electrolysis of fused halides or the aluminium reduction of oxides	
BARIUM	*Witherite*, $BaCO_3$ *Barytes*, $BaSO_4$		

Element	Occurrence	Extraction method	Notes
BORON	Borax, $Na_2B_4O_7.10H_2O$ Colemannite, $Ca_2B_6O_{11}.5H_2O$	Thermal reduction of B_2O_3 with Na, Mg, Al	Unique in Group III in occurring exclusively as an anionic constituent.
ALUMINIUM	Bauxite, $Al_2O_3.2H_2O$ Cryolite, Na_3AlF_6 Alumino-silicate rocks	Electrolytic reduction of Al_2O_3 dissolved in molten cryolite	
SCANDIUM, YTTRIUM and the heavy rare-earths, EUROPIUM to LUTETIUM	Thortveitite, $Sc_2Si_2O_7$ Gadolinite (basic silicate coloured black by iron) Xenotime (phosphate) Yttrotantalite, Samarskite Fergusonite (complex niobates and tantalates)	Electrolysis of fused chlorides	These metals of Group III A occur as silicates and phosphates. As a result of the lanthanide contraction, yttrium has the same ionic radius as the heavy lanthanoids.
CERIUM and the light rare-earths, LANTHANUM to SAMARIUM	Monazite, phosphate Cerite, hydrated silicate Orthite, complex silicate	Electrolysis of fused chlorides	
THORIUM	Monazite Thorite, ThO_2	Ca reduction of ThO_2	
URANIUM	Pitchblende, U_3O_8 Carnotite, $K_2O.2UO_3.V_2O_5$	Ca or Mg reduction of UF_4	
CARBON	Diamond, Graphite Dolomite, Chalk Limestone, Coal	Destructive distillation of coal. Carbon also obtained as a by-product of various industrial processes	The free element is found: also combined as carbonate.
SILICON	Quartz, SiO_2 Many silicates and alumino-silicates	Electrothermal reduction of SiO_2 Reduction of $SiCl_4$ by Zn or hydrogen	Next to oxygen, the most abundant element.
TITANIUM	Ilmenite, $TiO_2.FeO$ Rutile, TiO_2	Reduction of $TiCl_4$ by Mg (Kroll process) or Na	Group IV A metals with a very high affinity for oxygen. Necessary to convert to halides and reduce in an inert atmosphere. Hf is very similar to Zr because of the lanthanide contraction.
ZIRCONIUM	Baddeleyite, ZrO_2 Zircon, $ZrSiO_4$	Reduction of $ZrCl_4$ by Mg	
HAFNIUM	Accompanies Zr: Hf content usually 1–2% Zr content	As for Zr	

Element	Occurrence	Extraction method	Notes
VANADIUM	Vanadinite, $3Pb_3(VO_4)_2 \cdot PbCl_2$ Carnotite Patronite, sulphide	Aluminothermal reduction of V_2O_5	Group V A metals with smaller affinity for oxygen than the preceding group of transition metals. V is the only metal to be found as a sulphide in this group
NIOBIUM	Niobite, $Fe(NbO_3)_2$ containing Ta	Sodium reduction of K_2NbF_7 or K_2TaF_7	
TANTALUM	Tantalite, $Fe(TaO_3)_2$ containing Nb	Electrolysis of fused K_2TaF_7 Also: $Ta_2O_5 + 5TaC = 7Ta + 5CO$	
CHROMIUM	Chromite, $FeO \cdot Cr_2O_3$ Crocoisite, $PbCrO_4$	Reduction of Cr_2O_3 by Al or Si Also electrolysis of aqueous solutions of Cr^{III} salts	Group VI A metals which occur as mixed oxides or as part of an oxyanion.
MOLYBDENUM	Molybdenite, MoS_2 Wulfenite, $PbMoO_4$	Hydrogen reduction of MoO_3	MoS_2 is an exception.
TUNGSTEN	Wolframite, $FeWO_4/MnWO_4$ Scheelite, $CaWO_4$ Tungstite, WO_3	Hydrogen reduction of WO_3	Extraction method from oxide ores involves an initial roast with Na_2CO_3 to form the water-soluble Na salt.
MANGANESE	Pyrolusite, MnO_2 Hausmannite, Mn_3O_4	Reduction of Mn_3O_4 with Al or C	In Group VII A manganese is the only metal of commercial importance.
TECHNETIUM	Traces in nature; usually isolated from fission products	H_2 reduction of ammonium pertechnetate	
RHENIUM	Molybdenite contains up to 20 p.p.m. Re and is the chief source of this metal.	H_2 reduction of ammonium perrhenate	
IRON	Magnetite, Fe_3O_4 Haematite, Fe_2O_3 Pyrites, FeS_2	Reduction of oxides by CO in the blast furnace	The extraction of iron and its conversion to steels is the most important metallurgical process.
COBALT	Associated with Cu and Ni as sulphide and arsenide Smaltite, $CoAs_2$	Reduction of oxides by carbon or water-gas	Cobalt and nickel are much less common than iron and are usually found as low-grade ores.

Element	Occurrence	Extraction method	Notes
NICKEL	Occurs in *pentlandite*, largely iron sulphide containing up to 3% Ni. *Garnierite*, silicate of Mg and Ni produced by weathering *Millerite*, NiS	Carbon reduction of oxide followed by electrolytic refining Also by Mond carbonyl process $Ni(CO)_4 \underset{333K}{\overset{453K}{\rightleftharpoons}} Ni + 4CO$	
RUTHENIUM RHODIUM PALLADIUM OSMIUM IRIDIUM PLATINUM	Native, *e.g.* the alloy *Osmiridium*. 0.5 p.p.m. in nickel-containing iron sulphide (main source) Rare ores: *Braggite*, PdS *Sperrylite*, PtAs$_2$	Residues from nickel carbonyl process are worked up and pure compounds of the individual elements prepared. These are then thermally decomposed: *e.g.* $PdCl_2(NH_3)_2$ to Pd . and $(NH_4)_2PtCl_6$ to Pt	These are the platinum metals which have low free energies of formation for their compounds. They occur native or as easily reduced compounds.
COPPER	*Copper pyrites*, CuFeS$_2$ *Cuprite*, Cu$_2$O *Malachite*, CuCO$_3$.Cu(OH)$_2$ Also native	Partial oxidation of sulphide ore: $2Cu_2O + Cu_2S$ $= 6Cu + SO_2$ Or acid leach with H_2SO_4 followed by electrolysis	The native metals are found. The extraction from sulphide ores can be carried out by pyro- or hydro-metallurgy.
SILVER	*Argentite*, sulphide. *Horn Silver*, AgCl Native	Sodium cyanide leach of sulphide ore. This forms $Ag(CN)_2^-$ from which Ag is precipitated by Zn	
GOLD	Native. Small amounts in many ores such as pyrites	Similar cyanide leach as for Ag	
ZINC	*Zinc blende, wurtzite*, ZnS *Calamine*, ZnCO$_3$	For Zn and Cd, the sulphides roasted to the oxides which are then reduced by C. Electrolysis of ZnSO$_4$ for Zn preparation	In Group II B, sulphide ores again predominate. The metals are easily extracted from their ores.
CADMIUM	Small amounts in most zinc ores		
MERCURY	*Cinnabar*, HgS	Thermal decomposition: $HgS + O_2 = Hg + SO_2$	
GALLIUM	Small amounts in *zinc blende* and *bauxite*	By-product in the extraction of zinc or obtained by electrolysis of alkaline leachings of bauxite.	No concentrated ores of the Group III B metals are known.
INDIUM	Small amounts in *zinc blende* and *cassiterite*	In and Tl are recovered from the flue dusts of pyrites burners by electrolysis or chemical reduction methods	
THALLIUM	Found in *pyrites*		

Element	Occurrence	Extraction method	Notes
GERMANIUM	Found in *zinc blende*, Rare ores are complex sulphides like *argyrodite*, $4Ag_2S.GeS_2$	H_2 reduction of GeO_2	Group IV B metals are found generally as sulphides. Tin is unusual. Electrolytic refining of tin using aqueous solution is important.
TIN	*Cassiterite*, SnO_2	Carbon reduction of SnO_2	
LEAD	*Galena*, PbS	Sulphide roasted to oxide which is then reduced by carbon	
PHOSPHORUS	*Apatite*, $CaF_2.3Ca_3(PO_4)_2$ *Chlorapatite*, $CaCl_2.Ca_3(PO_4)_2$	Electric arc reduction by carbon in presence of SiO_2 (to form calcium silicate)	Sulphide ores are again important in the heavier elements of this group.
ARSENIC	*Nickel glance*, NiAsS *Mispickel*, FeAsS	Roasting of ore in absence of air	
ANTIMONY	*Stibnite*, Sb_2S_3	Iron reduction of sulphide	The only important example of the extraction of an element by direct reduction of its sulphide.
BISMUTH	*Bismuth glance*, Bi_2S_3 *Bismuthite*, Bi_2O_3	Carbon reduction of the oxide	
SULPHUR	*Native* Also combined as sulphides and sulphates	Native S recovered by melting undergound deposits and forcing the liquid to the surface—Frasch Process	Sulphur is very abundant but selenium and tellurium are trace elements. They are recovered as by-products from other processes; for example, Se and Te are found in anode sludges from the electrolytic refining of copper.
SELENIUM TELLURIUM	Found in sulphur-bearing ores	Reduction of their compounds by SO_2	
FLUORINE	*Fluorspar*, CaF_2 *Cryolite*, Na_3AlF_6	Electrolysis of molten KF/HF mixtures	The halogens are always found as anionic constituents. Fluorine, being highly electronegative and reactive, is obtained only by electrolysis of fused, anhydrous salts. Extraction of the halogens becomes progressively easier towards the heavier end of the group.
CHLORINE	Present as chloride ion in sea-water; also solid chloride deposits	Electrolysis of brine	
BROMINE	Also found in sea-water and salt deposits	Displacement by chlorine: $MgBr_2 + Cl_2 = MgCl_2 + Br_2$	
IODINE	Less than 0·1 p.p.m. in sea-water, but concentrated in seaweed As iodate, $NaIO_3$, in *Chilean nitrate*	Reduction of iodate with bisulphite	

Extraction of the metal

The concentrated ore must usually be converted to a compound which is suitable for reduction to the metal. *Pyrometallurgical processes*, involving the use of high temperatures, are most widely employed for the production of the crude metal. Thermodynamic factors are of chief importance in pyro-metallurgy. The kinetic aspects usually need not be considered because the use of high temperatures ensures that chemical equilibrium is rapidly reached.

Many oxide ores are directly reduced (smelted) to the metal. A variety of reducing agents is used, carbon being the one in most general use. Al, Si and hydrogen are other reducing agents and these may be preferred to carbon when, for instance, the metal to be isolated forms a carbide. The bulk of the impurities is removed by the addition of suitable fluxes to form a slag.

A sulphide ore, however, is almost invariably roasted to convert it to the oxide and this is then reduced to the metal. The roasting process also removes volatile impurities such as arsenic. For thermodynamic reasons, an oxide rather than a sulphide is used for reduction.

The highly electropositive metals, such as the alkali and alkaline-earth metals, are isolated by the electrolytic reduction of their fused halides. No suitable chemical reducing agent is available and preparation from aqueous solution is impossible because of the reactive nature of these metals. Less electropositive metals like Cr, Cu and Zn can be prepared by the electrolysis of concentrated aqueous solutions of their salts. One advantage of the electro-lytic over the oxide-reduction method for extraction is the high purity of the electrolytically prepared metal.

When neither oxide reduction nor electrolysis is suitable for chemical reasons, reduction of a metal halide by reactive metals such as Mg, Na and Ca is employed. Titanium is an outstanding example of a metal which cannot be prepared by conventional methods because of the enormous affinity it has for oxygen. This, coupled with the adverse effect on the mechanical properties of the metal of quite small amounts of dissolved oxygen, led to the develop-ment of the Kroll process in which highly purified $TiCl_4$ is reduced by mag-nesium in an inert atmosphere. In this way, oxygen is entirely excluded from the reactants before reduction takes place and so will not be present in the metal produced. There are many technical difficulties to be overcome in the production of titanium and it is not surprising that despite the abundance of the element in the Earth's crust (*Table 7.2*) the pure metal was not prepared for the first time until 1925 (by van Arkel's method). In contrast, metals such as copper and lead are much more rare than titanium but they have been well known for centuries simply because the metals are easily extracted from accessible ores.

Factors other than purely chemical ones are usually of crucial importance in determining the choice of method of extraction. Thus it is essential to make the most economic use of all the products of an extraction process. Again, the nature of the ore has an important bearing on the least expensive route to the pure metal. For example, copper can be made from its sulphide ores by conversion of the molten sulphide to the metal by an air blast. Alternatively, particularly in the case of low-grade ores, concentration is brought about by hydrometallurgical reactions and the metal is deposited electrolytically from solution. The proposed use of the metal can also influence the choice of extraction process. This is illustrated by the cases of Cr, W and Fe. Chromium is used most extensively as a constituent of steel and, when destined for this, it is quite sufficient to prepare ferrochromium by the direct reduction of chromite $(Cr_2O_3 . FeO)$ with silicon or aluminium. An important use of tungsten is in the manufacture of 'hard metal', tungsten carbide. Since only comparatively small quantities of tungsten are required, the pure metal is conveniently obtained by the reduction of WO_3 with hydrogen. In contrast, the vast amounts of iron required for structural materials must be made by a process capable of application on the largest possible scale, namely carbon reduction of the oxide.

Purification of the metal

Various refining processes are used and these will depend on the chemical nature of the metal and the impurities and on the previous history of the raw metal. Among the most important are the removal of impurities by oxidation, by electrolysis or by conversion to a volatile compound which is more readily purified. In addition, several new methods have been introduced to refine reactive and refractory metals.

Oxidative Refining. — This is usually necessary where a metal has been prepared by the reduction of its oxide. At the same time the oxides of other elements present have been reduced. Oxidative refining, using a limited amount of oxygen, is feasible when these other elements have a higher affinity for oxygen than the metal being refined. Then only the impurities are oxidised and the oxides formed are effectively separated from the metal by incorporation in a slag.

For example, when iron oxide is reduced by carbon in the blast furnace to produce pig iron, much of the impurity is removed as fusible slag but the product still contains Si, Mn, P and C. The metal is brittle because of the presence of carbon and refining is necessary to convert it to malleable iron. Oxidative refining is carried out in the Bessemer converter by blowing air through the molten iron. This oxidises Si, Mn and C. Phosphorus is removed in this process by the addition to the converter of basic compounds such as lime to produce 'basic slag'. The carbon content of the iron is reduced by air-blasting to a figure below that required for technical purposes. In addition, some of the iron has been reoxidised. Ferromanganese, of high carbon content, is therefore added with a two-fold purpose: to increase the carbon content and to reduce the iron oxide by the manganese. The manganese oxide formed is removed as a slag. Other deoxidants have been used in place of manganese, especially silicon which also has a high affinity for oxygen.

Electrolytic Refining. — The electrodeposition of pure metals from aqueous solution is important for refining a number of metals such as Cr, Cu, Sn, Ni, Zn and Ag. In the electrolytic refining of copper, for instance, castings of the crude metal are used as anodes in the electrolysis of acidified copper sulphate solution. On electrolysis, the anodes dissolve and pure copper is deposited on the cathode. The impurities which are not anodically dissolved collect at the bottom of the electrolytic cell. This is the anode sludge which is worked up for the extraction of the precious metals and for other elements such as selenium and tellurium.

Recently, fusion electrolysis has been introduced for titanium refining. Impure titanium anodes are used in an electrolyte of fused alkali metal chlorides such as the NaCl/KCl eutectic mixture, m.p. 923 K, which also contains some dissolved lower chlorides of titanium ($TiCl_2$ and $TiCl_3$). Pure titanium is deposited on the cathode as electrolysis proceeds.

Vapour-Phase Refining — This is exemplified by the Mond process for nickel. This involves the purification of nickel by forming its volatile carbonyl which is then thermally decomposed. Iron can also be prepared in a very pure state via its carbonyl.

The van Arkel process, based on the formation of a volatile metal iodide, is another example. Pure compact zirconium was first prepared in 1924 by van Arkel and de Boer by the reaction at 873 K between crude zirconium and iodine vapour to form zirconium tetraiodide, ZrI_4. The vapour of this compound was allowed to diffuse on to a tungsten filament maintained at 2073 K. Decomposition occurred and pure zirconium was deposited on the filament. This method has been most effective in the preparation of those metals which are very difficult to obtain in the pure state. Besides zirconium, other metals which have been made in this way include hafnium, silicon, titanium and beryllium. Recently, the van Arkel process has been adapted for industrial preparations.

New Methods — Brief mention must be made here of the important new methods which have been developed for the refining of refractory metals. The major problems to be solved are the attainment of a sufficiently high temperature to melt the impure metal and the prevention of chemical reactions involving the metal at this temperature. The use of a vacuum arc furnace for the refining of titanium, zirconium and molybdenum is now well established. The crude metal is compacted into an electrode and this is progressively melted in an arc furnace under vacuum. Volatile impurities are boiled off during the melting. An ingot of purified metal is obtained by chilling the molten metal in a copper crucible cooled externally by water. Another recent development in this field is the use of electron bombardment in which the metal to be melted is bombarded under high vacuum by electrons from a heated tungsten filament.

One method of purification worthy of special note is that usually referred to as *zone refining*. This involves the extraction of impurities by a solidifying front and can be used for the removal from a crystalline substance of any impurity which shows a difference in its solubility in the liquid and solid states of that substance. In some ways, therefore, it is similar to fractional crystallisation. The chief inorganic applications of zone refining are in the preparation of certain elements in an extremely high state of purity, particularly germanium, silicon and gallium for use as semi-conductors. Purification has reached such a stage that the impurity content is no longer detectable by normal analytical methods and must be estimated from the electrical resistivity of the material. Thus germanium has been prepared which contains as little as 10^{-7} p.p.m. of most other elements (with the exception of dissolved oxygen and hydrogen). Semi-conductors such as gallium arsenide and indium antimonide are also successfully zone refined.

THEORETICAL PRINCIPLES OF THE EXTRACTION OF METALS BY
PYROMETALLURGY

The reduction of a metal oxide by another element is, in effect, the competition
of both elements for the oxygen and the practical possibility of the process
depends on the relative affinities of the elements for oxygen.

A quantitative comparison of relative affinities may be made by considering
the free energies of formation of the oxides under standard conditions. For a
reaction to proceed spontaneously at a given temperature and pressure, there
must be a decrease in free energy, $-\Delta G^\circ$, in the system. ΔG is related to other
thermodynamic quantities by

$$\Delta G^\circ = \Delta H^\circ - T\Delta S^\circ \tag{1}$$

where ΔH° is the change in enthalpy and ΔS° is the change in entropy. ΔG°
can be expressed in any convenient energy units per unit of chemical change,
e.g. $kJ \, mol^{-1}$.

Standard free energy values, ΔG°, may be directly compared for a number
of reactions. These refer to the reaction between the metal and 1 mole of
gaseous oxygen at a partial pressure of one atmosphere and correspond with
the same amount of chemical change in each case. Although the experimental methods for evaluating ΔG are outside the scope of this book, it is
interesting to note the relationship with the equilibrium constant, K, of the
reaction, expressed in

$$\Delta G = -RT \ln \frac{K}{Q}$$

where R is the gas constant, T the absolute reaction temperature and Q a
constant which has the same form as K but in which the activities refer to the
values for the substances in the reaction. When all the reacting substances
are at unit activity, $Q = 1$ and $\Delta G = \Delta G^\circ$.

Equation (1) indicates that ΔG° is dependent on T and there is very nearly a
linear relationship between them. *Figure 7.1* demonstrates this for a number
of oxides. This is the graphical representation of free energy which was first
used by H. J. T. Ellingham and it provides a sound basis for discussing the
preparation of the elements by oxide reduction.

It is significant that most of the plots slope upwards from left to right, that is,
ΔG° becomes less negative as the temperature rises. The slope is given by

$$\left(\frac{\partial \Delta G^\circ}{\partial T} \right) = -\Delta S^\circ$$

and the change in ΔG° corresponds with the entropy loss resulting from the
disappearance of 1 mole of gaseous oxygen. Entropy is a measure of the
randomness of a given state: solids therefore have low entropies and the
entropy of any material increases on fusion or evaporation with the result
that gases have comparatively large entropies. For this reason a marked
increase in slope is often evident at the boiling point of the metal. Thus the
curve for calcium shows an abrupt change in slope at 1763 K (the boiling
point of the metal). Above this temperature the plot represents the free energy
change of the reaction between gaseous calcium and oxygen to form solid

calcium oxide. The production of 1 mole of oxide therefore involves the disappearance of 2 moles of gaseous reactants. Below 1763 K the same amount of calcium oxide is formed when only 1 mole of gaseous material (oxygen) is consumed. The entropy loss in the reaction is therefore greater above than below the boiling point of calcium. When other transitions such as phase

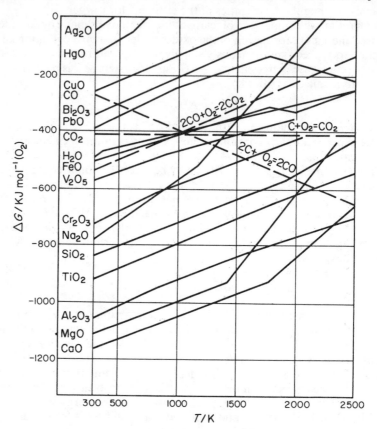

Figure 7.1. Ellingham diagram for oxide formation

changes or melting occur, the entropy changes are generally much smaller and the change in slope is correspondingly less.

The more negative the $\Delta G°$ value for oxide formation at a given temperature the more difficult it is to reduce the metal oxide concerned. When the plot for one metal lies below that for another, the first metal is thermodynamically capable of reducing the oxide of the second.

As a consequence of this the representative and A sub-group elements of Groups II, III and IV can reduce many metal oxides. Aluminium and silicon are used extensively for such purposes.

The particular utility of carbon as a reducing agent can be understood by reference to the change in $\Delta G°$ as T increases for the oxidation of carbon to carbon monoxide. $\Delta G°$ becomes more negative, that is, the trend is opposite to that shown generally by metal oxides. This is due to the increase in entropy which accompanies the reaction

$$2C(\text{solid}) + O_2(\text{gas}) \rightleftharpoons 2CO(\text{gas})$$

wherein 2 moles of gaseous product are obtained for every mole of gaseous oxygen consumed. The plot shows that at temperatures around 2000 K carbon is thermodynamically capable of reducing almost all metal oxides.

The alternative reaction possible between carbon and oxygen is

$$C + O_2 \rightleftharpoons CO_2$$

which becomes the more important at lower temperatures. CO_2 is the main

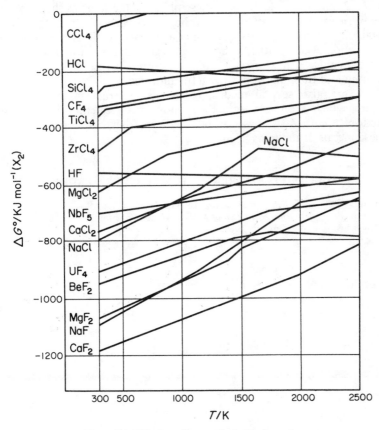

Figure 7.2. Ellingham diagram for halide formation

product below 983 K, CO above this temperature. In the region of 983 K, the equilibrium represented by

$$C + CO_2 \rightleftharpoons 2CO$$

determines the proportions of each gas present. The plot for

$$2CO + O_2 \rightleftharpoons 2CO_2$$

shows that carbon monoxide is an effective reducing agent only at temperatures below 973 K.

The same thermodynamic treatment is applicable to metal/sulphur reactions. The Ellingham plots for these have not been reproduced here because, with the sole exception of antimony, no metals are prepared industrially by the direct reduction of their sulphides. The free energies of formation of most metal sulphides are greater than those for H_2S and CS_2, carbon disulphide being in fact an endothermic compound, and so neither hydrogen nor carbon is a suitable reducing agent. Hence the common practice to roast sulphide ores to the corresponding oxides prior to reduction.

The $\Delta G°/T$ plots for a number of metal chlorides and fluorides are shown in *Figure 7.2*. The salient features of these plots are the same as those for oxide formation. In the case of chlorides, however, carbon cannot be used as reducing agent, and hydrogen has only a few applications, for example, in the reduction of $SiCl_4$. Recourse must generally be made to reactive electropositive metals such as magnesium or sodium, as in the production of zirconium and titanium from their tetrachlorides.

There are also several metals, in particular uranium, beryllium and niobium, which are made from their fluorides by calcium or magnesium reduction. *Figure 7.2* shows clearly the thermodynamic reason for choice of these metals as reducing agents.

SUGGESTED REFERENCES FOR FURTHER READING

DAY, F. H. *The Chemical Elements in Nature*, Harrap, London, 1963.
IVES, D. J. G. *Principles of the Extraction of Metals*, Monographs for Teachers No. 3, London, Royal Institute of Chemistry, 1960.

8   Solvent extraction and ion exchange

These are two modern techniques which have proved to be very effective in the accomplishment of separations which are much more difficult to realise by classical methods such as fractional crystallisation or precipitation. Both solvent extraction and ion-exchange procedures are widely applied on a laboratory scale for analytical purposes and on the plant scale for the purification of materials of industrial importance.

The advantages which these techniques possess over classical methods include (i) a high degree of selectivity in suitable circumstances, (ii) ease and simplicity of operation, (iii) the ability to concentrate minute amounts of material and to separate minor from major constituents and (iv) speed of operation (this applies more particularly to solvent extraction).

The main features of each technique will now be outlined, together with examples of some of the more important applications.

SOLVENT EXTRACTION SYSTEMS

The theoretical basis of solvent extraction is the *Nernst Partition Law*. This states that at equilibrium and at a constant temperature, the ratio of the concentration of a solute in two immiscible liquids is a constant. Although strictly true only for ideal dilute solutions, many substances obey this law, for example, iodine distributed between carbon tetrachloride and water.

Ionic compounds, as a general rule, are not soluble in organic solvents but unless the lattice enthalpy is high, show appreciable solubility in water and other solvents of high dielectric constant. Solution in water is accompanied by hydration of the cations and anions.

In contrast, covalent compounds show preferential solubility in organic solvents. Hence it is found that solvent extraction from an aqueous solution into an immiscible organic solvent can take place provided that an uncharged species is formed. This may either be a neutral molecule or an ion pair.

Several types of solvent extraction systems can be recognised and these are divisible into three broad classes:

1. Metal chelates

When an organic solvent containing a chelating reagent HR is shaken with an aqueous solution of a metal salt, the metal M^{n+} may be partly or completely extracted into the organic phase if an uncharged chelated complex MR_n is formed. The equilibrium reactions established in the two-phase system are summarised in *Figure 8.1*. It should be noted that the reagent and its metal complex are usually much more soluble in the organic phase than in the aqueous solution: also, as *Figure 8.1* indicates, the extent to which the reagent dissociates and hence the proportion of metal which is complexed depends on the pH.

This extraction system can be described as follows when certain assumptions are made:

(i) HR and MR_n are the only species present in the organic phase;

(ii) the metal ion forms no other complexes (such as hydroxo or polymeric species) in the aqueous phase;

(iii) concentrations may be used in place of activities.

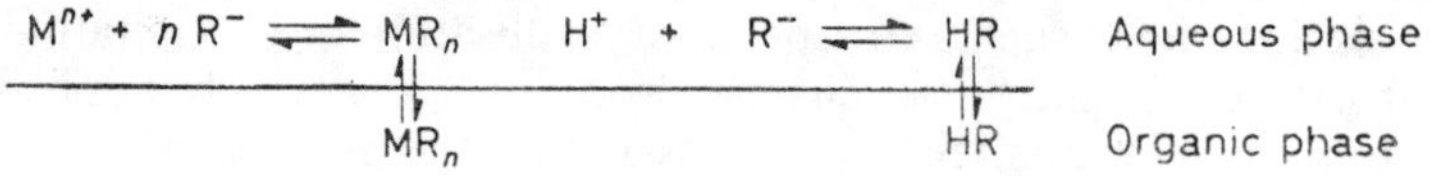

$$M^{n+} + n\,R^- \rightleftharpoons MR_n \qquad H^+ + R^- \rightleftharpoons HR \qquad \text{Aqueous phase}$$
$$\qquad\qquad\qquad MR_n \qquad\qquad\qquad\qquad HR \qquad \text{Organic phase}$$

Figure 8.1 The extraction of a metal chelate

The four equilibria can be represented thus:

$$K_{MR_n} = \frac{[MR_n]_{aq.}}{[M^{n+}].[R^-]^n} \qquad K_{HR} = \frac{[HR]_{aq.}}{[H^+].[R^-]}$$

$$p_{MR_n} = \frac{[MR_n]_0}{[MR_n]_{aq.}} \qquad p_{HR} = \frac{[HR]_0}{[HR]_{aq.}}$$

The distribution ratio, D, is defined as the ratio of the concentration of metal in the organic to metal in the aqueous phase. When the volumes of the two phases are equal,

$$D = \frac{[MR_n]_0}{[M^{n+}]} = p_{MR_n} . K_{MR_n}[R^-]^n$$

$$= \frac{p_{MR_n} . K_{MR_n} . [HR]_0^n}{K_{HR_n} . [H^+]^n . p_{HR}{}^n}$$

$$= K . \frac{[HR]_0^n}{[H^+]^n}$$

D thus varies as the nth power of the concentration at equilibrium of the reagent in the organic phase and inversely as the nth power of the equilibrium H^+ concentration in the aqueous phase. A plot of log D against pH at constant $[HR]_0$ or against log $[HR]_0$ at constant pH should yield a straight line of slope n when the above assumptions are correct.

Experimental results are conveniently represented by graphical plots of E, the percentage of metal extracted, as a function of pH.

$$E = \frac{[MR_n]_0}{[M^{n+}] + [MR_n]_0} \times 100 = \frac{100\,D}{1+D}$$

Figures 8.2 and *8.3* illustrate the extraction behaviour of various metal ions using 8-hydroxyquinoline in chloroform as the chelating agent. Other chelating agents used extensively in analytical work include acetylacetone, thenoyltrifluoroacetone, dithizone, dimethylglyoxime and many derivatives of 8-hydroxyquinoline.

2. Ion association

(a) The metal may be incorporated into a large cation or anion. This, in association with an ion of opposite charge, constitutes an ion pair which may

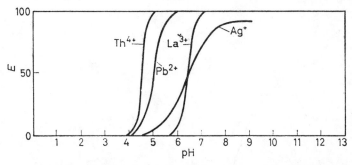

Figure 8.2. Effect of pH *on the extraction of various metal ions by 0·10* mol dm^{-3} *8-hydroxyquinoline in chloroform.* (From J. Stary, *The Solvent Extraction of Metal Chelates*, Pergamon Press, Oxford, 1964)

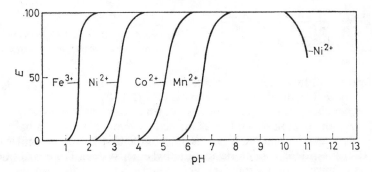

Figure 8.3. Effect of pH *on the extraction of various metal ions by 0·01* mol dm^{-3} *8-hydroxyquinoline in chloroform.* (From J. Stary, *The Solvent Extraction of Metal Chelates*, Pergamon Press, Oxford, 1964)

be readily extracted by an organic solvent. Examples of ion pairs are tetraphenylarsonium perrhenate, $[(C_6H_5)_4As]^+.ReO_4^-$, and permanganate, $[(C_6H_5)_4As]^+.MnO_4^-$. These are formed by reaction between the appropriate metal compound in aqueous solution and tetraphenylarsonium chloride dissolved in a non-reactive solvent such as chloroform.

(b) A very important type is that in which the organic solvent plays a major role in the extraction processes. For instance, displacement by solvent molecules of co-ordinated water from the cation and anion occurs during these extractions. It has been known for many years that uranyl nitrate, $UO_2(NO_3)_2$, is extracted from nitric acid solutions by diethyl ether. The hydrated species, $UO_2(NO_3)_2.6H_2O$, in the aqueous phase becomes $UO_2(NO_3)_2.2[(C_2H_5)_2O].2H_2O$ in the organic phase. Other solvents beside diethyl ether have been used, including derivatives of phosphoric acid such as tri-n-butyl phosphate $(C_4H_9O)_3P{=}O$.

An ion-association system of this class is more complicated than either 1 or 2(a) because more than one complex may be formed and extracted. Thus in

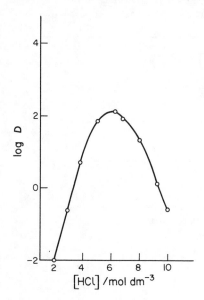

Figure 8.4. Log D versus HCl concentration for the extraction of iron (III) into diethyl ether

the uranyl nitrate system, the trinitrato-anionic complex $[UO_2(NO_3)_3]^-$ is known as well as the dinitrato complex, $UO_2(NO_3)_2$.

Another class of ion association system which has been extensively studied comprises the complex metal acids formed by many metals with halide and pseudohalide (CN^- and SCN^-) ions. One of the best-known examples is the iron (III) chloride/hydrochloric acid/diethyl ether system. The distribution of iron as a function of hydrochloric acid concentration in the aqueous phase is shown in Figure 8.4. A maximum occurs at a hydrochloric acid concentration of about 6 mol dm^{-3}.

Spectrophotometric studies have shown that the extracting species is the solvated ion-pair $H^+FeCl_4^-$. The spectrum of the iron species does not vary when different ethers are used and other investigations have confirmed that the iron is tetrahedrally surrounded by chloride ions and has no primary solvation. The solvation of the extracting species is thus limited to the secondary co-ordination sphere around the iron and to the proton.

The increase in D up to HCl concentration of 6 mol dm^{-3} is due to the increasing concentration of $FeCl_4^-$ which accompanies the rise in chloride ion concentration. The decrease beyond 6 mol dm^{-3} HCl is caused by the transfer (which occurs to a marked extent for high initial concentrations of HCl) of ether to the aqueous phase to solvate the protons there. This causes a significant increase in volume of this phase so that the equilibrium concentration of HCl is much lower than its initial concentration and the extraction of iron is correspondingly less.

The distribution ratio for the extraction of iron(III) chloride is given by:

$$D = \frac{[FeCl_3]_0 + [H^+FeCl_4^-]_0}{[Fe^{3+}] + [FeCl^{2+}] + [FeCl_2^+] + [FeCl_3] + [FeCl_4^-]}$$

Other more complex ion associations besides $H^+[FeCl_4]^-$ must be considered for high concentrations of iron in the organic phase. Moreover, the presence of HCl in the aqueous phase results in the formation of polymeric ions consisting of alternate $[FeCl_4]^-$ (tetrahedral) and $[FeCl_4(H_2O)_2]^-$ (octahedral) units with adjacent units sharing a chloride ion. Evidently the physical chemistry of this type of extraction system is very much more complicated than the metal chelate system, where only one complex is significant.

Many other metal nitrates, halides and other salts can be solvent-extracted under suitable conditions and a number of important applications are known.

(c) Recent developments have been associated with the use, in an inert solvent such as xylene or kerosene, of organic reagents of high molecular weight which contain either an acidic or a basic group in the molecule. For example, di(2-ethylhexyl) hydrogen phosphate contains one acidic hydrogen atom and it also possesses two long hydrocarbon chains which promote solution in organic solvents. This reagent in xylene effectively extracts uranium from aqueous sulphate, chloride or phosphate solutions and is akin to cation-exchange materials in its behaviour. Basic properties are associated with high molecular weight amines such as tris(6-methylheptyl)amine. In a suitable solvent, this compound will extract simple or complex anions from acid aqueous solutions and thus acts as a liquid anion-exchanger.

3. Covalent compounds

This class includes a diversity of substances such as the halogens (chlorine, bromine and iodine), sulphur dioxide, the tetraoxides of osmium and ruthenium, mercuric chloride, germanium tetrachloride, etc. These are extracted by hydrocarbon or halogenated hydrocarbon solvents. The process of extraction is closer to physical solution than in the classes *1* and *2*.

APPLICATIONS OF SOLVENT EXTRACTION

The purification of nuclear fuels

The most important large-scale uses of this technique are in the processing of fuels for nuclear reactors.

Metallic uranium is to date the most widely used nuclear fuel. This contains the fissile isotope ^{235}U. In order to utilise fully the ^{235}U content of uranium it is necessary periodically to remove the irradiated fuel elements from the reactor, to separate the fission products and ^{239}Pu (formed from ^{238}U) and to re-constitute the fuel rods with purified uranium. The fission products are valuable as a source of radioactive isotopes and ^{239}Pu is itself an important nuclear fuel.

The approximate composition of natural uranium after an irradiation corresponding to the energy release of 86·4 GJ kg^{-1} is: U, 99·8 per cent; Pu, 0·08 per cent; fission products, 0·08 per cent. Solvent extraction procedures have proved to be the most successful for the isolation of pure uranium and plutonium from such a mixture.

The process carried out at Windscale (Cumberland) by the U.K.A.E.A. is based on the preferential extraction of plutonyl and uranyl nitrates, PuO_2(NO_3)$_2$ and UO_2(NO_3)$_2$ respectively, by an organic solvent. The chief stages are:

(i) Dissolution of the irradiated fuel in nitric acid. This oxidising solvent produces U^{VI} and Pu^{VI}.

(ii) Extraction of U^{VI} and Pu^{VI} by an immiscible organic solvent. Dibutyl carbitol ('Butex') was originally used but it has been superseded by tri-n-butyl phosphate ('TBP'). Uranyl and plutonyl nitrates, together with some nitric acid, are extracted from aqueous solution, leaving almost all the fission products behind:

$$UO_2^{2+}(aq) + 2NO_3^-(aq) + 2TBP(org) = [UO_2(NO_3)_2.2TBP](org)$$

Nitric acid is co-extracted as the 1:1 solvate $-$ HNO$_3$.TBP.

(iii) Neutralisation of nitric acid in the TBP phase and reduction of Pu^{VI} to Pu^{III} by ferrous sulphamate. U^{VI} is unaffected.

(iv) Extraction of Pu^{III} from the TBP phase by aqueous 8·0 mol dm^{-3} ammonium nitrate (containing a small amount of nitric acid).

(v) Oxidation of the aqueous phase containing Pu^{III} with sodium dichromate. Then the Pu^{VI} formed is extracted by fresh TBP from nitric acid solution as before, leaving behind the remainder of the fission products.

(vi) Back-extraction of the separate TBP solutions of UO_2(NO_3)$_2$ and PuO_2(NO_3)$_2$ into dilute aqueous acid. This gives two aqueous solutions containing purified uranium and plutonium respectively. These are then worked up to obtain the metals themselves.

The whole of the above process is operated on a continuous basis with constant recirculation of solvents and solutions.

The separation of zirconium and hafnium

These two elements invariably occur together in nature and are noted for their great chemical resemblance to each other. Many attempts have been made to achieve a separation and these have received fresh impetus from the demand of the atomic energy industry for hafnium-free zirconium. This metal and its alloys have good corrosion resistance and a low absorption for neutrons. Both these properties are desirable when the material is to be used inside a nuclear reactor, either for structural purposes or to contain the radioactive fuel elements. Hafnium, in contrast, is a powerful neutron absorber.

Several solvent extraction processes are used on a large scale to effect the separation. In one method, hafnium thiocyanate, $Hf(SCN)_4$, is preferentially extracted from HCl solutions by isobutyl methyl ketone. Another process depends on the preferential extraction of zirconium from nitrate solutions using TBP.

The separation of niobium and tantalum

Niobium and tantalum are another pair of metals which are very difficult to separate by classical methods. Successful solvent extraction procedures have, however, been developed. For example, tantalum is extracted to a greater extent than niobium from HF/HCl solutions by isobutyl methyl ketone. Both niobium and tantalum form fluoro-complexes: in aqueous solutions containing moderate concentrations of HF (1 to 3 mol dm^{-3}) Ta^V is present largely as $[TaF_7]^{2-}$ and Nb^V is present as $[NbOF_5]^{2-}$. $[NbF_7]^{2-}$ is formed to an appreciable extent only at high HF concentrations. It is probable that the preferential extraction of tantalum is related to the greater ease of formation and extraction of the acido-complex H_2TaF_7 compared with H_2NbF_7.

LIQUID METAL SYSTEMS

Silver is distributed between immiscible molten lead and zinc phases with a partition coefficient of about 300 in favour of the zinc phase. This solvent extraction system was the basis of the Parkes process (1850) for the desilverisation of lead which was in use for many years in this country.

Recent investigations into the use of a liquid metal as an extractant have been with the aim of devising a more direct method for the re-processing of irradiated uranium than the aqueous/organic solvent process. A liquid-metal extraction process would have the chief advantage of the preservation of the metallic state throughout. The two most promising metals for the extraction of plutonium from irradiated uranium are silver and magnesium. Silver is the better extractant but the subsequent separation of the silver and plutonium presents difficulties owing to the high boiling point of silver. Magnesium extracts plutonium to a lesser extent than does silver but it can be readily removed afterwards because of its relatively high volatility. Neither metal, however, is very selective for the separation of fission product activity from either the uranium or plutonium. Another problem which must be solved is the provision of a suitable containing material for the liquid metals. Tantalum metal, coated with suitable ceramic material to reduce corrosion, appears to be the most promising to date.

Molten salt/liquid metal systems occur very widely in metallurgical processes. Impurities distribute themselves between the slag (molten salt) and the liquid metal.

Certain liquid metals have been used in nuclear fuels and solvent extraction techniques have been applied in their processing. For example, the Liquid Metal Fuel Reactor, developed at Brookhaven National Laboratory, in the United States, has a solution of uranium (1000 p.p.m.) in liquid bismuth as fuel. Equilibration with molten magnesium chloride serves to remove many of the fission products. The extraction process involves a chemical reaction. Thus lanthanum is distributed between the liquid metal phase (m) and the liquid salt phase (s) in accordance with the equilibrium:

$$2La(m) + 3MgCl_2(s) \rightleftharpoons 2LaCl_3(s) + 3Mg(m)$$

ION EXCHANGE

An ion-exchange material is an insoluble substance containing anions or cations which are exchangeable with ions in a surrounding solution without physical change in the structure of the ion-exchange material.

Certain naturally occurring alumino-silicates, the zeolites, show the property of ion exchange. Synthetic zeolites, whose main use for many years was in water softening, have also been made. Dissolved calcium and magnesium salts are responsible for the hardness of water. The softening process involves the removal of the metal ions. This can be carried out by allowing the hard water to percolate through a bed of zeolitic exchanger in the sodium form. Replacement of the calcium ions in water by sodium ions takes place according to:

$$Ca^{2+} + 2Na^+(Z^-) \rightleftharpoons 2Na^+ + Ca^{2+}(Z^-)_2$$

where (Z^-) represents the zeolite. Magnesium ions are replaced by a similar reaction. If percolation is continued, a stage is reached when all the sodium ions on the exchanger have been replaced by calcium. The exchange material may then be regenerated for further water softening by washing with a concentrated solution of sodium chloride. This procedure reverses the above equation.

The first synthetic organic cation-exchangers were produced in 1935 by the sulphonation of various coals and soon afterwards ion-exchange resins were synthesised by condensation reactions like that between phenol and formaldehyde. These resins incorporate acidic groups like sulphonic, $-SO_3H$, in strongly acid resins, and carboxyl, $-COOH$, in weakly acid resins.

Many of the ion-exchange resins now used are made with styrene as starting material. Polymerisation of styrene, $C_6H_5.CH=CH_2$, gives a linear polymer, polystyrene, which is soluble in organic solvents and which can be sulphonated to give a water-soluble polymer. If a small quantity of divinylbenzene $CH_2=CH.C_6H_4.CH=CH_2$ is present when polymerisation is carried out, a three-dimensional cross-linked matrix is produced. The extent of cross-linking and hence the dimensions of the holes in the matrix is determined primarily by the divinylbenzene content of the reaction mixture and also by the nature of the solvent used during polymerisation. A low degree of cross-linking allows large molecules or ions to enter the holes of the resin matrix and vice versa. The cross-linked polymer is an insoluble material of high thermal and chemical stability. Sulphonation with sulphuric or chlorosulphonic acid introduces a sulphonic acid group into almost all the benzene rings and produces a strong acid resin with cation-exchange properties.

$$RSO_3H + NaCl \rightleftharpoons RSO_3Na + HCl$$
$$2RSO_3Na + Ca^{2+} \rightleftharpoons (RSO_3)_2Ca + 2Na^+$$

Commercially available exchangers of this type include Zeo-Karb 225, Dowex 50 and Amberlite IR-120.

Weak acid cation-exchangers contain carboxylic instead of sulphonic acid groups. They are copolymers of methacrylic acid, $CH_3.(C{=}CH_2).COOH$, and divinylbenzene and are exemplified by Zeo-Karb 226 and Amberlite IRC-50.

Anion-exchange resins contain amine or quaternary ammonium groups. The basic groups can be introduced in various ways, for example, by the

Table 8.1. INORGANIC ION-EXCHANGE MATERIALS

Type	Structure	Properties	Uses
(A) Alumino-silicates (i) Kaolin clays Lamellar Zeolites	Layer lattices	Stable to heat and radiation. Cation or anion exchanger depending on pH of solution	Fixation of radioactive waste (removal of Cs^+)
(ii) Zeolites	Three-dimensional lattices	Ionic sieves. Sorption of ions of a certain size only	E.g. use of synthetic ultramarine to separate ^{223}Fr from ^{227}Ac
(B) Salts of heteropoly acids Ammonium molybdophosphate (A.M.P.)	Insoluble, crystalline and ionic	Cation exchanger. Stable up to 573 K and towards radiation. Requires support (asbestos)	Fixation of radioactive waste
(C) Zirconium phosphate	Insoluble and polymeric	Stable to heat and radiation. PO_4^{3-} ion leached slowly so of limited usefulness	Separation of uranium and fission products. Also alkali from alkaline-earth cations
(D) Hydrous zirconium oxide	Microcrystalline gel	Cation exchanger in basic solutions; anion exchanger in acid solutions	High selectivity for polyvalent anions

chloromethylation of polystyrene followed by treatment with base. Weak base resins are obtained from ammonia or primary or secondary amines. Strong base resins containing quaternary ammonium groups are derived from tertiary amines.

Examples of the anion-exchange process are:

$$R.N(CH_3)_3OH + NaCl \rightleftharpoons R.N(CH_3)_3Cl + NaOH$$
$$2R.N(CH_3)_3Cl + SO_4^{2-} \rightleftharpoons [R.N(CH_3)_3]_2SO_4 + 2Cl^-$$

Typical strong base anion-exchangers commercially available are De-Acidite FF, Amberlite IRA-400 and Dowex 1.

A variety of inorganic ion-exchange materials have been studied in recent years. The properties of these are summarised in *Table 8.1*.

The great success of these materials in effecting difficult separations is due to their different affinities for various ions. For example, the order of affinities, obtained in aqueous solutions where the chloride ion concentration is $0 \cdot 1$ mol dm^{-3} at 298 K, of some simple ions for a sulphonic cation-exchanger is:

$$Li^+ < H^+ < Na^+ < K^+ < Cd^{2+} < Rb^+ < Cs^+ < Mg^{2+} < Cu^{2+} < Co^{2+}$$
$$< Ca^{2+} < Sr^{2+} < Ba^{2+} < Al^{3+} < Th^{4+}$$

The greater the charge, the more firmly bound is the ion and the greater the tendency for it to displace one of lower charge from the resin. For two ions of equal charge, the larger one is the more strongly held by the resin.

Ion-exchange materials are applied to an ever-increasing extent both in the laboratory and on an industrial scale and some of the more important uses are described below.

1. The demineralization of water

Complete removal of dissolved salts can be brought about by the use of both types of exchanger. Thus the aqueous solution can be passed first through a bed of cation-exchanger in the hydrogen form to convert the salts to the corresponding free acids. The acids may then be removed on an anion-exchanger. Alternatively, the process can be carried out by passing the solution through a 'mixed-bed' of resin which contains both anion- and cation-exchanger.

2. Separations by ion-exchange chromatography

These are most conveniently achieved by sorption of the mixture of ions to be separated on a column of resin material followed by the elution (removal) from the column by washing with an aqueous solution of suitable reagents.

(a) *The rare earths* — The history of the discovery of the rare earths is characterised by the application of improved techniques of identification and separation resulting in the further subdivision of what had hitherto been considered as homogeneous substances. Spectroscopic analysis was of major importance in the characterisation of the rare-earth metals. The close similarity of the rare earths to one another posed difficult problems in their separation and up to 25 years ago they could only be separated by the repetition of the time-consuming procedures of fractional precipitation or crystallisation of their salts. In a few cases it was possible to utilise valency states different from the group valency of 3. Thus Ce^{III} can be oxidised to Ce^{IV} and then resembles a Group IV rather than a Group III metal; Eu^{III}, Sm^{III} and Yb^{III} can all be reduced to the divalent state in which they show resemblances to the alkaline earths.

Arising from intensive research on atomic energy projects during World War II, a chromatographic separation of the lanthanoids was worked out using ion-exchange resins. The tripositive ions are strongly sorbed by a cation-exchange resin: La^{3+}, the largest, is most strongly held; Lu^{3+}, the smallest, is least strongly held. Elution by HCl therefore removes Lu^{3+} first and La^{3+} last. The separation is not good, however, particularly between the heavier lanthanoids, but is markedly improved by the addition of a complexing agent such as a citrate to the eluting solution. The citrate ion complexes most strongly with the heaviest metals and least strongly with the lightest. When elution is carried out using an ammonium citrate/citric acid buffer, the partial separation effected by the resin is greatly enhanced by complex-formation in solution. Other complexing agents used in the aqueous phase include ethylenediaminetetraacetic acid (EDTA). As much as 80 per cent of an element is obtained in a high state of purity from a single passage through the exchange column. *Figure 8.5* illustrates the chromatographic

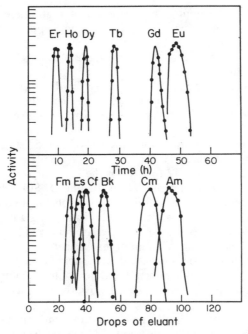

Figure 8.5. The elution of tripositive lanthanoid ions (upper diagram) and tripositive actinoid ions (lower diagram) from Dowex-50 cation-exchange resin using ammonium citrate/citric acid eluant of pH *3·35 at 360* K (Diagram from Katz, J. J. and Seaborg, G. T. *The Chemistry of the Actinide Elements*, Methuen, London, 1957)

separation of the lanthanoids. In this way it has proved possible for the first time to prepare and study the properties of kilogram quantities of many of the rare-earth metals in a pure state.

(b) *The transuranic elements* — Ion-exchange resins have also been of great value in the isolation of the artificially prepared transuranic elements. For example, berkelium (97) is produced by the bombardment of americium (95) with α-particles from a cyclotron. Curium (96) and fission products are obtained at the same time. The most effective means of purifying berkelium is by elution from a cation-exchange resin, using a citrate buffer. *Figure 8.5* demonstrates that complete separation from Cm and Am is achieved.

Many other separation procedures involving cation-exchange resins have been devised. A number of other separations are possible using an anion-exchange resin — usually a strongly basic one — which sorbs metals as complex anions. These are then successively eluted with a mixture of aqueous HF and HCl. In this way, niobium and tantalum, and zirconium and hafnium have been separated.

3. The production of uranium concentrates

Uranium is widely distributed but deposits of high-grade ore are rare. At the present time the major part of the uranium supply comes from ores which have a uranium content as low as 0·1 per cent. Ion exchange is very important in the working-up of these low-grade ores for the preparation of uranium concentrates.

Physical methods of concentration are used for the initial removal of unwanted materials. Then the ore is chemically treated either by an acid leach using dilute sulphuric acid or by an alkaline leach using sodium carbonate solution.

The acid leach is carried out in the presence of an oxidising agent such as MnO_2 or $NaClO_3$ to ensure that the uranium originally present (usually as U_3O_8) is converted completely to U^{VI}. In the leach liquor, the uranium is present in the form of complex anions such as $[UO_2(SO_4)_2]^{2-}$ and $[UO_2(SO_4)_3]^{4-}$. These will exchange readily for other ions on an anion-exchange resin. Uranium is one of the few elements to form strong anionic complexes in sulphate solutions and so many impurities are removed by the sorption of uranium from such solutions by an anion-exchanger.

The carbonate leach is more selective than the acid extraction because most metal carbonates are insoluble in water and do not form soluble complexes. High temperatures or pressures are needed, however, in order to speed up the solution process.

A typical ion-exchange column for use in uranium concentration is 2 m in diameter, 4 m high and contains about 7 Mg of resin. For sorption from acid leach solutions a three-column system is employed. Two columns are used to remove uranium from the leach liquors, whilst the third is being eluted with aqueous NH_4NO_3 ($1·0$ mol dm^{-3}). After removal of uranium from the third column, the resin is regenerated with sulphuric acid and is re-used in the next sorption cycle. The prime advantage of the ion-exchange process is that concentration of the uranium in solution is achieved at the same time as the removal of impurities.

4. Analytical applications

There are many analytical applications of ion exchange. These depend on one of the following processes:

(i) Replacement of ions in solution by ions in the resin. For example, the substitution of metal ions by hydrogen and subsequent titration of the eluted acid with alkali.

(ii) Concentration of ions by sorption on a resin from very dilute solution followed by elution with a small volume of suitable eluting agent. This facilitates the analysis of trace quantities of elements.

(iii) Removal of interfering ions by selective sorption; thus cobalt (II) and nickel (II) in strong hydrochloric acid solution are separated by an anion-exchange resin; cobalt forms complex chloro-anions which are sorbed by the resin, but nickel does not and is not retained on the resin.

SUGGESTED REFERENCES FOR FURTHER READING

AMPHLETT, C. B. *Inorganic Ion Exchanges*, Elsevier, Amsterdam, 1964.

ARDEN, T. V. *Water Purification by Ion-Exchange*, Butterworth, London,

DAWSON, J. K. and LONG, G. *Chemistry of Nuclear Power*, Newnes, London, 1959.

HELFFERICH, F. *Ion-Exchange*, McGraw-Hill, New York, 1962.

IRVING, H. M. N. H. 'Solvent extraction', *Quart. Rev. chem. Soc., Lond.*, 5 (1951) 200.

MARTIN, F. S. and HOLT, R. J. W. 'Liquid-liquid extraction in inorganic chemistry', *Quart. Rev. chem. Soc., Lond.*, 13 (1959) 327.

The comparative chemistry of the representative elements

HYDROGEN

The hydrogen atom has the electron configuration $1s^1$. The properties of the element may be compared either with those of the alkali metals, since H^+ is formed like Li^+, Na^+, etc. by the loss of one electron, or else with the halogens, since like them the hydrogen atom can attain the configuration of an inert gas by the acquisition of one electron to form the hydride ion H^- or by forming one covalent bond. Although certain analogies are possible with both groups, there are many unique aspects of hydrogen chemistry and so the element is generally considered on its own.

As indicated in *Table 9.1*, three isotopes exist: hydrogen (H), deuterium (D) and tritium (T). Of these, only the first two are important in nature, the

Table 9.1. THE PROPERTIES OF HYDROGEN

Electron configuration	$1s^1$		
Ionisation energy $\Delta U_0/\text{kJ mol}^{-1}$	$1312 \cdot 1$		
Electron affinity $\Delta H_E^{\ominus}/\text{kJ mol}^{-1}$	$-72 \cdot 4$		
Electronegativity	$2 \cdot 1$		
Ionic radius r/pm H^+	10^{-3}		
$\qquad\qquad\qquad\; H^-$	210		
Isotopes	^1_1H	^2_1H	^3_1H
Relative atomic mass	$1 \cdot 007\ 82$	$2 \cdot 014\ 10$	$3 \cdot 016\ 05$

relative abundance ratio being of the order of 6000:1 in favour of the lighter isotope. The third isotope occurs to only about one part in 10^{17} of ordinary hydrogen and is generally prepared by nuclear bombardment of the light elements.

The diatomic molecule (H_2) is found in two forms existing in equilibrium, namely *ortho-* and *para*-hydrogen; these differ from each other in the direction of nuclear spin. In the former, spins are parallel but in the latter they are opposed. At room temperature and above, the ratio of *ortho* to *para* is 3:1 whereas at low temperatures (about 20 K) the equilibrium lies almost entirely in favour of the *para* form. Conversion of *ortho* to *para* may be catalysed by absorption on to activated charcoal. The two forms have slightly different physical properties. A similar situation arises in the case of the deuterium molecule.

Chemically, hydrogen is characterised by the formation of binary covalent hydrides and the hydride ion. The unipositive oxidation state is shown in combination with the more electronegative elements such as chlorine and bromine to form the polar molecules HCl and HBr. The hydride ion is formed only in combination with the most electropositive elements such as lithium and sodium. In solution chemistry, molecules like HCl are ionised and the solvated proton is formed by the attraction of the extremely small H^+ ion, with its associated high charge density, for the solvent molecules. In aqueous solution the oxonium ion is formed. This is conventionally written as H_3O^+ though there are almost certainly more water molecules attached to the proton than this simple formulation would suggest.

The covalency of hydrogen is limited to one because only two electrons can be accommodated in the first quantum shell. There are, however, certain instances where a hydrogen atom is to be found between two atoms being covalently bound to the first atom and joined by hydrogen bonding to the second. Hydrogen bonding, as previously mentioned, is only observed where hydrogen is in combination with the most electronegative elements. This bond is responsible for association in hydrogen fluoride, water and liquid ammonia and dictates the structure of crystals of ice, oxo-acids, ammonium salts and many minerals.

Compounds

Hydrides

This is the general name assigned to compounds formed between hydrogen and other elements of the Periodic Table, though to be dogmatic this should only be applied to compounds formed with elements of lower electronegativity. From a consideration of the type of bonding that occurs in these compounds, three classes may be distinguished.

(a) *The Ionic Hydrides* — These are typified by the compounds formed with the Group I A and II A elements, except magnesium and beryllium, at elevated temperatures. The presence of the hydride ion H^- is demonstrated by electrolysis of the molten hydride when hydrogen is formed at the anode. This is best accomplished with LiH since the thermal stability decreases from LiH to CsH and CaH_2 to BaH_2.

As ionic compounds, these hydrides are characterised by high melting points and boiling points and are electrically conducting in the fused state. The alkali metal hydrides possess the sodium chloride structure but the structures of the hydrides of the alkaline-earth metals are more complex.

Vigorous hydrolysis occurs with the formation of hydrogen and the corresponding hydroxide:

$$H^- + H_2O = H_2 + OH^-$$

Reducing properties are exhibited at high temperatures; thus sodium hydride will reduce Fe_3O_4 to metallic iron and carbon dioxide to sodium formate:

$$NaH + CO_2 = NaCOOH$$

At room temperature, however, this reducing property is lost; it appears, therefore, that thermal dissociation of the hydride must be a necessary step in these reactions.

(b) *The Covalent Hydrides* — The representative elements of Groups IV, V, VI and VII form mononuclear hydrides where the element exhibits the group valency. These compounds, except water and a few others, are gaseous under normal conditions. The stability of the hydrides within any one group decreases as the atomic number increases, those formed by the more metallic elements in particular are extremely unstable although there is evidence for the existence of hydrides such as PbH_4 and BiH_3.

With the members of Group III B, the simplest hydrides formed are of the polynuclear type, e.g. B_2H_6, Ga_2H_6 and $(AlH_3)_n$. In these compounds, which are electron deficient, multicentre bonds are present.

A number of polynuclear hydrides are also formed by certain other elements: these include the lighter elements of Group IV B, which give rise to hydrocarbons, silanes, germanes; and nitrogen and oxygen, which form hydrazine and hydrogen peroxide respectively.

The preparation of the hydrides may be effected by a variety of methods:

(i) Direct combination of the elements, e.g. H_2O, NH_3 and HF.
(ii) Reduction of certain compounds in the presence of hydrogen, e.g. AsH_3 and SbH_3.
(iii) Electrolytic reduction, e.g. SnH_4.
(iv) Hydrolysis of metal borides, carbides and similar compounds.
(v) Reduction of halides by lithium aluminium hydride, e.g. SiH_4, GeH_4 and SnH_4.

(c) *Interstitial Hydrides* — The uptake of hydrogen by palladium metal is a well-known characteristic of the metal. This is accomplished by absorption of the hydrogen into the interstices of the metallic lattice, the resultant composition approximating to $PdH_{0.7}$. Non-stoichiometric formulae are found for most compounds of this group, thus zirconium forms $ZrH_{1.92}$ and titanium forms a compound $TiH_{1.73}$, though the use of different pressures and temperatures of reaction may cause variations in the exact proportion of hydrogen. A close similarity is found between the hydrides and the parent metal; thus they are involatile and metallic in character. The absorption of hydrogen causes an expansion of the metal lattice as indicated by the lower density of the hydride.

Although the more basic of the lanthanide elements form hydrides of similar non-stoichiometric constitution, e.g. $LaH_{2.76}$, $CeH_{2.7}$ and $PrH_{2.9}$, their heats of formation are comparable with the salt-like hydrides. Again, their densities are less than that of the parent metal.

Certain metallic hydrides do possess stoichiometric formulae, e.g. a hydride of copper, obtained by the reduction of Cu^{2+} by the phosphinate ion at 343 K, has the formula CuH, though accurate measurements show there is a very slight hydrogen deficiency. Hydrides of composition NiH_2, CoH_2, FeH_2 and CrH_3 may be prepared by reaction between the anhydrous metal chloride in ether with phenyl magnesium bromide and hydrogen.

DEUTERIUM

Evidence for the existence of this isotope was first obtained when the examination of residues from large quantities of liquid hydrogen by spectral techniques showed shifts in the spectral lines of the Balmer series. These shifts could only be explained by the assumption that an isotope of hydrogen with mass two was present (Urey, Brickwedde and Murphy, 1931).

This is the more accessible of the heavy isotopes and is obtained by a process of fractional electrolysis of water. This involves repeated electrolyses until at the later stages deuterium is obtained at the cathode in fairly high proportions. The cathode gases are burned and returned to the cell and eventually a residue of heavy water, D_2O, is left. Deuterium may then be obtained by any one of the processes used to prepare hydrogen from water.

Chemically, deuterium is indistinguishable from hydrogen but there are significant differences in the physical properties, as shown in *Table 9.2*.

Table 9.2. COMPARISON OF PHYSICAL PROPERTIES OF HYDROGEN AND DEUTERIUM

	m.p./K	b.p./K	ΔH_{fusion}/kJ mol^{-1} at m.p.	Vapour pressure p/Nm^{-2} at b.p. of H$_2$
Hydrogen (H$_2$)	13·95	20·55	0·12	101 325
Deuterium (D$_2$)	18·75	23·75	0·22	333 306

Compounds of deuterium may be prepared by reactions involving D_2O; thus ND_3 may be obtained by reacting magnesium nitride with D_2O, and SO_3 reacts with heavy water to form D_2SO_4.

Rapid exchange takes place between ionisable hydrogen and ionisable deuterium, but covalently bound deuterium does not appear to take part in exchange reactions unless hydrogenation catalysts are present. Because of these exchange properties deuterium finds considerable application in the study of mechanisms of reactions and as an isotopic indicator.

TRITIUM

This isotope was discovered during mass spectrometric investigations of deuterium-enriched hydrogen; its occurrence in ordinary water is minute and generally it is prepared by one of a number of nuclear reactions.

Deuteron bombardment of deuterium compounds leads to the following reaction:

$$^2_1H + {}^2_1H \longrightarrow {}^3_1H + {}^1_1H$$

Another method of production involves the bombardment of beryllium by deuterons:

$$^2_1H + {}^9_4Be \longrightarrow {}^3_1H + 2{}^4_2He$$

Neutron bombardment of certain lithium or boron isotopes produces tritium:

$$^1_0n + {}^{10}_5B \longrightarrow {}^3_1H + 2{}^4_2He$$
$$^1_0n + {}^6_3Li \longrightarrow {}^3_1H + {}^4_2He$$

This isotope of hydrogen is radioactive, being a weak β-emitter, with a half-life of approximately 12·5 y.

GROUP 0—THE NOBLE GASES

The elements constituting this group of the Periodic Table are helium (He), neon (Ne), argon (Ar), krypton (Kr), xenon (Xe) and radon (Rn).

All are monatomic gases characterised by high volatility and a short liquid range. With the exception of helium, which has the electron configuration $1s^2$, all the other elements of the group have a completely filled p shell of electrons. These complete shells of electrons, which have high ionisation potentials, have appeared to be chemically inert until quite recently, hence the name inert gases. Because, however, a number of xenon compounds have now been prepared, together with two krypton fluorides, the term *noble gases* is to be preferred. Extraction of all except helium and radon is carried out by fractional distillation of liquid air. Isotopes of radon are formed during

Table 9.3. THE PROPERTIES OF THE NOBLE GASES

	He	Ne	Ar	Kr	Xe	Rn
Atomic number	2	10	18	36	54	86
Electron configuration	$1s^2$	$2s^2\,2p^6$	$3s^2\,3p^6$	$4s^2\,4p^6$	$5s^2\,5p^6$	$6s^2\,6p^6$
m.p. T/K	1·05*	24·55	83·75	116·55	161·65	202·15
b.p. T/K	4·17	27·15	87·25	120·25	166·05	208·15
Ionisation energy $\Delta U_0/\text{kJ mol}^{-1}$	2 371·5	2 080·3	1 520·5	1 350·6	1 170·3	1 037·2
Atomic radius r/pm	93	112	154	169	190	

*At 2·533 MPa

certain radioactive disintegrations and helium is a product of radioactive decay—the alpha particles take up two electrons to form the helium atom. Helium gas is thus associated with minerals that contain α-emitters, e.g. pitchblende; production is effected in the U.S.A. from natural gases by liquefaction of all except helium.

Some physical properties of the noble gases are given in *Table 9.3*.

Compounds

Until very recently, no chemical compounds of the noble gases were known. In discharge tubes species such as He_2^+ and HeH^+ have been observed: under these excited conditions the s electrons in helium are unpaired and one is excited to a higher energy state.

Physical trapping of atoms of the noble gases may occur in crystal lattices. Thus argon, krypton and xenon atoms become trapped in quinol when the latter substance is crystallised from water under considerable pressure of the noble gas. The empirical formulae of these solids, known as *clathrate compounds*, are (3 quinol. NG), the noble gas being encaged by three quinol molecules linked by hydrogen bonding.

Hydrates and deuterates have been obtained for the heaviest of the noble gases, e.g. $Xe \cdot 6H_2O$; with the lighter elements these are formed only under pressure.

More recently, a number of stoichiometric compounds, such as $XePtF_6$ and XeF_4, have been made for the first time and currently the synthesis of further noble gas compounds is of great interest. The preparation of the first true compound of xenon was achieved in 1962 by Bartlett, who, having utilised the strong oxidising properties of PtF_6 to prepare dioxygen$(1+)$hexafluoro-platinate (V), $O_2^+PtF_6^-$, suggested that an analogous reaction might be expected to occur with xenon because of the similarity of the first ionisation potential of molecular oxygen and of atomic xenon (approximately 1150 kJ mol^{-1}). The preparation of xenon hexafluoroplatinate (V) and some other xenon metal fluorides was soon achieved by the reaction at room temperature between xenon and the appropriate metal (VI) fluoride.

$$Xe + MF_6 = Xe^+MF_6^- \qquad (M = Pt, Pu, Ru, Rh)$$

Three xenon fluorides have been synthesised: XeF_6, by heating xenon with excess fluorine under pressure at 573 K; XeF_4, by heating xenon with fluorine at 673 K; and XeF_2 by irradiating a mixture of xenon and fluorine with light from a high pressure mercury arc. At room temperature these compounds are white crystalline solids, XeF_6 being the most volatile. Reduction by hydrogen occurs according to

$$XeF_n + n/2\,H_2 = Xe + nHF$$

a reaction that has proved useful in the analysis of these fluorides. Hydrolysis of the fluorides leads to a further series of xenon compounds. Thus, $XeOF_4$ is a colourless liquid formed by the incomplete hydrolysis of XeF_6 and XeO_3 is the end product of slow hydrolysis of XeF_6. Salts of $Xe(OH)_6$ result from alkaline hydrolysis of the hexafluoride.

The structures of xenon compounds, which may be predicted by simple electron-pair repulsion, are

(cf. ICl_2^-)   (cf. ICl_4^-)   (cf. IF_6^-)   (cf. IO_3^-)

The linear and square-planar arrangements of XeF_2 and XeF_4 respectively have been confirmed experimentally. The experimental data on XeF_6 are consistent with a symmetrical, non-octahedral structure, presumably arising from sp^3d^3 orbitals. The XeO_3 structure is supported by crystallographic and infra-red data. Infra-red and Raman data support a square-pyramidal structure for $XeOF_4$ with oxygen replacing a lone-pair in the XeF_4 arrangement.

Fluorides of two other noble gases have also been reported. The preparation of KrF_4, which is much less stable than XeF_4, can be realised by the passage of an electric discharge through mixtures of krypton and fluorine at liquid nitrogen temperatures, whereas the photolysis of a solid mixture of krypton and fluorine in an argon matrix at 20 K yields KrF_2. The existence of a radon fluoride has been established. Compounds of helium, neon and argon have yet to be prepared. The existence of compounds between the noble gases and electronegative elements is not really surprising when ionisation potential data are considered: thus the first ionisation potential of xenon is very close to that of bromine and is appreciably less than that of hydrogen.

Because of their unreactive character, these elements, particularly argon, are useful as inert atmospheres where experimentation is not possible in the presence of air. Both neon and argon find application in the electronics field and liquid helium is used as a coolant.

GROUP I—REPRESENTATIVE ELEMENTS

The elements of this group, lithium (Li), sodium (Na), potassium (K), rubidium (Rb), caesium (Cs) and francium (Fr), the first series of representative elements, have one s electron in the new quantum shell after the inert-gas core. A consideration of their ionisation potentials shows that this electron is but loosely held and the group, as a whole, therefore, is characterised by the great tendency to form monovalent cations. The electropositive character is the highest of the known elements and the metals are thus extremely reactive. The ease of loss to the outermost electron determines the chemical properties of the metals to a large extent: thus the reaction with water increases with vigour from lithium to caesium. The metals exhibit a high degree of conductivity and are soft and readily fusible. In the vapour state a small proportion of diatomic molecules is present. The atoms are held together by a weak covalent bond formed by the overlap of the s orbitals.

Table 9.4. THE PROPERTIES OF THE ALKALI METALS

	Li	Na	K	Rb	Cs	Fr
Atomic number	3	11	19	37	55	87
Electron configuration	$2s^1$	$3s^1$	$4s^1$	$5s^1$	$6s^1$	$7s^1$
m.p. T/K	452·3	370·7	336·7	312·2	301·7	
b.p. T/K	1 613	1 158	1 048	963	943	
Ionisation energy $\Delta U_0/kJ\ mol^{-1}$	520·1	495·4	418·4	402·9	373·6	
Atomic radius r/pm	123	157	203	216	235	
Ionic radius r/pm	74	102	138	149	188	
Electric mobility of ion $10^5\ u_i/cm^2\ s^{-1}\ V^{-1}$	40·1	51·9	76·2	79·2	79·6	

Compounds

Hydrides

All the metals of the group form hydrides of the type MH by direct combination which requires increasingly higher temperatures from lithium to caesium. These are ionic compounds, the most stable being lithium hydride, and all have the sodium chloride lattice.

Hydrolysis results in the formation of the corresponding hydroxide with liberation of hydrogen. Electrolysis in the fused state yields hydrogen at the anode as expected for the presence of H^-.

In addition to these simple hydrides a number of complex hydrides are also known, for example $LiBH_4$ and $LiAlH_4$. Like the simple hydrides, these are good reducing agents. $LiBH_4$ has the constitution of a salt with Li^+ and BH_4^- ions in the lattice.

Nitrides

The ease of formation of the nitrides falls with increasing atomic number and only lithium nitride is capable of formation by heating the metal in nitrogen. The remaining alkali metals form nitrides by reaction with activated nitrogen but the products are frequently contaminated by azide formation.

Extreme sensitivity is shown to moisture, with hydrolysis to ammonia and the hydroxide.

Oxides

Lithium and sodium show distinct differences from the other metals of the group in their reaction with oxygen.

Burning of lithium in excess oxygen leads to the formation of the oxide Li_2O, whereas under the same conditions sodium yields mainly the peroxide Na_2O_2. With the remaining metals this reaction results in the formation of the hyperoxides MO_2. Small amounts of the hyperoxide are often found as impurity in commercial samples of sodium peroxide and this imparts a yellow colour to the otherwise white solid. Normally, sodium hyperoxide is obtained by reacting the peroxide with oxygen under pressure and at elevated temperatures.

Special methods are employed for the preparation of the less stable oxides. Thus K_2O_2 is obtained by passing oxygen into solution of the metal in liquid ammonia at about 213 K. The oxides of the heavier metals are obtained by heating the metal with the nitrate:

$$\text{e.g.} \quad 10K + 2KNO_3 = 6K_2O + N_2$$

The hyperoxides, all highly coloured, were for some time thought to have the dimeric formula M_2O_4. Such a formulation would require them to be diamagnetic, whereas paramagnetism is exhibited corresponding to the presence of one unpaired electron as required by the ion O_2^-. This ion may be formulated as below, with one three-electron bond as well as the single electron-pair bond; or better from the molecular-orbital approach, as O_2 plus one electron in a π^*2p orbital.

$$:O\!\!-\!\cdots\!\!-\!O:^-$$

The presence of O_2^- is supported by x-ray data on the solids which are found to have a distorted NaCl structure (O_2^- replacing Cl^-).

Rapid reaction with water occurs with all the oxides and results in the formation of the hydroxides:

$$2Na_2O_2 + 2H_2O = 4NaOH + O_2$$
$$2KO_2 + 2H_2O = 2KOH + H_2O_2 + O_2$$

In addition to these oxides, rubidium and caesium are reported to form sesquioxides Rb_2O_3 and Cs_2O_3. Investigations have shown that these are correctly formulated as mixed hyperoxides–peroxides, e.g. $Rb_2O_2 . 2RbO_2$.

Furthermore the reaction of ozonised oxygen with the hydroxides in the solid state or with metals in liquid ammonia yields the ozonides MO_3. The stability of these compounds increases with increasing atomic number and the formation of LiO_3 is extremely doubtful. Paramagnetism is exhibited by the ozonides and they are characterised by high colouration and great reactivity. Slow decomposition to the hyperoxide and oxygen occurs on standing and rapid hydrolysis is effected by moisture.

Halides

With the exception of certain lithium halides, the alkali metal halides are essentially ionic in their constitution. The thermal stability of the halides increases with increasing atomic number of the metal atom and decreasing atomic number of the halogen atom.

Except for CsCl, CsBr and CsI, the crystal structure is that of sodium chloride with 6 co-ordination of the ions. For the caesium halides, 8 co-ordination is exhibited.

A general survey of the salts of the alkali metals shows them to be more soluble in water and other polar solvents than the corresponding salts of other metals. The degree of solvation in solution diminishes from lithium to caesium as demonstrated by the ionic mobilities of the individual ions (*Table 9.4*); lithium, because of the high positive charge density in the ion, shows the greatest tendency to hydrate. Sodium chloride shows less tendency to hydrate, although a dihydrate has been prepared by crystallisation of a saturated solution of the salt at less than 273 K. Hydration persists in the solid state where some 75 per cent of lithium and sodium salts are found to be hydrated. For the heavier metals this percentage drops to 25 per cent for potassium salts and for rubidium and caesium only one salt is found to be hydrated, this being the ferrocyanide. Evidence suggests that for salts of potassium, rubidium and caesium the hydration is associated with the anion rather than the cation.

A number of alkali metal complexes have been obtained with organic molecules, for instance salicylaldehyde will complex with all the group members. In these complexes the co-ordination of lithium is invariably 4 but the other alkali metals exhibit both 4 and 6 co-ordination.

Peculiarities of lithium

Because of its small ionic size, comparable with that of magnesium, lithium shows distinct differences from the lower group members and in fact strongly

resembles magnesium in its chemical behaviour. Several instances of this are:
   (a) Lithium is the only member of the group to form a nitride by direct
       combination with nitrogen. Similarly, the ready combination with
       carbon to form a carbide is an analogy with magnesium.
   (b) The carbonate is relatively unstable thermally and the hydrogen-
       carbonate cannot be isolated from aqueous solution.
   (c) The solubilities in water of the compounds of lithium as a whole
       resemble those of magnesium; thus the chloride, bromide and iodide
       all have high solubility but the fluoride, carbonate, phosphate and
       oxalate are only slightly soluble.

Francium

Until 1946 the gap in the Periodic Table at element 87 had defied filling, but it
was then established that the element arose from a branch chain decay of
^{227}Ac by alpha emission, the main decay (98·8 per cent) of which is by electron
emission to ^{227}Th. Characterisation of the chemical properties of francium
is made difficult by the problem of the short half-life (21 min) of the longest-
lived isotope ^{223}Fr which decays to ^{223}Ra, but it appears to fit into the alkali
metal group and in its precipitation reactions shows a strong resemblance
to caesium.
   Several other isotopes are known and these all have shorter half-lives than
^{223}Fr; the three longest-lived after ^{223}Fr are ^{212}Fr, ^{222}Fr and ^{221}Fr, with
respective half-lives of 19·3, 14·8 and 4·8 min.

GROUP II—REPRESENTATIVE ELEMENTS

This group contains beryllium (Be), magnesium (Mg); the alkaline earths
calcium (Ca), strontium (Sr) and barium (Ba); the radioactive element
radium (Ra).
   With the elements of this group there are two valency electrons contained
in a completed s quantum shell. The loss of these electrons with the formation
of the dipositive oxidation state becomes easier with increasing atomic
number, but, unlike the members of Group I A, the chemistry of this family

Table 9.5. THE PROPERTIES OF THE GROUP II REPRESENTATIVE ELEMENTS

		Be	Mg	Ca	Sr	Ba	Ra
Atomic number		4	12	20	38	56	88
Electron configuration		$2s^2$	$3s^2$	$4s^2$	$5s^2$	$6s^2$	$7s^2$
m.p. T/K		1 553	923	1 123	1 033	983	
b.p. T/K		1 773	1 373	1 763	1 653	1 913	
Ionisation energy							
$\Delta U_0/kJ\ mol^{-1}$	1st	899·1	737·6	589·5	548·9	502·5	502·5
	2nd	1 756·8	1 450·2	1 145·2	1 064·4	964·8	
Atomic radius r/pm		89	136	174	191	198	
Ionic radius r/pm M^{2+}		35	72	100	116	136	

is not completely dominated by the chemistry of the cations. Covalent character is predominant in the compounds of the first member, beryllium, which, because of its small size and high nuclear charge, shows distinct anomalies in its behaviour and in many instances shows a diagonal relationship to aluminium. The ease of loss of the valency electrons is again the governing factor in the chemistry of the group. The reaction with cold water is vigorous for the alkaline-earth metals but for magnesium attack is only readily effected by hot water and beryllium remains inert even at elevated temperatures.

The general physical properties of the elements are summarised in *Table 9.5*.

Compounds

Hydrides

As the group is descended there is an increase in reactivity towards hydrogen. Beryllium and magnesium show only a slight tendency to react with hydrogen but calcium, strontium and barium react to form the saline hydrides MH_2 at high temperatures. Compared with the alkali metal hydrides, those of calcium, strontium and barium are more thermally stable, but they are similar in behaviour on hydrolysis and electrolysis and hence contain the H^- ion.

Nitrides

On heating, all the group members react with nitrogen forming the nitrides of general formula M_3N_2. Heating the metals in ammonia results in the formation of the same compounds. The stability of the nitrides falls with increasing atomic number. Hydrolysis results in the formation of ammonia and the corresponding hydroxide.

Oxides

Reaction between oxygen and the metals takes place on heating to form the normal oxides, MO. In the case of calcium, strontium and barium these may be obtained also by the thermal decomposition of the corresponding carbonates at very high temperatures.

The oxides, except BeO, have the ionic sodium chloride lattice; beryllium oxide, mainly covalent, has the crystal structure of wurtzite.

Peroxides, MO_2, of the alkaline-earth metals are known, the most stable being that of barium which is obtained in the reaction between barium and oxygen at high temperatures and under pressure. The calcium and strontium peroxides may be precipitated as octahydrates by adding hydrogen peroxide to alkaline solutions of calcium and strontium salts. Acidification of the peroxides yields H_2O_2 and this reaction has been used in the preparation of hydrogen peroxide from BaO_2.

Beryllium oxide is inert towards hydrolysis but the remaining oxides of the group hydrolyse with the formation of the corresponding hydroxides which are strong bases. Compared with the alkali metal hydroxides, these are relatively much less soluble. Thermal stability increases from magnesium to barium; barium hydroxide is sufficiently stable to be heated to fusion without decomposition to the oxide.

Halides

Beryllium halides are essentially covalent in their behaviour, having low melting points, appreciable volatility and showing no electrical conductivity in the fused anhydrous state. The compounds with any particular halogen show an increase in ionic character with an increase in cationic radius and the halides of any particular metal show an increase in ionic character for a decrease in anionic radius.

The fluorides of the group have higher lattice enthalpies than the other halides and are consequently less soluble in water. The solubility of the other halides decreases from magnesium to barium. Hydrate formation is characteristic of the halides and attempts to dehydrate by heating often result in the formation of the basic halide by the loss of hydrogen halide; this tendency decreases from magnesium to barium.

Complexes

Beryllium has a great tendency to increase its covalency to the maximum of four by sp^3 hybridisation in the formation of tetrahedral complexes such as BeF_4^{2-}. Organic complexes of beryllium are well characterised, in particular compounds with β-diketones and monobasic acids, e.g. acetic acid, are typical of the complexes with oxygen as donor atom.

$$
\begin{array}{ccccc}
H_3C & & & & CH_3 \\
\diagdown & & & & \diagup \\
& C=O & & O-C & \\
\diagup & & \diagdown \diagup & & \diagdown \\
HC & & Be & & CH \\
\diagdown & & \diagup \diagup & & \diagup \\
& C-O & & O=C & \\
\diagup & & & & \diagdown \\
H_3C & & & & CH_3 \\
\end{array}
$$

Acetylacetone complex of Be^{II}

With the monobasic carboxylic acids, 'basic' complexes are formed of which the best known is basic beryllium acetate $Be_4O(CH_3COO)_6$. In this compound tetrahedral symmetry is exhibited as shown in *Figure 9.1*. A central oxygen atom is tetrahedrally surrounded by four beryllium atoms and six acid groups are along the edges of the beryllium tetrahedron thus giving a set of 4 co-ordinate beryllium atoms.

For the remainder of the group, weak β-diketone and β-keto ester complexes are formed. Complexes with ethylenediaminetetra-acetic acid are, however, far more stable and better characterised; the formation of such complexes is used in the quantitative estimation of the metal ions.

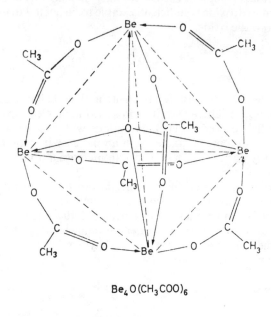

$$Be_4O(CH_3COO)_6$$

Figure 9.1. Structure of basic beryllium acetate

GROUP III—REPRESENTATIVE ELEMENTS

This group comprises the elements boron (B), aluminium (Al), gallium (Ga), indium (In) and thallium (Tl). The electron configuration is ns^2np^1 and the group valency is three. In addition, lower valency states are obtained and the univalent state in particular becomes of increasing importance as the group is descended; with thallium it is the most important state.

The chemistry of the group may be conveniently divided into that of boron and aluminium on the one hand and that of gallium, indium and thallium on the other.

BORON AND ALUMINIUM

Some general properties of these two elements are given in *Table 9.6*. The high ionisation potentials for the formation of the tripositive cations means that bonding in the three-valent state will be mainly covalent. The formation of three covalent bonds leaves the atoms two short of a completed octet. A great tendency to accept two more electrons and hence increase the number of electrons to eight is shown by both elements.

Table 9.6. THE PROPERTIES OF BORON AND ALUMINIUM

		B	Al
Atomic number		5	13
Electron configuration		$2s^2 2p^1$	$3s^2 3p^1$
m.p. T/K		2 303	933
b.p. T/K		4 203	2 723
Ionisation energy $\Delta U_0/kJ\ mol^{-1}$	1st	800·4	577·4
	2nd	2 426·7	1 815·9
	3rd	3 658·5	2 743·9
Atomic radius r/pm		80	125
Ionic radius $r/pm\ M^{3+}$		23	53

Compounds

Hydrides

Both boron and aluminium form hydrides of the covalent type.

In the case of boron, several hydrides have been prepared (*Table 9.7*). These were originally investigated by Stock, who obtained a mixture of them by the acid hydrolysis of magnesium boride, Mg_3B_2. This results mainly in the formation of B_4H_{10} and small quantities of the others. At least six volatile

Table 9.7. PHYSICAL PROPERTIES OF THE BORON HYDRIDES

Formula	B_2H_6	B_4H_{10}	B_5H_9	B_5H_{11}	B_6H_{10}	$B_{10}H_{14}$
Name	Diborane (6)	Tetra-borane (10)	Penta-borane (9)	Penta-borane (11)	Hexa-borane (10)	Deca-borane (14)
m.p. T/K	107·7	153·2	226·6	150·2	208·2	372·9
b.p. T/K	180·7	291·2	321·2	336·2	—	486·2

hydrides are known and these have attracted much attention because of their reactions and the difficulty experienced in devising suitable structural formulae for them.

Diborane, B_2H_6, is the simplest of the boron hydrides and several methods are available for its preparation:

(i) The reaction between boron trichloride or trifluoride and lithium aluminium hydride in ethereal solution.

(ii) Sparking of boron trichloride with hydrogen.

and (iii) The reduction of boron trifluoride with lithium hydride in ether

$$6LiH + 8BF_3 . Et_2O = 6\ LiBF_4 + B_2H_6 + 8\ Et_2O$$

Higher hydrides may be obtained by pyrolysis of diborane. At 373–393 K the main products of pyrolysis are B_4H_{10} and B_5H_{11}.

Diborane is spontaneously inflammable, especially in damp air, and decomposes to boric acid and hydrogen in the presence of alkali.

With hydrogen chloride, B_2H_5Cl is formed which slowly decomposes to BCl_3 and diborane.

The reaction with excess ammonia at high temperatures yields borazine, $B_3N_3H_6$, a compound somewhat similar to benzene, which has a ring structure of alternate boron and nitrogen atoms.

Structure of the boron hydrides

The lowest of the boron hydrides is a dimer, B_2H_6, and this immediately poses problems as to the electronic structure of the molecule. The least number of electron-pair bonds necessary to hold eight atoms together is seven, but in diborane the number of valency electrons is only twelve. Diborane is thus said to be electron-deficient.

Physical evidence suggests that the two boron atoms are bridged by two hydrogens. Various attempts have been made to explain this bridging via hydrogen. Pitzer has attempted to describe it in terms of a protonated double bond in which the two bridge hydrogens are situated at the points of

$$
\begin{array}{ccc}
H & & H \\
\diagdown & \!\!_{-}H^{+}\!_{-} & \diagup \\
& B\!\!=\!\!\!=\!\!B & \\
\diagup & H^{+} & \diagdown \\
H & & H
\end{array}
$$

greatest electron density in a double bond similar to that in ethylene. Although this is a reasonably simple approach, it has the drawback of suggesting that the bridge hydrogens should have acidic properties whereas the chemistry of diborane suggests that they react as H^-.

A more convincing approach envisages the boron atom as sp^3 hybridised and forming two molecular orbitals over the boron atoms and the bridge hydrogen atoms. These three-atom orbitals are bent and are frequently referred to as 'banana' bonds; each contains two electrons, one from hydrogen and one from boron:

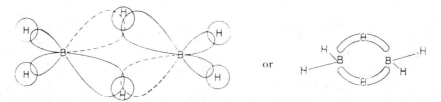

The structures of the higher hydrides also pose similar difficulties and for more details of the structures of these molecules the student is referred to more advanced texts.

Aluminium forms a solid polymeric hydride of formula $(AlH_3)_n$ by treatment of lithium hydride with aluminium chloride. The presence of excess lithium hydride leads to the formation of lithium tetrahydroaluminate (III), $LiAlH_4$. No dimeric hydride has been obtained and the structure of the polymer is not exactly known.

Besides the hydroaluminates, hydroborates are known. Those of the alkali metals are ionic in character but aluminium tetrahydroborate is covalent, being a colourless liquid boiling at 318 K. Both the AlH_4^- and BH_4^- ions are unstable in the presence of water and the alkali metal salts of both are useful reducing agents.

Oxides

Boron oxide, B_2O_3, is the product of combustion of boron in oxygen though it also results as the product of dehydration of orthoboric acid, H_3BO_3, via the intermediate formation of metaboric acid $(HBO_2)_n$ and tetraboric acid, $H_2B_4O_7$. Of these acids, the salts most commonly encountered are those of $(HBO_2)_n$ and $H_2B_4O_7$. The orthoborate ion has a planar structure like NO_3^- and the metaborate ion in $NaBO_2$ has the structure

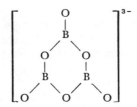

Boron oxide shows a resemblance to silica in physical properties and has the property of forming complex glasses with other metal oxides.

Aluminium oxide, Al_2O_3, is obtained either by heating the metal in air or by dehydration of the hydrated oxides. This oxide is amphoteric in behaviour and exists in several crystalline modifications. When formed at very high temperatures, a variety of Al_2O_3 is obtained that is extremely resistant to attack, needing fusion with $NaHSO_4$ to dissolve it.

Halides

Boron forms the halides BF_3, BCl_3, BBr_3 and BI_3. The method of formation of the last three is by reduction of the oxide with carbon in the presence of the appropriate halogen at high temperatures. The action of HF on B_2O_3 is used to prepare the fluoride.

All four trihalides are covalent molecules existing as triangular monomers in the vapour phase. Hydrolysis occurs to orthoboric acid:

$$4BF_3 + 3H_2O = H_3BO_3 + 3HBF_4$$

$$BX_3 + 3H_2O = H_3BO_3 + 3HX \ (X = Cl, Br \ or \ I)$$

All the trihalides behave as acceptor molecules, in particular nitrogen and oxygen are common donor atoms; typical complexes formed are

$$F_3B \longleftarrow NH_3 \quad and \quad F_3B \longleftarrow OEt_2$$

Aluminium similarly forms all four trihalides and, of these, the fluoride is ionic. The chloride, bromide and iodide exist as dimeric molecules both in the vapour phase and in solution of solvents such as benzene. This dimerisation occurs by bridging through the halogen atoms as shown. The dimeric molecules are easily broken down by reaction with donor molecules such as ethers:

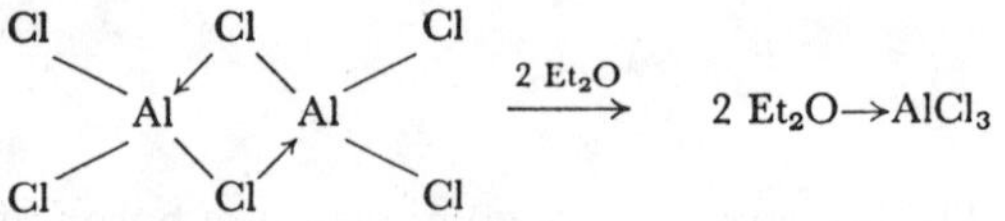

$$2\ Et_2O \longrightarrow 2\ Et_2O \rightarrow AlCl_3$$

Complex fluoroborates and fluoroaluminates are well known, e.g. KBF_4 and Na_3AlF_6.

Complexes

Both boron (III) and aluminium (III) form complexes with β-diketones such as acetylacetone (Hacac). Aluminium gives the trichelate complex $Al(acac)_3$, whereas boron trihalides react in a different manner to form such complexes as $B(acac)F_2$ (shown below) and $[B(acac)_2]^+\ Cl^-$.

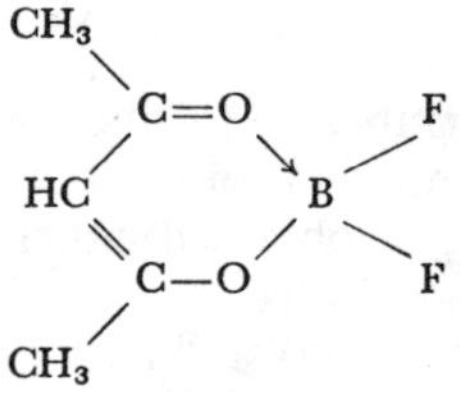

Aluminium (III) also forms complexes with such reagents as oxalic acid, and 8-hydroxyquinoline.

GALLIUM, INDIUM AND THALLIUM

All three elements have three electrons in the outermost shell and exhibit the group valency. Indium also shows a valency of one, though this is unstable with respect to oxidation to indium (III). Thallium (III) compounds are strong oxidants and rapidly decompose to the corresponding thallium (I) compounds. Reasons for the occurrence of both the $+1$ and the $+3$ oxidation states have been proposed by Drago for the chlorides from a thermodynamic standpoint.

A consideration of ionisation energies alone would suggest that the attainment of the $+3$ oxidation state would be most difficult for gallium, which is contrary to the observed facts.

	Ga	In	Tl
Sum of $\Delta U_0(M^+) + \Delta U_0(M^{2+})$/kJ mol^{-1}	4 931·3	4 508·7	4 835·9

The trend in the thermodynamic instability of the higher state with increasing atomic number is indicated by the data for the reaction

$$MCl_3(s) = MCl(s) + Cl_2(g) \tag{1}$$

	Al	Ga	In	Tl
$\Delta H^\circ/kJ\ mol^{-1}$	674·9	561·9	461·9	284·1

Two main factors may account for this instability:
(i) Decrease in bond energy of the compounds of the higher oxidation state.
(ii) Increase in bond energy of the compounds of the lower oxidation state.
Consideration of the reactions

$$M(g) + n/2Cl_2(g) = MCl_n(g) \tag{2}$$

and $$MCl_{n-2}(g) + Cl_2(g) = MCl_n(g) \tag{3}$$

and the fact that $GaCl_3$ and $TlCl_3$ have similar sublimation energies (and therefore probably similar lattice stability) means that the data for reaction (2) may be used as a measure of the bond strength of the higher chlorides. Reaction (3), which is the reverse of the decomposition reaction of the higher chloride, may be used as a comparison of the bond strengths of the chlorides of the higher oxidation state relative to the bond strengths of the oxidation state two below.

	B	Al	Ga	In	Tl
Reaction (2) $\Delta H^\circ/kJ\ mol^{-1}$	−802·1	−892·0	−721·3	−617·6	−457·7
Reaction (3) $\Delta H^\circ/kJ\ mol^{-1}$	−502·5	−557·7	−482·8	−298·7	−204·6

Comparison of the figures for (2) and (3) indicates the instability of thallium (III) which may be attributed to a decrease in the strength of the bonds holding the molecule together.

The higher oxidation state, which is attained by the promotion of an s electron to a p level, may not be easily reached if the energy released in the formation of two further covalent bonds is insufficient to promote the' s electron to the p level. Values of the promotion energies are

	Al	Ga	In	Tl
$M_{s^2p}(g) \rightarrow M_{sp^2}(g)\ \Delta H^0/kJ\ mol^{-1}$	350·1	458·1	421·7	451·9 .

Where the energy is insufficient the lower oxidation state may be formed by utilising only the p electron in bond formation.

The characteristic properties of the three elements are given in *Table 9.8*.

The compounds of gallium are similar to those of aluminium; the oxide is less acidic than Al_2O_3.

In the case of indium, the oxide is even more basic than Ga_2O_3, but compounds of indium (III), on the whole, tend to resemble those of gallium (III).

Both gallium and indium form compounds of empirical formula MCl_2, where the metal is apparently in the divalent state. Physical evidence points to the formulation of these chlorides as $M^I(M^{III}Cl_4)$, since if M^{2+} were present paramagnetism would be expected, and the compounds are in fact diamagnetic.

Thallium forms two series of compounds. The thallic, Tl^{III}, compounds show certain resemblances to Ga^{III} and In^{III}. The oxide is completely basic in behaviour. Thallous, Tl^I, compounds show a strong resemblance to both

Table 9.8. THE PROPERTIES OF GALLIUM, INDIUM AND THALLIUM

		Ga	In	Tl
Atomic number		31	49	81
Electron configuration		$3d^{10} 4s^2 4p^1$	$4d^{10} 5s^2 5p^1$	$5d^{10} 6s^2 6p^1$
m.p. T/K		303·0	429·4	576·8
b.p. T/K		2 513	2 323	1 743
Ionisation energy $\Delta U_0/kJ\ mol^{-1}$	1st	579·1	558·1	589·1
	2nd	1 979	1 816·1	1 970·2
	3rd	2 952·2	2 692·0	2 865·6
Atomic radius r/pm		125	150	155
Ionic radius r/pm M^{3+}		62	79	88
M^+				150

lead (II), with which Tl^I is isoelectronic, and to silver (I). In addition, there are certain resemblances to the alkali metals; thus, the oxide and hydroxide are strong bases.

GROUP IV—REPRESENTATIVE ELEMENTS

This group shows a pronounced change from non-metallic to metallic character as the atomic number increases. Carbon (C) has non-metallic properties, silicon (Si) and germanium (Ge) are semi-conductors, whilst tin (Sn) and lead (Pb) are usually regarded as true metals. Allotropy is shown by carbon and tin. Physical data of the elements are summarised in *Table 9.9*.

The outermost electron shell is four short of the next noble gas configuration. The gain of four electrons by one atom appears to be energetically

Table 9.9. THE PROPERTIES OF THE GROUP IV REPRESENTATIVE ELEMENTS

	C	Si	Ge	Sn	Pb
Atomic number	6	14	32	50	82
Electron configuration	$2s^2 2p^2$	$3s^2 3p^2$	$4s^2 4p^2$	$5s^2 5p^2$	$6s^2 6p^2$
m.p. T/K		1 683	1 210	505	600
b.p. T/K	4 123 (subl)	2 953	3 103	2 960	2 024
Electronegativity (Pauling)	2·5	1·8	1·8	1·8	1·8
Atomic radius r/pm	77	117	122	141	154
Ionic radius r/pm M^{2+}			93	112	120
M^{4+}		26	54	69	77

impossible except with carbon in the ionic carbides of strongly electropositive metals. Alternatively, four covalent bonds are formed, as in the hydrides MH_4, and the tetrachlorides MCl_4, which are known for all five elements. These molecules are tetrahedral and the central atom is therefore sp^3 hybridised. Covalent bond formation of this type is the dominant feature in carbon and silicon chemistry but with the heavier elements account must also be taken of the formation of positive ions. Thus the tetrafluorides of carbon, silicon and germanium are molecular but both SnF_4 and PbF_4 have salt-like properties.

For Ge, Sn and Pb both the $+2$ and $+4$ oxidation states are well known. A consideration of the thermodynamic data for the chlorides of Group IV B, similar to that given to Group III B, again indicates the instability of the higher oxidation state as the atomic number increases. The $+2$ state of Ge and Sn are reducing but dipositive Pb is stable. Pb in the $+4$ state has strong oxidising properties and readily reverts to the $+2$ state. There is, in fact, a well-defined increase in the stability of the $+2$ state in going from Ge to Pb and a corresponding decrease in the stability of the tetravalent state. In combination with very electronegative elements, for example, in the dioxides SnO_2 and PbO_2 which have the rutile structure, the tetrapositive ions Sn^{4+} and Pb^{4+} respectively are found.

Carbon, as a first short period element, shows marked differences from the remaining members of the group.

Firstly, the carbon atom is covalently saturated; in other words the second quantum shell is completely filled whenever it is joined by four covalent bonds to other atoms. The maximum covalency of Si and the other elements is six because here d orbitals are available for bonding. In this way, we can account for the inertness of the carbon tetrahalides towards water compared with the ready hydrolysis shown by the tetrahalides of the succeeding elements. The first stage in hydrolysis is the co-ordination of a water molecule followed by the elimination of hydrogen halide from the complex:

and it is apparent that the carbon tetrahalides, with no vacant orbital on the carbon, cannot react in this way. Silicon halides are hydrolysed to SiO_2 irreversibly but hydrolysis is incomplete with the halides of the heavier elements of the group and may be suppressed by the addition of the appropriate hydrogen halide:

$$SiCl_4 + 2H_2O \rightarrow SiO_2 + 4HCl$$
$$GeCl_4 + 2H_2O \rightleftharpoons GeO_2 + 4HCl$$
$$SnCl_4 + 2H_2O \rightleftharpoons SnO_2 + 4HCl$$

Secondly, carbon–carbon bonds are stronger than bonds between like atoms formed by the other members of the group. Numerous carbon compounds are known in which two or more carbon atoms are covalently bound together and this property of forming links with itself (catenation), coupled with the high stability of carbon–hydrogen bonds, is responsible for the immense range of organic compounds. Silicon forms much stronger bonds with fluorine, oxygen and chlorine than with itself or hydrogen (*Table 9.10*) and in contrast to the hydrocarbons the silicon compounds which contain Si—Si or Si—H bonds are generally very reactive. Thus only a few hydrides of silicon, the silanes, are known (SiH_4 to S_6H_{14}) and these are spontaneously inflammable in air and rapidly decomposed by water. The silanes are analogous to the straight-chain paraffins and, from Si_2H_6 onward, contain Si—Si bonds. Catenation is also found to a limited extent in the germanes (GeH_4 to Ge_5H_{12}) but not in compounds of tin and lead.

Lastly, carbon is the only element of the group to form stable multiple bonds with itself (*Table 9.10*) and other elements. As a result, there are many

Table 9.10. ENTHALPIES OF BONDS INVOLVING EITHER CARBON OR SILICON

Bond	$\Delta H°/\text{kJ mol}^{-1}$	Bond	$\Delta H°/\text{kJ mol}^{-1}$
C—H	413	Si—H	318
C—C	346	Si—Si	222
C=C	610		
C≡C	835		
C—O	358	Si—O	452
C—F	485	Si—F	565
C—Cl	339	Si—Cl	381
C—Br	285	Si—Br	310
C—I	218	Si—I	234

carbon compounds which have no analogy elsewhere in the group. Examples are the cyanides (—CN), cyanates (—CNO), thiocyanates (—SCN) and unsaturated organic compounds: also the molecular compounds carbon suboxide, C_3O_2, carbon monoxide, CO, and carbon disulphide, CS_2.

Compounds

Hydrides

The formation of volatile hydrides is a property associated mainly with non-metals and carbon is the supreme example. Compared with the enormous number of hydrocarbons, only a few silanes and germanes are known, and the weakly electropositive metals tin and lead form only stannane (SnH_4) and plumbane (PbH_4).

For details of the preparation of hydrocarbons, reference must be made to a textbook of organic chemistry. A mixture of the silanes is obtained when magnesium silicide, Mg_2Si, is hydrolysed with aqueous acid or reacted with ammonium bromide/liquid ammonia. SiH_4 alone is prepared by the reduction of $SiCl_4$ with $LiAlH_4$ in diethyl ether at 273 K. GeH_4 and SnH_4 are made by similar reactions. The formation of PbH_4 has been demonstrated by the addition of a dilute acid to an alloy of magnesium and thorium-B (radioactive ^{212}Pb). The production of a volatile hydride of lead is shown by the transfer of radioactivity to the vapour phase.

The thermal stability of the tetrahydrides falls off steadily with increasing molecular weight. The stability towards water, however, follows the sequence:

$$CH_4 \quad > \quad GeH_4 \quad > \quad SnH_4 \quad > \quad SiH_4$$

stable	unattacked by 33% alkali	unattacked by 15% alkali	quantitative decomposition by water

This and other chemical evidence has led other workers to dispute Pauling's electronegativity values for Group IV as it is very unlikely that such great differences in reactivity as are shown by SiH_4, GeH_4 and SnH_4 are consistent with the identical values of electronegativity assigned by him to the three elements concerned (*Table 9.9*). Allred and Rochow have suggested an alternation of electronegativities in this group:

$$C = 2 \cdot 6; \quad Si = 1 \cdot 9; \quad Ge = 2 \cdot 0; \quad Sn = 1 \cdot 93; \quad Pb = 2 \cdot 45$$

These values apply to compounds in which the element is sp^3 hybridised and forms 4 covalent bonds. The order of increasing ease of hydrolysis of the tetrahydrides is seen to be that of decreasing electronegativity. The hydrolysis of SiH_4, for example, occurs easily and is catalysed by OH^- ion. It may therefore proceed by co-ordination of OH^- to the silicon (this carries a fractional positive charge because it is less electronegative than hydrogen) followed by the formation of molecular hydrogen from the proton of water and one of the negatively charged hydrogen atoms in the SiH_4 molecule. This process is less likely to occur when the central atom is any of the other elements of this group because of their greater electronegativities and hence their smaller attraction for the OH^- ion.

Oxides

(a) *Dioxides* — Some properties of these are summarised in *Table 9.11*.

CO_2 is conveniently displaced from a carbonate by treatment with dilute acid. It is slightly soluble in water to give a solution of carbonic acid:

$$H_2O + CO_2 \rightleftharpoons H_2CO_3 \overset{H_2O}{\rightleftharpoons} H_3O^+ + HCO_3^-$$

The transition in structure type from the molecular CO_2 to the three-

Table 9.11. THE DIOXIDES OF THE GROUP IV REPRESENTATIVE ELEMENTS

Compound	m.p. T/K	b.p. T/K	Structure	Properties
CO_2		194·7(subl)	Molecular	Acidic
SiO_2	1 983·2	2 863·2	3 dimensional lattice (quartz, tridymite and cristobalite)	Acidic
GeO_2	1 389·3	1 473·2	Ionic: rutile (quartz) above 1303K	Amphoteric
SnO_2		2 173·2(subl)	Ionic: rutile	Amphoteric
PbO_2	1 025·2 (decomp. under O_2 at 101·325 k Pa)		Ionic: rutile	Amphoteric

dimensional SiO_2 is clearly illustrated by the boiling points of the two compounds. The electronic description of the CO_2 molecule has been given earlier.

All three forms of SiO_2 — quartz, tridymite and cristobalite — occur in nature. Like CO_2, silica has acidic properties and reacts with basic oxides and carbonates at high temperature to give silicates.

The hydrolysis of $SiCl_4$ or the addition of acids to alkali metal silicate solutions gives hydrated silica. This can be dried to an amorphous powder (silica gel), containing about 4 per cent water. Its main use is as a drying agent and as a catalyst support.

GeO_2 and SnO_2 are prepared by treatment of the element with concentrated nitric acid: PbO_2 is made by the electrolytic oxidation of Pb^{II} salts or the reaction of Cl_2 or Br_2 with an alkaline solution of lead acetate. PbO_2 is the chief constituent of the anode of the charged lead accumulator. It is an effective oxidising agent, liberating Cl_2 from HCl and oxidising Mn^{2+} in acid solution to MnO_4^-. It cannot be classed as a peroxide because H_2O_2 is not liberated from PbO_2 by acid.

All five dioxides have acidic properties and the derived oxo-acids or their salts are known. In the case of C, Si and Ge, only the salts have been isolated. These are the carbonates (containing CO_3^{2-}), the silicates and the germanates (both meta-, GeO_3^{2-}, and ortho-salts, GeO_4^{4-}). SnO_2 on fusion with alkali gives rise to a water-soluble stannate, for example Na_2SnO_3. The addition of acid causes precipitation of hydrated stannic oxide, α-stannic acid. This dissolves easily in more concentrated acid, but it ages on standing to an unreactive insoluble form, β-stannic acid. The loss in reactivity is believed to be due to a·progressive increase in the size of colloidal particles. PbO_2 similarly interacts with alkali to form plumbates. These and the stannates have been shown by x-ray analysis to contain 6 co-ordinated metal atoms. For instance, the hydrated salt potassium 'metastannate', $K_2SnO_3 . 3H_2O$, should more properly be formulated as $K_2Sn(OH)_6$.

(b) *Monoxides* — CO can be made by the interaction of carbon and carbon dioxide at high temperatures. It behaves as a neutral compound. The metallic carbonyls are important derivatives. The electronic structure of CO has been discussed earlier.

The evidence for a monoxide of silicon is not conclusive. SiO appears to be formed around 1573 K when a mixture of SiO_2 and Si is vaporised under reduced pressure.

GeO is made by the reduction of Ge^{IV} in solution using phosphinic acid, HPH_2O_2. The preparation must be carried out in an inert atmosphere because rapid oxidation back to the tetrapositive state would otherwise occur. SnO, obtained when the metal is heated with a limited supply of air, is similarly oxidised to SnO_2. PbO (litharge) is formed when lead is heated in air.

SnO and PbO have layer structures in which each metal atom has four equidistant neighbouring oxygen atoms on one side of it. A plan view of one layer of SnO (or red, tetragonal PbO) is shown in *Figure 9.2*. In the tin compound, the Sn—Sn distance between adjacent layers is 370 pm and this is short enough for there to be some metal–metal bonding present. The yellow, orthorhombic form of lead monoxide has a puckered layer structure which has been described aptly as a 'crumpled' version of the structure of red PbO.

The monoxides of Ge, Sn and Pb are amphoteric with basic properties increasing in importance as the atomic number of the metal increases. They are the parent compounds both of the normal salts of Ge^{II}, Sn^{II} and Pb^{II} and of the oxo-acid salts known as the germanates (II) (GeO_2^{2-}), stannates (II) (SnO_2^{2-}) and plumbates (II) (PbO_2^{2-}).

(c) *Other oxides* — Carbon is unique in forming a suboxide C_3O_2. This is a gaseous compound formed when malonic acid is heated with P_2O_5. It poly-merises readily on heating and behaves towards water as the anhydride of

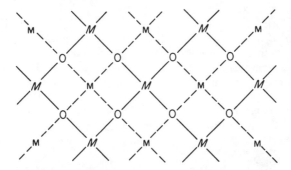

Figure 9.2. Plan of one layer of SnO *(or* PbO*) structure: O in plane; M above plane; M below plane (M = Sn or Pb). From* Progress in Stereochemistry, *vol. 4 (1969) edited by Aylett and Harris*

malonic acid. The molecule is linear, a chain of three carbon atoms being terminated at each end by an oxygen atom. The molecular structure is usually represented in terms of multiple bonds between the atoms with non-localised π-bonds at an important feature (cf. the structure of CO_2, p. 76).

Two other oxides of lead are known. Pb_3O_4 (red lead) is made by heating litharge in air to 723 K. This is formulated as $(Pb^{II})_2Pb^{IV}O_4$. Pb_2O_3, made when a soluble lead (II) salt is treated with a plumbate, also contains lead in two valency states — $Pb^{II} Pb^{IV}O_3$.

Halides

(a) *Tetrahalides* (MX_4) — All possible ones have been made except the tetrabromide and tetraiodide of lead. The non-existence of these is attributed to the strong oxidising properties of Pb^{IV} which would at once convert a heavy halide ion to free halogen. Preparative methods are illustrated by the tetrachlorides:

CCl_4 is made by chlorination of CS_2. $SiCl_4$ and $GeCl_4$ are prepared by heating the dioxide and carbon in chlorine. $SnCl_4$ is obtained by direct chlorination of the metal. $PbCl_4$ is formed as an unstable yellow liquid by the reaction between PbO_2 and concentrated HCl.

The tetrafluorides of C, Si and Ge are gaseous at ordinary temperatures: SnF_4 and PbF_4 are involatile solids, subliming at 978 K and melting at 873 K respectively. The other tetrahalides are generally volatile liquids or easily melted solids. CBr_4 and CI_4 are unstable: this is related to the steric difficulty of forming four bonds between the small carbon atom and the large halogen atoms.

A hexafluoro-complex of Si is known, namely $[SiF_6]^{2-}$. This is produced when SiF_4 is hydrolysed:

$$SiF_4 + 2H_2O = SiO_2 + 4HF$$

then
$$SiF_4 + 2F^- \rightleftharpoons [SiF_6]^{2-}$$

Other complex hexahalide ions are $[GeF_6]^{2-}$, $[GeCl_6]^{2-}$, $[SnF_6]^{2-}$ $[SnCl_6]^{2-}$, $[SnBr_6]^{2-}$, $[SnI_6]^{2-}$, $[PbF_6]^{2-}$, $[PbCl_6]^{2-}$ and $[PbBr_6]^{4-}$.

(b) *Catenated halides* — Carbon forms a large number of fluorocarbons, compounds which are noted for their chemical stability. They are made by the electrolytic generation of fluorine in the presence of hydrocarbons. Under these conditions, replacement of H by F takes place. Several chlorocarbons are also known, for example C_2Cl_6, C_3Cl_8 and C_4Cl_{10}, but these are much less stable than the corresponding fluorine compounds.

Silicon also forms a number of catenated halides:

$$Si_2F_6,\ Si_2Cl_6,\ Si_2Br_6\ and\ Si_2I_6;$$
$$Si_3Cl_8,\ Si_4Cl_{10},\ Si_5Cl_{12},\ Si_6Cl_{14}\ and\ Si_{10}Cl_{22}.$$

Here the chloro-compounds are the most numerous and the most stable. Direct reaction between the halogen and a Ca–Si alloy at elevated temperatures serves as a preparative method. These halides are easily soluble in organic solvents and hydrolysis results in the removal of all halogens to give, as one of the chief products, silico-oxalic acid, $H_2Si_2O_4$.

Ge_2Cl_6 appears to be the only similar compound formed by the elements following Si.

(c) *Dihalides* (MX_2) — These are formed by Ge, Sn and Pb.

$GeCl_2$ is obtained when $GeCl_4$ is heated with Ge. $SnCl_2$ is prepared by heating the metal in HCl. $PbCl_2$ has a low solubility in water and is readily made by precipitation from aqueous solution.

$GeCl_2$ has also been made by the low-temperature distillation of $GeHCl_3$ (prepared by the reaction of GeS with dry HCl gas). Even at 253 K, $GeCl_2$ made in this way begins to decompose to form subchlorides in which the germanium: chlorine ratio is lower than 1:2. These subchlorides are polymeric materials. In contrast, $SnCl_2$ is a well-characterised compound, melting at 515 K, with very much greater stability. It is used extensively in catalysts for polymerisation processes, as a vulcanisation accelerator, etc.

Compounds of tin (II) have structures which are generally intermediate between three-dimensional lattices and discrete molecular crystals and this reflects the nature of the bonding, which is intermediate between ionic and covalent. Thus in the solid state, $SnCl_2$ is composed of infinite Sn–Cl chains, weakly bound to adjacent chains, each tin atom being co-ordinated by a distorted octahedron of chlorines (one Cl at 267, two at 278, one at 306, one at 318 and one at 328 pm). The dichloride is an angular molecule (bond angle 95°) in the anhydrous vapour state. The tin atom has two bonding pairs and one lone pair of electrons. The dihydrate, $SnCl_2 . 2H_2O$, contains one molecule of water which is easily lost on heating and a second water molecule more strongly bound by direct co-ordination to the metal. The groups around the tin atom in $SnCl_2 . H_2O$ form a pyramidal arrangement with the fourth (non-bonding) pair of electrons on the tin atom directed towards the fourth corner of the tetrahedron.

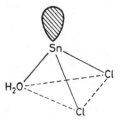

Lead (II) chloride has a characteristic structure in which each metal atom has a co-ordination number of 9. Six chlorines are arranged at the apices of an irregular trigonal prism and the remaining three are above the rectangular faces of the prism.

Tin (II) fluoride has a structure similar to that of $SnCl_2$: there are two Sn, F distances of 215, one of 245 and three longer than 280 pm. Two forms of PbF_2 are known. One, stable at high temperatures, has the fluorite structure; the other, a low-temperature modification, has the lead chloride structure. Germanium (II) fluoride is a chain-like polymer with germanium atoms linked by fluorine bridges. Each germanium is effectively surrounded by two near and two more distant fluorines and its co-ordination can be regarded as that of a distorted trigonal bipyramidal arrangement in which the fifth (one of the equatorial) position is occupied by a lone pair of electrons.

The di-iodides of Ge, Sn and Pb all have the cadmium iodide structure. This is very characteristic of compounds in which the bonding has appreciable covalent nature.

Cyanides and cyanogen

Hydrogen cyanide is a volatile, poisonous liquid (b.p. 298 K) formed when a metal cyanide is treated with sulphuric acid. It is an associated liquid due to hydrogen bonding. In aqueous solution, HCN acts as a weak acid. The metal cyanides are either ionic (NaCN, KCN, TlCN) or covalent ($Cu(CN)_2$, AgCN). These ionic cyanides have the NaCl or CsCl structure because the cyanide ion assumes pseudo-spherical symmetry by rotation. Aqueous solutions of ionic cyanides have a high pH because the weakness of HCN as an acid favours reaction to the right in

$$CN^- + H_2O = HCN + OH^-$$

Cyanide is an important ligand in many metal complexes. It has a marked stabilising effect on certain oxidation states of transition metal ions, as in $[Cu^I(CN)]_2^-$ and $[Co^{III}(CN)_6]^{3-}$. When the carbon and nitrogen atoms of one cyanide group are simultaneously co-ordinated to two different metal atoms, polymeric structures such as $(C_3H_7)_2AuCN$ are established.

Dicyanogen, $(CN)_2$, is a poisonous, colourless gas formed by heating a noble metal cyanide such as $Hg(CN)_2$. It is readily soluble in water and undergoes slow hydrolysis to a variety of products including oxamide

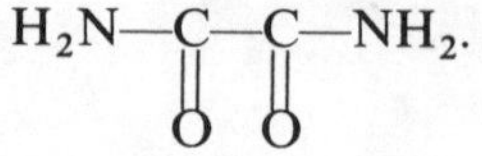

$$H_2N-\underset{\underset{O}{\|}}{C}-\underset{\underset{O}{\|}}{C}-NH_2.$$

This product indicates the presence of a C—C bond in the $(CN)_2$ molecule. Structural studies confirm a linear and symmetrical arrangement of atoms $N\overset{116}{\rule{1em}{0.4pt}}C\overset{137}{\rule{1em}{0.4pt}}C\overset{116}{\rule{1em}{0.4pt}}N$. The bond lengths are such that the molecule must be regarded as a resonance hybrid with $N\equiv C-C\equiv N$ as a major contributing form.

On heating to 673–773 K, dicyanogen polymerises to solid paracyanogen, $(CN)_n$:

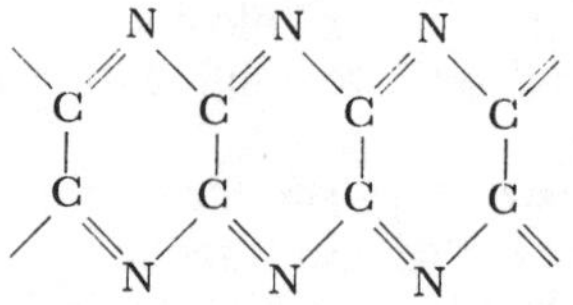

Many of the properties of dicyanogen, hydrogen cyanide and cyanide ion are similar to properties shown by the halogens. For this reason dicyanogen is often called a pseudohalogen. Several other pseudohalogens are known including dithiocyanogen $(SCN)_2$ and diselenocyanogen $(SeCN)_2$. The corresponding pseudohalogen ions are CN^-, SCN^- and $SeCN^-$, and cyanate, OCN^-, is also known.

Silicones

These are polymeric organosilicon compounds which, because of their special properties, have a number of important industrial applications. They are made from the products of hydrolysis of organochlorosilanes, $SiRCl_3$, SiR_2Cl_2 and SiR_3Cl, where R is methyl, phenyl or hydrogen.

Organochlorosilanes are themselves prepared by the reaction between alkyl or aryl halides and a copper–silicon alloy at about 570 K. When methyl chloride is used the major product is $(CH_3)_2SiCl_2$, together with smaller amounts of CH_3SiCl_3, $(CH_3)_3SiCl$ and CH_3HSiCl_2.

The next stage in the preparation of silicones is the hydrolysis of dichlorodimethylsilane, $(CH_3)_2SiCl_2$. In acid solution this is converted firstly to $(CH_3)_2SiCl(OH)$ and then to a mixture of products including linear polymers of general formula $HO[(CH_3)_2SiO]_n.(CH_3)_2SiOH$ (n varies between 3 and 9) and cyclic polymers such as $[(CH_3)_2SiO]_4$, containing non-planar rings composed of alternate Si and O atoms.

The hydrolysis products are converted to silicone oils by further polymerisation processes catalysed by acids and alkalis. The linear polymers undergo condensation, the terminal hydroxo groups on two molecules forming a new O—Si—O link by the elimination of water.

$$
\begin{array}{ccccc}
& CH_3 & & CH_3 & CH_3 \\
& | & & | & | \\
-O-Si-OH + HO-Si-O- & = & -O-Si-O-Si-O- + H_2O \\
& | & & | & | \\
& CH_3 & & CH_3 & CH_3
\end{array}
$$

The cyclic polymers break down and are converted to long-chain polymers.

A characteristic property of silicone oils is their low-temperature co-efficient of viscosity compared with hydrocarbon oils. Their main industrial use is to confer water-repellent properties on surfaces such as textiles.

Silicone rubbers are long-chain polymers with some cross-linking between the chains. They are made by the hydrolysis of $(CH_3)_2SiCl_2$ or $[(CH_3)_2SiO]_4$ followed by polymerisation. The addition of a small quantity of $(CH_3)_3$ $SiOSi(CH_3)_3$ serves to control the length of the polymer chains because it acts as a chain terminator by way of the $(CH_3)_3Si$-group. Silicone rubbers are thermally more stable than hydrocarbon rubbers and remain flexible over a wide range of temperature. These properties, together with their chemical inertness, have led to many applications.

If the hydrolysis of $(CH_3)_2SiCl_2$ is carried out in the presence of CH_3SiCl_3, extensive cross-linking accompanies polymerisation. This results in rigid polymers, known as silicone resins. These are used in the manufacture of electrically-insulating material and for rendering masonry water-repellent.

GROUP V—REPRESENTATIVE ELEMENTS

Nitrogen (N) and phosphorus (P) are non-metals. Metallic properties first become significant when the underlying shell of 18 electrons is present, that is, with arsenic (As) and antimony (Sb) and bismuth (Bi), which are increasingly metallic. Three elements of this group show allotropy and the variations in melting and boiling points reflect the changes in the structural units present in the elements (*Table 9.12*).

Each atom is three electrons short of the configuration of the nearest inert gas. Nitrogen gains three electrons in forming the nitride ion N^{3-}, found in the ionic nitrides of lithium and Group II metals. The electronegativity of the

Table 9.12. THE PROPERTIES OF THE GROUP V REPRESENTATIVE ELEMENTS

	N	P	As	Sb	Bi
Atomic number	7	15	33	51	83
Electron configuration	$2s^2\,2p^3$	$3s^2\,3p^3$	$4s^2\,4p^3$	$5s^2\,5p^3$	$6s^2\,6p^3$
m.p. T/K	63·2	317·3 (white)	1 090·2 (under press.)	903·7	544·2
b.p. T/K	77·4	553·7	883 (subl)	1 653·2	1 723·2
Electronegativity	3·0	2·1	2·0	1·9	1·9
Atomic radius r/pm	74	110	121	141	152
Ionic radius r/pm M^{3+}			58	100	102

remaining elements of the group is insufficient for them to form triply charged anions. Thus the bonding in metallic phosphides and arsenides is largely covalent.

Alternatively, the electronic shell is completed when the element forms three covalent bonds, as in the hydrides MH_3, and in most of the trihalides. The compound containing the most electropositive Group V element and the most electronegative halogen, BiF_3, has salt-like properties and hence contains the Bi^{3+} ion. The tripositive cation is probably also present in the salts of strong acids such as $Bi(ClO_4)_3.5H_2O$ and $Sb_2(SO_4)_3$, but does not appear to be stable in the presence of water because hydrolysis occurs to bismuth oxide and antimony oxide salts.

The first member (nitrogen) shows differences from the other elements. The atom is restricted to a maximum covalency of four (as in the ammonium ion) because only four orbitals are available for bonding. The heavier elements all have vacant d orbitals which can be used in bonding and show an extra valency of 5 and a maximum covalency of six. As examples, we may quote the electronic configurations of the phosphorus atom by itself and in various chemical combinations:

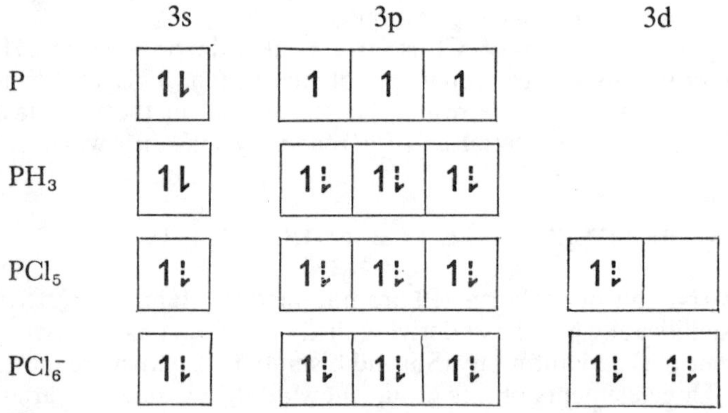

	3s	3p			3d	
P	⇅	↑	↑	↑		
PH_3	⇅	⇅	⇅	⇅		
PCl_5	⇅	⇅	⇅	⇅	⇅	
PCl_6^-	⇅	⇅	⇅	⇅	⇅	⇅

There are conflicting views on the extent to which the d orbitals of second and subsequent row elements take part in bonding in compounds where the elements shows a high covalency. Further reference is made to this point in the discussion of bonding in sulphur compounds.

The later Group V elements (P onwards) can thus show oxidation states of -3 (in MH_3), $+3$ (in MCl_3) and $+5$ (in MCl_5). Additional states arise in the case of N and P, frequently because of the formation of multiple bonds. The covalency maximum of 6 is shown in ions such as $[AsF_6]^-$, $[SbCl_6]^-$ and $[Sb(OH)_6]^-$.

Nitrogen

Four-fifths of the atmosphere consists of nitrogen but the element is not of great abundance in the combined state in the Earth's crust. This is a striking illustration of the stability of molecular nitrogen and many nitrogen compounds are in fact endothermic with respect to nitrogen itself. Nitrogen forms diatomic molecules in which the atoms are held together by a triple bond (N≡N), the strength of which is shown by the magnitude of the dissociation energy, 944·8 kJ mol^{-1}. Molecular nitrogen is therefore chemically rather inert at ordinary temperatures.

Direct reaction with hydrogen to form ammonia is favoured by high pressures and relatively low temperatures:

$$\tfrac{1}{2}N_2 + \tfrac{3}{2}H_2 \rightleftharpoons NH_3; \ \Delta H_f^\circ = -46\cdot2\,\text{kJ mol}^{-1}$$

This reaction is the basis of the Haber process.

Combination with oxygen to produce NO, nitric oxide, takes place to a slight extent at high temperatures, although even at 3473 K the yield of NO is only 4·4 per cent.

Nitrogen reacts directly with many other elements at high temperatures. Ionic nitrides are formed by Li, by all Group II metals and by Th (Th_3N_4). These react readily with water to form metal hydroxides and ammonia and their composition and properties are consistent with the presence of N^{3-} ions. Volatile covalent nitrogen compounds are formed with the non-metals H, C, F, Cl and O. Also non-volatile covalent substances are known for the elements of the boron group (except Tl) and for Si and P. Boron nitride, BN, is of interest because it has the same hexagonal structure as graphite. A number of transition metal nitrides are known—these are of an interstitial nature.

Compounds

Hydrides (MH$_3$)

These are formed by all five elements.

Ammonia and its derivatives are of great chemical importance. NH_3 acts as a weak base in aqueous solution. It combines with protonic acids to yield ammonium salts. These salts strongly resemble the alkali metal salts in their solubilities and structures (the radius of NH_4^+ is 143 pm compared with that of Rb^+ of 149 pm). Ammonium compounds are thermally unstable and the products of decomposition are dependent on the nature of the anion present. If it cannot be reduced, ammonia is formed:

$$NH_4Cl \rightleftharpoons NH_3 + HCl$$

If the anion has oxidising properties, an oxidation product of NH_3 is obtained:

$$NH_4NO_2 = N_2 + 2H_2O$$
$$NH_4NO_3 = N_2O + 2H_2O$$

Many complexes are known in which NH_3 acts as a ligand molecule. Metal amides, containing NH_2^- ions, are formed primarily by the representative elements of Groups I and II. Preparation is effected by heating the metal in gaseous NH_3.

Ammonia is relatively stable towards many oxidising agents. It reacts with atmospheric oxygen above 773 K in the presence of a platinum catalyst to form NO. This is the first stage in the production of nitric acid from ammonia (Ostwald process).

Hydrazine, N_2H_4, and hydroxylamine, NH_2OH, are two important derivatives of NH_3. Hydrazine is prepared by the oxidation of NH_3:

$$NH_3 \quad + NaOCl \qquad = NH_2Cl \quad + NaOH$$
$$NH_2Cl + NH_3 + NaOH = H_2N.NH_2 + NaCl + H_2O$$

Hydroxylamine can be made by the electrolytic reduction of low concentrations of nitrate ion in 50 per cent H_2SO_4 at an amalgamated lead cathode:

$$HO.NO_2 + 6e^- + 6H^+ = HO.NH_2 + 2H_2O$$

Both compounds are bases but they are weaker bases than ammonia. The following equilibrium constants illustrate this:

$$NH_3 \quad + H_3O^+ \overset{K_1}{\rightleftharpoons} NH_4^+ \quad + H_2O \quad K_1 = 1 \cdot 8 \times 10^{-5}$$

$$N_2H_4 \quad + H_3O^+ \overset{K_2}{\rightleftharpoons} N_2H_5^+ \quad + H_2O \quad K_2 = 8 \cdot 5 \times 10^{-7}$$

$$NH_2OH + H_3O^+ \overset{K_3}{\rightleftharpoons} NH_3OH^+ + H_2O \quad K_3 = 6 \cdot 6 \times 10^{-9}$$

Phosphine, PH_3, is made by reaction of phosphonium iodide, PH_4I, with alkali:

$$PH_4I + H_2O = PH_3 + H_3O^+ + I^-$$

PH_4I itself is prepared by the interaction of white phosphorus and iodine in carbon disulphide solution, followed by evaporation of the solvent and treatment of the residue with small amounts of water. PH_3 and its derivatives are generally less stable than NH_3 and its derivatives. Phosphine is also a weaker base and a stronger reducing agent than NH_3.

PH_3 and substituted phosphines form a large number of co-ordination compounds, especially with the platinum group of metals. In these complexes the lone pair on the phosphorus atom is donated to the metal. An additional process which increases the stability of the complex is π-bonding involving the d orbitals of the phosphorus atom. PH_3 is thus acting as an electron donor as well as an electron acceptor whereas NH_3 can act only as an electron donor because N has no vacant d orbitals.

Diphosphane, P_2H_4, is the analogue of N_2H_4, and is formed in small amounts when phosphine is prepared.

AsH_3, SbH_3 and BiH_3 are increasingly unstable and have strong reducing properties. The existence of BiH_3 (bismuthine) can be demonstrated by using a radioactive isotope of the metal. This is incorporated in a Bi–Mg alloy, which forms BiH_3, a volatile compound detectable by its radioactivity, on treatment with acid.

The ammonia molecule is pyramidal with an H—N—H bond angle of 106°45′. In the hydrides of the other elements of this group the H—M—H angle is much smaller: PH_3, 93°50′; AsH_3, 91°35′; SbH_3, 91°30′. This decrease is associated with a decrease in the electronegativity of the central atom. In passing from NH_3 to PH_3 for example, the bonding pairs of electrons are less strongly attracted by the phosphorus than by the nitrogen atom. As a result the repulsion between the lone pair on the central atom and the bonding pairs becomes relatively more important and the bond angle closes up. It is interesting to note that for AsH_3 and SbH_3 the bond angle is very close to that expected (90°) for bonds involving pure p orbitals. An alternative interpretation of the observed decrease in bond angle in the trihydrides is therefore that the nitrogen atom in NH_3 is sp^3 hybridised whilst the orbitals used in bonding by the heavier atoms show an increasing amount of p character.

The thermal stability and basic strength of the hydrides is greatly increased by the replacement of hydrogen by groups of lower electronegativity, such as alkyl groups. For example, tetra-alkyl ammonium hydroxides have similar properties to the caustic alkalis. The organic derivatives of phosphine and the hydrides of As and Sb are similarly more stable than the parent hydrides.

Oxides and oxo-acids

The binary compounds between nitrogen and oxygen provide more examples of multiple bond formation. The five well-characterised oxides are: N_2O,

Table 9.13. THE OXIDES OF NITROGEN AND PHOSPHORUS

Compound	Physical state (at 293·2K)	Structure	Properties
N_2O	gas	molecular	neutral oxide
NO	gas	molecular	neutral oxide
N_2O_3	gas	molecular	acid anhydride
N_2O_4	liquid (b.p. 294·3 K)	molecular	acid anhydride
N_2O_5	solid	ionic	acid anhydride
P_4O_6	solid (m.p. 297 K)	molecular	acid anhydride
P_4O_{10}	solid	molecular; hexagonal	acid anhydride
$(P_4O_{10})_x$	solid	polymeric; orthorhombic 1	acid anhydride
$(P_4O_{10})_x$	solid	polymeric; orthorhombic 2	acid anhydride
$(PO_2)_n$	solid	—	acid anhydride

NO, N_2O_3, N_2O_4 and N_2O_5. These correspond with the formal oxidation states for nitrogen of $+1$, $+2$, $+3$, $+4$ and $+5$ respectively. Physical properties are summarised in *Table 9.13*.

Dinitrogen oxide, N_2O, is prepared by the thermal decomposition of NH_4NO_3. It is formally the anhydride of hyponitrous acid, $H_2N_2O_2$, but it does not react with water to form this. It is a linear molecule and its structure may be described in terms of the resonating forms

$$\overset{-}{N}=\overset{+}{N}=O \quad \text{and} \quad N\equiv\overset{+}{N}-\overset{-}{O}$$

Since N_2O is isoelectronic with CO_2, the electronic descriptions of the two are similar.

Nitric oxide, NO, is synthesised directly from the elements. It is a neutral molecule which has remarkable stability considering that it contains an odd number of electrons. NO dimerises to the diamagnetic N_2O_2 in the liquid state. The molecular-orbital description of the NO molecule has been given earlier.

Dinitrogen trioxide, N_2O_3, is prepared by the reduction of nitric acid with As_2O_3. It reacts with water to form nitrous acid, HNO_2, and is accordingly classed as an acid anhydride. It behaves at room temperature as a mixture of NO and NO_2; association to the diamagnetic N_2O_3 molecules is more extensive the lower the temperature. Resonance structures of the type

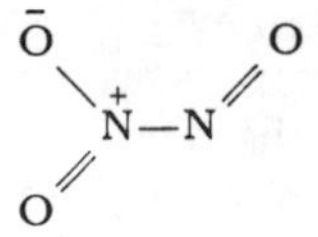

have been suggested.

Dinitrogen tetraoxide, N_2O_4, the mixed anhydride of HNO_2 and HNO_3, is prepared by reaction of NO with O_2 or by the heating of heavy metal nitrates:

$$2Pb(NO_3)_2 = 2PbO + 4NO_2 + O_2$$

Increasing use is made of this oxide as a solvent medium for inorganic reactions. In the vapour state, dissociation to the brown NO_2 monomer takes place to an extent which depends on the temperature. Like NO, NO_2 is paramagnetic (it contains 17 valency electrons) and its structure has been described in terms of resonance between

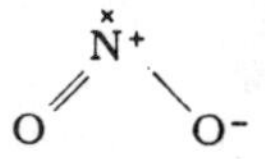

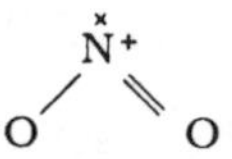

or of localised σ-bonds and a delocalised π-bond.

Dinitrogen pentaoxide, N_2O_5, is the anhydride of nitric acid, from which it may be prepared by dehydration using P_4O_{10}. It is molecular in the vapour state but the solid has been shown to have an ionic structure, being composed of nitryl, NO_2^+, and nitrate, NO_3^-, ions.

The three best-known oxo-acids of nitrogen are given in *Table 9.14*, together with their characteristic properties.

Salts of the unstable hyponitrous acid, $H_2N_2O_2$, may be obtained by reacting hydroxylamine hydrochloride with ethyl nitrite and ethanolic sodium ethoxide

$$NH_2OH + C_2H_5NO_2 + 2C_2H_5ONa = Na_2N_2O_2 + 3C_2H_5OH$$

and by reduction of a nitrite with sodium amalgam

$$2NO_2^- + 4Na + 2H_2O = N_2O_2^{2-} + 4Na^+ + 4OH^-$$

Treatment of the silver salt, which is insoluble in water, with ethereal hydrochloric acid yields a solution from which crystals of the acid may be prepared. Decomposition of the acid both in the solid state and in aqueous solution yields dinitrogen oxide.

From spectroscopic evidence the structure of the anion is

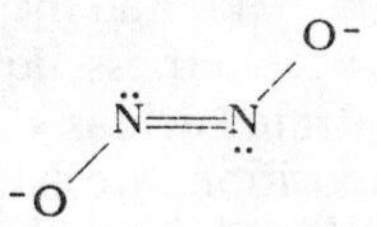

$$\overset{-O}{\underset{-O}{\diagup}}\ddot{N}=\!\!=\!\ddot{N}\diagup^{O^-}$$

Aqueous solutions of nitrous acid, HNO_2, are unstable, rapidly decomposing to NO and HNO_3, especially on heating. Salts may be prepared either by thermal decomposition of the corresponding nitrate or thermal reduction of the nitrate with a reagent such as lead or carbon. The angular structure of

Table 9.14. THE OXO-ACIDS OF NITROGEN AND PHOSPHORUS

Acid	Formula	Oxidation state	Properties
Hyponitrous	$H_2N_2O_2$	+1	Unstable, weak, dibasic and reducing
Nitrous	HNO_2	+3	Unstable, weak, monobasic and oxidizing
Nitric	HNO_3	+5	Strong, monobasic and oxidizing
Phosphinic	HPH_2O_2	+1	Strong, monobasic and strongly reducing
Phosphonic	H_2PHO_3	+3	Moderately strong, dibasic and strongly reducing
Diphosphonic (IV)	$(HO)_2OP\text{--}PO(OH)_2$	+4	Moderately strong, tetrabasic. No reducing properties and resists oxidation
Orthophosphoric	H_3PO_4	+5	Moderately strong and tribasic. Stable

Also known for phosphorus are the condensed acids:
Diphosphonic $H_2P_2H_2O_5$;
Pyrophosphoric ($H_4P_2O_7$) and metaphosphoric $(HPO_3)_n$, where $n = 3, 4, 6$ etc.
Three per-acids are known:
Peroxonitric acid HNO_4, peroxomonophosphoric acid, H_3PO_5, and peroxodiphosphoric acid, $H_4P_2O_8$.

the ion may be interpreted as follows. In NO_2^- the nitrogen atom is sp^2 hybridised, two of the hybrids forming a σ-bond with each oxygen and the third containing the lone pair. The valency octet around nitrogen is completed by the formation of one non-localised π-bond:

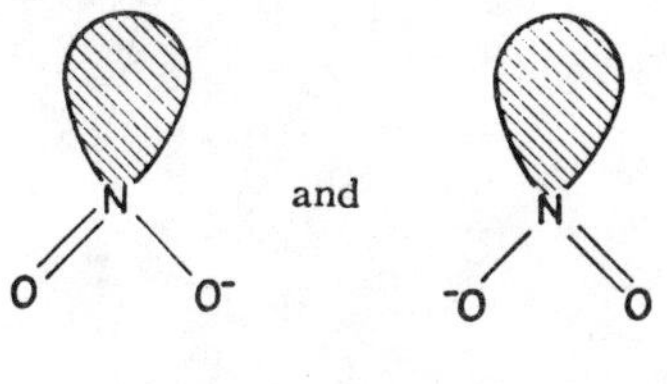

Nitric acid, HNO_3, prepared by synthetic methods involving the catalytic oxidation of ammonia to nitric oxide and then absorbing the oxide in water in the presence of oxygen, is a colourless liquid in the anhydrous form. In aqueous solution the concentrated acid is generally yellow in colour (arising from photochemical decomposition into NO_2, O_2 and H_2O).

In the vapour phase there is considerable hydrogen bonding between the nitric acid molecules. The molecular structure is

and the nitrogen atom can be regarded as sp^2 hybridised. In the anion, which is planar and symmetrical, resonance forms of the type

may be postulated where the nitrogen is again sp^2 hybridised and forms three σ-bonds and one non-localised π-bond.

It is interesting to note that the anions BO_3^{3-}, CO_3^{2-} and NO_3^-, which contain a first short period element as central atom, are all planar. The oxo-anions of the second period elements are tetrahedral (SiO_4^{4-}, PO_4^{3-}, SO_4^{2-} and ClO_4^-). In NO_3^-, the π-bond must involve a p electron from the nitrogen and so the number of σ-bonds is restricted to three. In PO_4^{3-}, d orbitals are available for π-bonding and four tetrahedrally distributed σ-bonds are formed. Another difference between the oxo-anions of the first and second short periods is that carbonates and nitrates are found only as mononuclear ions, but condensed anions are easily formed by silicates and phosphates.

The oxides of phosphorus show very little resemblance in their physical properties to those of nitrogen. This is because in its compounds with oxygen phosphorus tends to form polymeric structures.

Phosphorus (III) oxide, P_4O_6, is the chief product when the element is burned in a limited supply of air. It is the anhydride of phosphonic acid, H_2PHO_3. The structural unit is a molecule with four P atoms at the apices of a tetrahedron, each P atom being joined to the other three by oxygen bridges. The P—O bond length is 165 pm.

P_4O_6

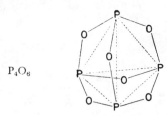

Phosphorus (V) oxide, P_4O_{10}, is the product of combustion of the element in excess air. It is the anhydride of phosphoric acid, H_3PO_4. This oxide is polymorphic and at least three crystalline and two amorphous forms have been recognised. One of the crystalline polymorphs contains discrete P_4O_{10} molecules (this molecule is also present in the vapour of the oxide). These are related to P_4O_6 by the addition of an extra oxygen to each phosphorus. The four additional oxygen atoms are bound more firmly than the other six. The evidence for this comes from the shorter P—O distance of 139 pm compared with 162 pm for the bond between P and the bridging O. The shorter bonds therefore appear to be multiple and so in P_4O_{10} the valency shell of the phosphorus atom must be expanded beyond 8 electrons.

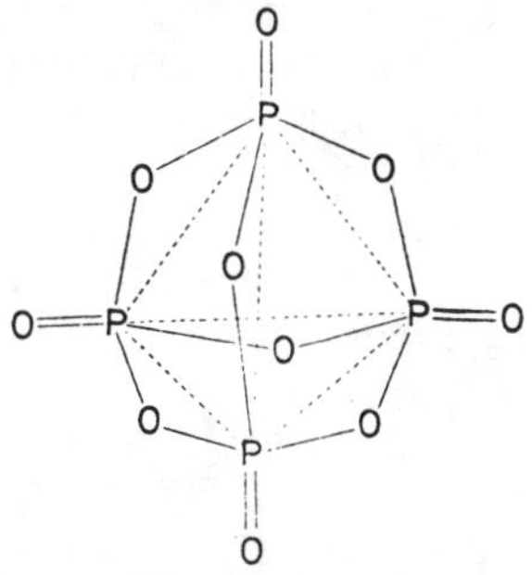

P_4O_{10}

The other two crystalline forms are orthorhombic and have polymeric structures. Each P atom is tetrahedrally surrounded by four O atoms and three of these are shared with adjacent tetrahedra to give sheet polymers of infinite extent. One of these polymorphs has interlocking rings containing 6 P and 6 O atoms: the second contains larger rings of 10 P and 10 O atoms. The P—O bonds in such crystals are largely covalent.

$(PO_2)_n$, commonly called phosphorus tetroxide, is made by heating P_4O_6 *in vacuo* above 483 K. Its properties suggest that it also has a polymeric structure.

Several oxo-acids of phosphorus are known. Their properties are summarised in *Table 9.14*.

In a systematic approach to these compounds it is of value to consider how they arise by the replacement of successive hydrogen atoms with oxygen in the phosphonium cation, PH_4^+.

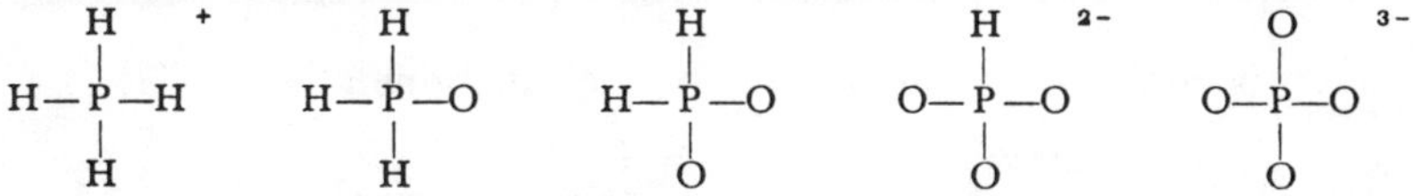

As oxygen replaces hydrogen around the phosphorus atom there is a change from basic to acidic properties. The number of hydrogen atoms combined with oxygen determines the basicity of the acid for the P—H group shows no tendency to lose a proton. This is in accordance with the much lower electronegativity of P compared with O. Also it is evident that reducing properties are possessed by the acid if it contains a P—H bond.

Both phosphinic acid, HPH_2O_2, and phosphonic acid, H_2PHO_3, are colourless, crystalline solids prepared by the following reactions:

$$8P + 3Ba(OH)_2 + 6H_2O = 3Ba(PH_2O_2)_2 + 2PH_3$$
$$\downarrow H_2SO_4$$
$$2HPH_2O_2 + BaSO_4$$
$$PCl_3 + 3H_2O = H_2PHO_3 + 3HCl$$

They have strong reducing properties associated with the P—H bonds and are oxidised to form phosphate. The P—H bonds in both acids are non-ionisable as shown by deuterium exchange experiments.

Orthophosphoric acid, H_3PO_4, which may be prepared by hydrolysing P_4O_{10}, is a colourless crystalline solid in the pure state. Pyro- and polyphosphoric acids are also known; these are condensed forms of phosphoric acid, containing more than one phosphorus atom, linked via oxygen, and are prepared from the ortho acid by heating; loss of water ultimately occurs.

The structures of all the phosphoric acids are based on PO_4 tetrahedral units. Discrete PO_4^{3-} ions occur in orthophosphates, whilst in pyro- and triphosphate ions, two and three tetrahedra respectively are attached by shared oxygens.

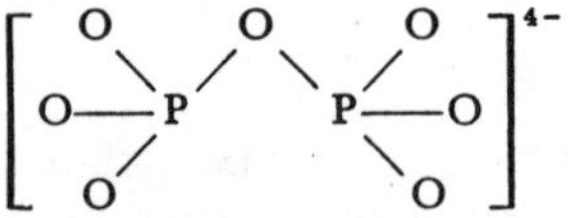

Pyrophosphate 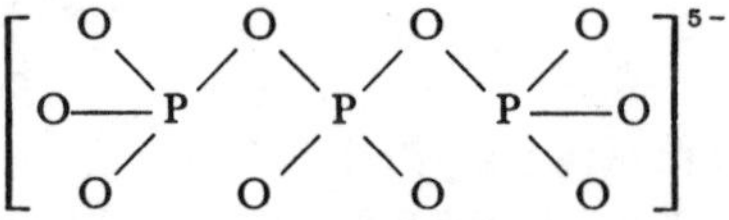   Triphosphate

In cyclotriphosphates, the process of sharing of oxygens has gone further with the formation of ring structures:

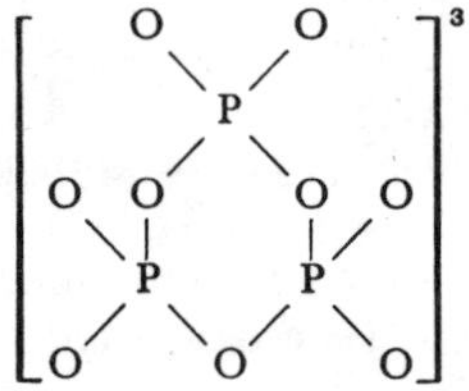

Cyclotriphosphate

These processes may be summarised by the equations:

$$2HPO_4^{2-} = P_2O_7^{4-} + H_2O$$
$$H_2PO_4^- + 2HPO_4^{2-} = P_3O_{10}^{5-} + 2H_2O$$
$$xH_2PO_4^- = (PO_3)_x^{x-} + xH_2O$$

The above reactions do not occur when orthophosphates are acidified but are brought about by heating. A wide range of polyphosphate ions, of either chain or ring structure, exists in the crystalline phosphates.

The oxides of As^{III} and Sb^{III} correspond with P_4O_6. As_4O_6, prepared by burning arsenic in air, has a similar structure, being composed of tetrahedral molecules. Sb_4O_6 is made by reaction between $SbCl_3$ and Na_2CO_3. It is dimorphic. One form consists of discrete Sb_4O_6 molecules: the other is composed of double chains:

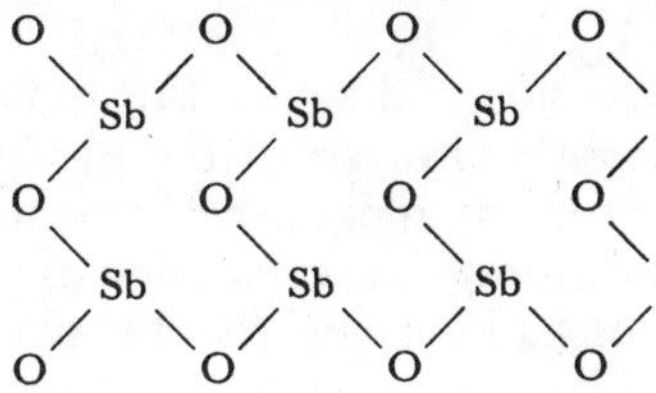

These two oxides are amphoteric giving rise to As^{III} and Sb^{III} compounds on reaction with strong acids and to arsenate (III) (AsO_3^{3-}) and antimonate (III) (SbO_3^{3-}) with alkalis.

Bismuth trioxide, Bi_2O_3, has basic properties only and crystallises in several forms. One of these is cubic and has an ionic structure. The trend of nonmetallic to metallic character of the elements is therefore accompanied by a change from molecular to polymeric to ionic structure for the lower oxides.

As_2O_5 and Sb_2O_5 are prepared by oxidation of the elements with nitric acid. Their structures are unknown. These are oxidising agents and they have pronounced acidic character, as shown by reactions with alkali to form arsenates (V) (AsO_4^{3-}) and antimonates (V) $[Sb(OH)_6]^-$ respectively. The change from tetrahedral to octahedral co-ordination by oxygen takes place on passing from As to Sb.

Bismuthates containing Bi^V are known and have powerful oxidising properties. The pentaoxide has not been obtained as a pure compound.

Halides

Two series of compounds, the tri- and penta-halides, are known for phosphorus and the heavier elements. Trihalides only are formed by nitrogen because its covalency maximum is four.

NF_3, prepared by the electrolysis of molten ammonium hydrogen difluoride, NH_4HF_2, is a colourless gas (b.p. 144 K) which is notable for its stability and resistance to hydrolysis.

NCl_3, made by electrolysis of saturated ammonium chloride solution, is an unstable yellow oil which is readily decomposed by water. The electronegative nitrogen atom acts as an electron-pair donor towards the hydrogen of water and hydrolysis proceeds with the successive replacement of Cl by H atoms:

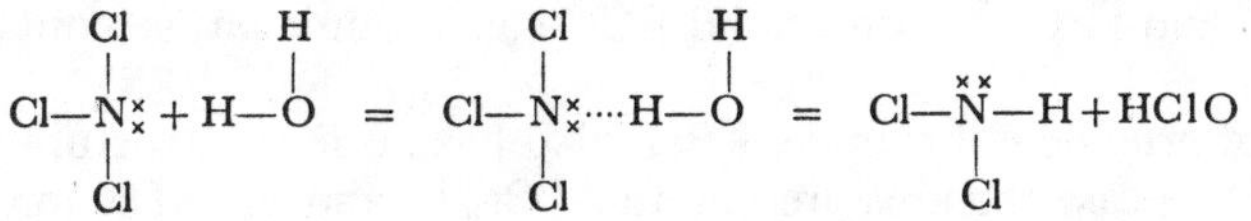

$$NHCl_2 + H_2O = NH_2Cl + HClO$$
$$NH_2Cl + H_2O = NH_3 + HClO$$

The tribromide and triiodide of nitrogen have been prepared only in the form of their very unstable ammoniates, such as $NBr_3.6NH_3$.

The remaining elements each form the four binary trihalides and a number of mixed trihalides ($PClF_2$, PCl_2F, $PBrF_2$, PBr_2F and SbI_2Br). The binary compounds, with the exception of the ionic BiF_3 (m.p. 1000 K), are all covalent molecular substances. Hydrolysis of the phosphorus and arsenic trihalides occurs with the production of the corresponding oxo-acid. Here, the P or As atom acts as an electron-pair acceptor and its valency shell is expanded to 10 electrons by accepting a lone pair from the oxygen atom of water. For example,

$$Cl-P^x_x\overset{Cl}{\underset{Cl}{|}}+{}^x_xO-H\ \ =\ \ Cl-P_{xx}\overset{Cl}{\underset{Cl}{|}}\overset{H}{\underset{}{\diagup}}O-H\ \ =\ \ P(OH)Cl_2+HCl$$

$$P(OH)Cl_2+H_2O = P(OH)_2Cl+HCl$$
$$P(OH)_2Cl+H_2O = P(OH)_3\ \ +HCl$$

The trihalides of antimony and bismuth (except BiF_3) are reversibly decomposed to form insoluble antimony oxide and bismuth oxide compounds

e.g. $$BiCl_3 + H_2O \rightleftharpoons ClBiO + 2HCl$$

The covalent trihalides form pyramidal molecules. The shape arises from a tetrahedral arrangement of three bonding and one non-bonding pairs of electrons around the central atom. The compounds are generally poor conductors of electricity in the liquid state in accordance with their covalent character. AsF_3 and SbF_3 are exceptional in that they show, in the fused state, a specific conductivity similar to that of BrF_3 and IF_5. This is attributed to self-ionisation:

$$2AsF_3 \rightleftharpoons AsF_2^+ + AsF_4^-$$

Phosphorus pentachloride is the best-known of the pentahalides. It is a solid at ordinary temperatures which undergoes thermal decomposition:

$$PCl_5 \rightleftharpoons PCl_3 + Cl_2$$

This dissociation has been thoroughly studied as an outstanding example of a reversible reaction. In the vapour state PCl_5 exists as a trigonal bipyramidal molecule and the phosphorus atom is accordingly sp^3d hybridised (PF_5, AsF_5, $SbCl_5$ and PCl_2F_3 also have the same molecular shape). In the fused state the compound shows very little electrical conductivity but greater conductivity is shown by its solutions in polar solvents. This is due to the presence of $[PCl_4]^+$ and $[PCl_6]^-$ ions. Solid PCl_5 has an ionic lattice containing these two ions.

Phosphorus pentabromide, PBr_5, also has an ionic structure in the solid state: in this case the ions present are $[PBr_4]^+$ and Br^-. The ion $[PBr_6]^-$ is known to be present in solutions of PBr_5 in a polar solvent such as methyl cyanide.

Antimony pentachloride, $SbCl_5$, is prepared by the chlorination of $SbCl_3$, but the corresponding pentachloride of arsenic is unknown. The bonding in a pentahalide involves a d orbital of the Group V atom, and the formation of such a compound requires the promotion of an electron from an s or a p orbital of lower energy. There is an appreciably larger energy difference between the 4d and the 4s and 4p orbitals in arsenic than between the corresponding orbitals in the third quantum shell in phosphorus or the fifth quantum shell in antimony. As a result, arsenic is more reluctant to show pentavalency than either phosphorus or antimony.

The cation $AsCl_4^+$ is known, being present in the compound $AsCl_5.PCl_5$ (correctly formulated as $[AsCl_4]^+[PCl_6]^-$). Also known are the ionic compounds $[AsCl_4]^+[SbCl_6]^-$ and $[AsCl_4]^+[AsF_6]^-$. Indeed, recent studies on the halides of Group V have shown how important ions of this type are in their chemistry.

The pentabromides and pentaiodides of As, Sb and Bi are not formed owing to the oxidising tendencies of the $+5$ state of these three elements.

Phosphonitrilic halides

These are the polymeric compounds containing P, N and a halogen; they show several unusual features in their properties and structure.

The chlorides are synthesised by the reaction between PCl_5 and NH_4Cl in an inert solvent such as *sym*-tetrachloroethane.

$$nPCl_5 + nNH_4Cl = (PNCl_2)_n + 4nHCl$$

Removal of the solvent by distillation *in vacuo* followed by fractional distillation under reduced pressure yields first of all cyclic $[PNCl_2]_3$ and then $[PNCl_2]_4$.

These are acid chlorides which undergo hydrolysis with the replacement of some or all of the chlorines by hydroxyl groups. As they are insoluble in water, hydrolysis is best carried out by shaking their ethereal solutions with water. $[PN(OH)_2]_3$ is obtained from $[PNCl_2]_3$ and $[PN(OH)_2]_4.2H_2O$ from $[PNCl_2]_4$.

The trimer has a planar ring composed of alternate P and N atoms. The P—N bonds are all equal in length and much shorter (165 pm) than the single P—N bond length observed in other compounds (178 pm). This suggests that resonance must occur between the structure

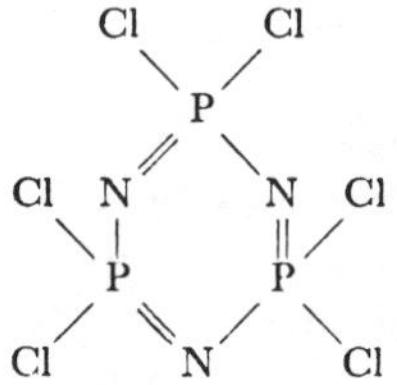

and a second structure in which the three double bonds are in the alternative positions. The $[PNCl_2]_3$ ring accordingly has 'aromatic' character and the chemical reactions of the trimer have been interpreted on this basis.

$[PNCl_2]_4$ is also a cyclic polymer containing alternate P and N atoms in a puckered 8-membered ring.

Other polymers of higher molecular weight have been isolated from the reaction mixture of PCl_5 and NH_4Cl. These appear to be linear. Further polymerisation takes place on heating phosphonitrilic chlorides to 523 K to 623 K with the production of a rubber-like polymer of high molecular weight. Practical applications of this are limited by the chemical reactivity of the chlorine atoms.

Phosphonitrilic fluorides and bromides have been synthesised but no iodides have been reported.

GROUP VI—REPRESENTATIVE ELEMENTS

These elements — oxygen (O), sulphur (S), selenium (Se), tellurium (Te) and the radioactive polonium (Po) — show systematic changes in chemical properties with increase in atomic number. Thus O and S are non-metals, Se and Te are semi-conductors and Po has metallic character. The structure of the elements changes strikingly from diatomic molecules, through ring and chain molecules to a metallic lattice.

The electronic configuration common to all is two short of that of the nearest noble gas. In many of their compounds the atoms attain a completed octet of electrons. There are several ways in which this can take place.

(a) The divalent anion X^{2-} is formed by gaining two electrons. This is an endothermic process, hence the positive sign for electron affinity in *Table 9.15*;

Table 9.15. THE PROPERTIES OF THE GROUP VI REPRESENTATIVE ELEMENTS

	O	S	Se	Te	Po
Atomic number	8	16	34	52	84
Electron configuration	$2s^2\,2p^4$	$3s^2\,3p^4$	$4s^2\,4p^4$	$5s^2\,5p^4$	$6s^2\,6p^4$
m.p. T/K	54·3	392·1*	490·6†	723·0	
b.p. T/K	90·2	717·8	958·0	1 663·2	
Electronegativity	3·5	2·5	2·4	2·1	
Electron affinity $(X(g) \rightarrow X^{2-}(g))\ \Delta H_E^{\ominus}/kJ\ mol^{-1}$	+639·7	+395·4	+422·6		
Atomic radius r/pm	74	104	117	137	152
Ionic radius r/pm X^{2-}	140	184	198	221	

*For monoclinic sulphur   †For grey selenium

whilst energy is released by the acquisition of the first electron, the addition of the second is opposed by Coulombic repulsion and requires the absorption of a greater amount of energy. Compounds containing X^{2-}, namely oxides, sulphides, etc., are considered later.

(b) The anions OH^- and SH^- are formed. The hydroxide ion, OH^-, is much more stable than the hydrogensulphide ion SH^- (for example, hydroxides are stable but hydrogensulphides are decomposed to H_2S by boiling their aqueous solution). Compounds containing SeH^- or TeH^- have not been isolated.

(c) Two covalent bonds are formed as in the hydrides H_2O, H_2S, H_2Se and H_2Te; hydrogen peroxide, H_2O_2, and persulphides, H_2S_2, etc.; the halides OF_2, SCl_2, etc.; and organic derivatives such as the ethers R_2O and thioethers R_2S. The allotropy shown by S and Se is attributable to ring or straight-chain structures wherein each atom is covalently bound to two neighbours. Sulphur in particular shows a marked tendency for catenation and there are many sulphur compounds which have no analogy elsewhere in the group. These include the polysulphur dichlorides, S_nCl_2 where $n = 3$ to 6, and the thionic acids, $H_2S_nO_6$ where $n = 2$ to 6. Selenium has a smaller tendency to form compounds with Se—Se bonds. Thus Se_2Cl_2 is less stable than S_2Cl_2 and there are no Se analogues of the thionic acids.

(d) The formation of one double bond. The property is shown by oxygen in many compounds. For instance,

phosgene,
$$\begin{array}{c} Cl \\ \diagdown \\ \diagup \\ Cl \end{array} C{=}O;$$
urea,
$$\begin{array}{c} NH_2 \\ \diagdown \\ \diagup \\ NH_2 \end{array} C{=}O;$$

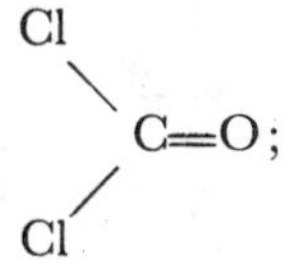

acetic acid,
$$CH_3 . C \begin{array}{c} {\diagup\!\!\diagup} O \\ \diagdown \\ OH \end{array};$$
the sulphoxides,
$$\begin{array}{c} R \\ \diagdown \\ \diagup \\ R \end{array} S{=}O;$$

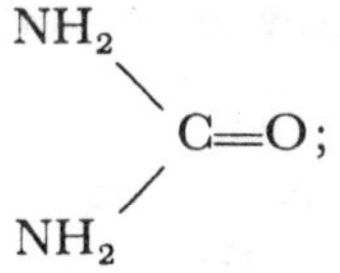

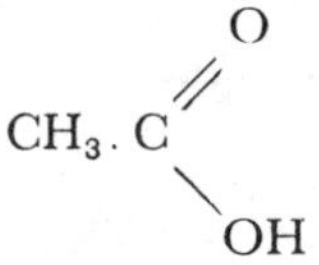

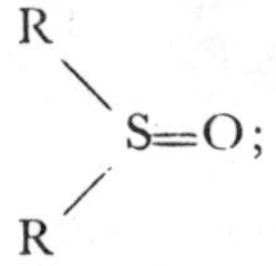

and the sulphones,
$$\begin{array}{c} R \\ \diagdown \\ \diagup \\ R \end{array} S \begin{array}{c} {\diagup\!\!\diagup} O \\ {\diagdown\!\!\diagdown} O \end{array}$$

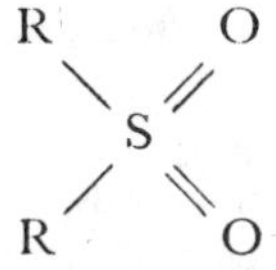

A number of sulphur compounds are also known, such as

thiourea,
$$\begin{array}{c} NH_2 \\ \diagdown \\ \diagup \\ NH_2 \end{array} C{=}S;$$

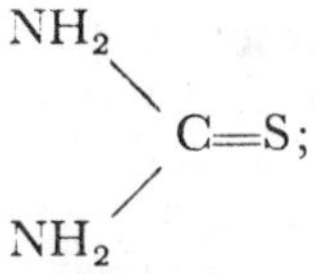

and ethanethionic acid,
$$CH_3 . C \begin{array}{c} {\diagup\!\!\diagup} S \\ \diagdown \\ OH \end{array}$$

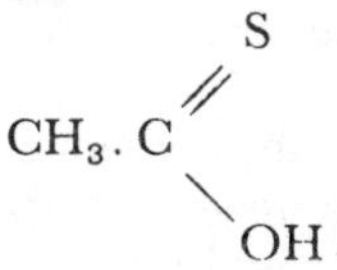

(e) The acceptance of a pair of electrons donated by another atom. The tendency for this to occur diminishes as the group is descended, that is, as the electronegativity decreases. With oxygen, this behaviour is found in the amine and phosphine oxides, $R_3N \rightarrow O$ and $R_3P \rightarrow O$ respectively. In aqueous solution alkali metal sulphides can react with sulphur to form polysulphide ions which may be regarded as formed by the donation of an electron pair from S^{2-} to an S atom to complete the valency octet of the latter. Poly-selenides and polytellurides are produced in a similar way and can be formu-lated in like manner.

Oxygen is a first short period element and hence is restricted to a covalency maximum of 4. Basic beryllium acetate, $Be_4O\,(CH_3COO)_6$, is an example of 4-covalent oxygen. More commonly, oxygen has a covalency of 3, as in the oxonium ion H_3O^+. Here the oxygen atom forms two covalent bonds with the hydrogens and uses a lone pair for co-ordination of the proton.

S, Se and Te can form up to 6 covalent bonds by the expansion of the valency shell to include d orbitals. Examples are the hexafluorides, SF_6, SeF_6 and TeF_6:

	3s	3p			3d	
Ground state of S	1↧	1↧	1	1		
Excited state	1	1	1	1	1	1
S in SF_6	1↧	1↧	1↧	1↧	1↧	1↧

These three elements can show a variety of oxidation states, namely -2, $+2$, $+4$ and $+6$. The best known compounds are those in which the Group VI element shows an oxidation state of $+4$ or $+6$ in combination with oxygen or the halogens.

On the grounds of orbital size and energy, it appears at first sight unlikely that the 3d orbitals in sulphur are suitable for participation in bond-forma-tion. In general, the atomic orbitals involved in bonds from a given atom are of similar sizes. In the free sulphur atom, however, the 3d orbitals are larger and more diffuse than the 3s and the 3p orbitals. Again, the promotion energy from the s^2p^4 ground state to the $s^1p^3d^2$ excited state is large and this would also appear to preclude the involvement of d orbitals. It appears that we should not include d orbitals in our description of the bonding in a mole-cule like SF_6 unless the properties of these orbitals are modified so much by the molecular environment as to allow them to take part.

As regards orbital size, the ligand field contraction theory provides a possible mechanism whereby d orbitals can contribute to the bonding. This theory is based on the fact that the high-valent states of second-row elements are found when they are combined with strongly electronegative elements or groups. Bonds have a marked polarity and there is effectively a withdrawal of electron density from the central atom leaving it with a formal positive charge. This positive charge causes all the atomic orbitals to contract, the more diffuse (d) orbitals contracting by a greater amount than the less diffuse (s and p). By this mechanism, d and s and p orbitals can approach one another in size and all can participate in bonding. A similar line of argument can be developed to conclude that the 3s, 3p and 3d orbitals are all rather closer in energy in the combined than in the free atom.

The same approach can be made to the bonding in high-covalent molecules of other second-row elements, for example, in PCl_5.

Some degree of π-bonding through a back-donation mechanism is to be expected in all compounds between second-row elements and highly electronegative atoms. Thus, in SF_6, the bonds are stronger and the bond lengths shorter than would be expected from σ-bonding alone.

There is some evidence for the formation of cations by the heavy elements of this group. Thus both Te^{4+} and Po^{4+} ions exist in the ionic dioxides, TeO_2 and PoO_2. Po is more electropositive than Te and tends to form a normal salt, for instance $Po(SO_4)_2$, whereas only the basic salt is known with Te, $TeO.SO_4$. The formation of tetrapositive ions by Te and Po shows the persistence of the inert pair effect into this Group.

Compounds

Hydrides

Four of general formula H_2X are known — H_2O, H_2S, H_2Se and H_2Te.

H_2S is prepared by the decomposition of a metallic sulphide with acid: H_2Se and H_2Te are formed when metallic selenides and tellurides respectively are hydrolysed.

The volatility increases sharply from H_2O to H_2S because of association in water by hydrogen bonding. In this group only oxygen is sufficiently electronegative to show this property in its compounds. The hydrides of the remaining elements show the increase in m.p. and b.p. with molecular weight, which is normally shown in any series of homologous compounds.

Standard molar enthalpies of formation are: H_2O, -286; H_2S, -20; H_2Se, $+77$; H_2Te, $+143\,kJ\,mol^{-1}$. Although these data might suggest otherwise, hydrogen selenide appears to be quite stable thermally provided it does not come into contact with air. Thus no decomposition occurs below 553 K until air is introduced, when the selenide decomposes rapidly to selenium. The acid strengths increase as the size of the anion XH^- increases and hence as its power of attraction for a proton decreases. Accordingly, H_2S is an acidic solute in water.

The molecules are V-shaped with the following inter-bond angles:

$$H_2O,\ 104°\ 30';\quad H_2S,\ 92°\ 20';\quad H_2Se,\ 91°$$

The decrease in angle as the electronegativity of the Group VI element decreases can be interpreted, as for the Group V hydrides, in terms of the repulsion between electron pairs which results in increasing deviation from the tetrahedral angle. Alternatively, it appears to be perfectly adequate to describe the bonding in H_2Se and H_2S as involving pure p orbitals of the Se and S atoms respectively.

Oxides, sulphides, selenides and tellurides are derivatives of the hydrides. The oxides of most metals are ionic and show basic properties. The oxides of electropositive metals such as the alkali and alkaline–earth elements react vigorously with water to give alkaline solutions. The oxide ion is very strongly basic, and has a powerful affinity for the proton. The oxides of metals showing higher valencies have high lattice enthalpies and as a result are often sparingly soluble in water. These compounds will usually dissolve in acidic solutions because the high concentration of protons necessary to effect solution is then available. For instance TiO_2, insoluble in water, is slowly dissolved by hot strong sulphuric acid. The oxides of the more electronegative elements are often characterised by amphoteric properties and insolubility in aqueous media. Fusion under alkaline or acid conditions is then necessary to render them soluble. Acidic properties predominate in the case of transition metal oxides in which the metal shows a high oxidation state and for the oxides of non-metals. A few 'neutral' oxides like CO, N_2O and NO are exceptional.

Sulphides of the alkali and alkaline-earth metals have ionic structures and are decomposed by water thus:

$$S^{2-} + H_2O \rightleftharpoons OH^- + SH^-$$

On boiling, $$SH^- + H_2O = OH^- + H_2S\uparrow$$

A number of other sulphides, for example Al_2S_3, Cr_2S_3 and SiS_2, are also readily hydrolysed. Sulphides of the heavy metals, in contrast, have extremely low solubilities in water and some are not even decomposed by mineral acid. Hence the great utility of H_2S as a reagent in qualitative analysis for the precipitation of the sulphides of Pb, Cu, Cd, Bi, Hg, Sb, Sn and As from acid solution.

Where a metal oxide has an ionic structure, the corresponding sulphide may often have a layer lattice due to the greater polarisability of S^{2-}. For example, TiO_2 and SnO_2 have the rutile structure but TiS_2 and SnS_2 have the cadmium iodide structure.

Alkali metal selenides and tellurides are ionic and resemble the corresponding sulphides in their properties. Heavy metal selenides and tellurides have many similarities to the sulphides of these elements.

O and S form other molecular hydrides. These are hydrogen peroxide (H_2O_2), hydrogen persulphide (H_2S_2) and a series of hydrogen polysulphides (H_2S_3 to H_2S_8) containing chains of sulphur atoms.

H_2O_2 is made by electrolytic oxidation of ammonium sulphate/sulphuric acid solution. Peroxodisulphuric acid, $H_2S_2O_8$, is first formed and this is hydrolysed to H_2O_2:

$$2SO_4^{2-} = S_2O_8^{2-} + 2e^-$$
$$H_2S_2O_8 + 2H_2O = 2H_2SO_4 + H_2O_2$$

Hydrogen peroxide has a variety of uses — as an oxidising agent in organic synthesis, a bleaching agent and a rocket fuel. The chief advantage of this over other oxidising agents is that its only by-product is water and so, in organic preparative work, the recovery of a product of high purity is facilitated. The H_2O_2 molecule is dihedral, that is, the two O—H bonds are not in the same plane (*Figure 9.3*). This is the most stable configuration because the repulsion between the lone pairs on the two oxygen atoms is a minimum. Here it is assumed that only oxygen p orbitals are used in bonding. Repulsion between the lone pairs is greatest when the p orbitals containing them, one on each oxygen atom, are parallel and least when the orbitals are at right angles.

Important derivatives of H_2O include the ionic peroxides, formed by the alkali and alkaline-earth metals and containing the peroxide ion, O_2^{2-}, and the peroxo-acids and their salts, formed by the representative elements and transition metals of Groups IV, V and VI; for instance, peroxonitric acid, HNO_4, peroxomonosulphuric acid, H_2SO_5, and peroxotitanic acid, H_4TiO_5. These compounds are chiefly of note for their strong oxidising properties.

H_2S_2 and the polysulphides are prepared by the addition of acid to an aqueous solution of sulphur in alkali metal sulphides. A water-insoluble

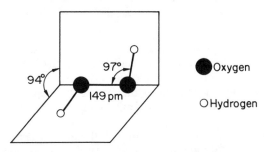

Figure 9.3. The hydrogen peroxide molecule

yellow oil is obtained and this may be fractionally distilled under reduced pressure to give individual sulphides. The H_2S_2 molecule is similar to H_2O_2 in its shape.

Oxides and oxo-acids

The binary compounds with oxygen are:

S_2O	—	TeO	PoO
S_2O_3	—	—	—
SO_2	SeO_2	TeO_2	PoO_2
SO_3	SeO_3	TeO_3	—
S_2O_7	—	—	—
SO_4	—	—	—

The most important are the dioxides and trioxides.

SO_2 is prepared by burning sulphur or sulphides in air. It is a reducing agent, itself being converted to SO_3. SO_2 is a molecular oxide and the S—O bonds are essentially double. The angular shape of the molecule —

—is attributable to sp^2 hybridisation of the sulphur. Two hybrid orbitals are used to form σ-bonds with each oxygen and the third contains a lone pair. Two π-bonds are also formed. The S atom therefore has ten electrons in its outermost shell.

SeO_2, made by burning Se in air, is a solid subliming at 588 K. It is easily reduced to elementary Se and so finds uses as an oxidising agent. In the solid state it has a chain-like structure:

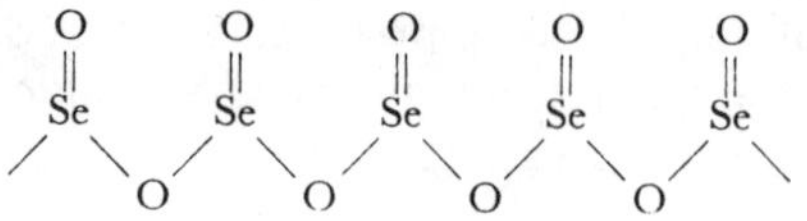

TeO_2, prepared by the combustion of Te, has oxidising properties. It is a solid which is much less volatile than SeO_2 and has the ionic rutile structure. The dioxides thus illustrate clearly the transition from a covalent to an ionic structure as the electronegativity of the Group VI element decreases. SO_2 and SeO_2 have acidic properties and TeO_2 is amphoteric, in accordance with this trend.

Of the three trioxides, SeO_3 is the most difficult to prepare and this compound is one example of the reluctance of Se to show its maximum oxidation state in combination with oxygen.

SO_3 is made by heating SO_2 in oxygen to 673 K in the presence of a catalyst. It condenses at 318 K and freezes at 290 K to a colourless ice-like solid. This is γ-SO_3 which appears to be a ring-shaped trimer, S_3O_9. Two other modifications are known, α- and β-SO_3, and these also appear to be polymeric. The structures of the oxide forms may be regarded as

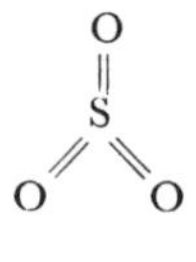

SO$_3$ vapour 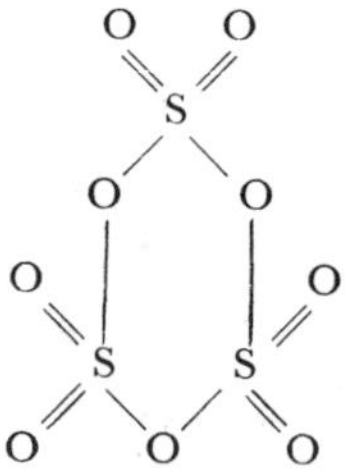 γ-SO_3 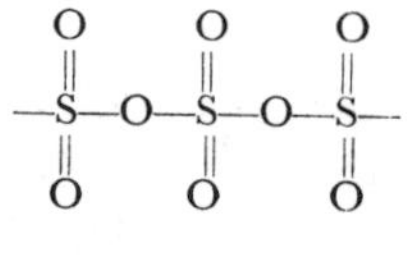 β-SO_3 (α-SO_3 is layer form)

SeO_3 is prepared by the oxidation of SeO_2 with oxygen in a high-frequency electrical discharge. It is a white solid, m.p. 391 K, resembling SO_3.

TeO_3 is obtained by dehydration of ortho telluric acid, H_6TeO_6. It is a non-volatile solid, thermally very stable. SeO_3 and TeO_3 structures are unknown.

The oxo-acids in Table 9.16 are derived from the above-mentioned oxides.

In the case of the oxides of S and Se, the oxo-acids are prepared by reaction of the parent oxide with water. Te oxides are sparingly soluble in water and the oxo-acids are best made by direct oxidation of the element.

The sulphite ion is pyramidal and the sulphate is tetrahedral. The bonding between sulphur and oxygen has been the subject of great speculation and to explain the observed equivalence of the S—O bond lengths in both ions, resonance between a number of different structures has been proposed. The S—O bond length in SO_2 and SO_3 is 144 pm and in these molecules the bonds are regarded as essentially double in character. (A single S—O bond, as found in potassium ethyl sulphate, $K.SO_3.O.C_2H_5$, is appreciably longer— 160 pm.) In the sulphate ion in hydrazinium sulphate, $N_2H_4.H_2SO_4$, the

Table 9.16

Parent oxide	Oxo-acid	Basicity	Properties
SO_2	Sulphurous, H_2SO_3	Weak, dibasic	Reducing
SeO_2	Selenious, H_2SeO_3	Weak, dibasic	Oxidizing
TeO_2	Tellurous, H_2TeO_3	Weak, dibasic	Oxidizing. Tends to polymerise
SO_3	Sulphuric, H_2SO_4	Strong, dibasic	Weak oxidising agent
SeO_3	Selenic, H_2SeO_4	Strong, dibasic	Very strongly oxidising
TeO_3	Orthotelluric, H_6TeO_6	Very weak, dibasic. Orthotellurates such as Ag_6TeO_6 are known	Strongly oxidising. Tends to polymerise

S—O bond length is 149 pm. The bond thus is intermediate between a single and a double bond and this is consistent with Pauling's formulation of SO_4^{2-} as a resonance hybrid between such structures as:

$$^-O-\overset{\displaystyle O^-}{\underset{\displaystyle O^-}{\overset{+}{S}}}=O \qquad O=\overset{\displaystyle O^-}{\underset{\displaystyle O^-}{S}}=O \qquad O=\overset{\displaystyle O^-}{\underset{\displaystyle O}{\overset{-}{S}}}=O \qquad O=\overset{\displaystyle O}{\underset{\displaystyle O}{\overset{\|^{2-}}{S}}}=O$$

The molecular-orbital description is in terms of 4 σ-bonds (hence the tetrahedral shape) and two non-localised π-bonds. Similar descriptions have been put forward for the sulphite ion, either a resonance hybrid or the sulphur atom, in a state of sp^3 hybridisation, forming 3 σ-bonds, 1 non-localised π-bond and carrying one lone pair of electrons:

In addition to H_2SO_3 and H_2SO_4, a number of other sulphur oxo-acids are known. In some cases the free acid has been made; in others, only the salts are known. Some interrelations of these acids are summarised in *Figure 9.4*.

Among the other oxo-acids of sulphur that are of some importance are dithionous, thiosulphuric and disulphurous acid. As free acids they are unstable but their salts are well known.

Dithionites, $S_2O_4^{2-}$, may be obtained by the reduction of sulphite (in solution) with zinc dust and excess sulphur dioxide. Acid solutions of the ion rapidly decompose to SO_2 and elemental sulphur, though in contrast to thiosulphate which decomposes likewise, the dithionite ion is a strong reducing agent. Thiosulphate, $S_2O_3^{2-}$, the sodium salt of which is widely used in quantitative analysis, may be obtained by boiling sulphur with solutions of sulphite. The presence of two chemically non-equivalent sulphur atoms in this ion has been demonstrated by radioisotopic substitution. Disulphites, $S_2O_5^{2-}$, arise from the interaction of SO_2 with sulphites in aqueous solution; they differ from other diacids in general by not possessing an oxygen bridge in their structure.

The structures of all three ions contain a S—S bond:

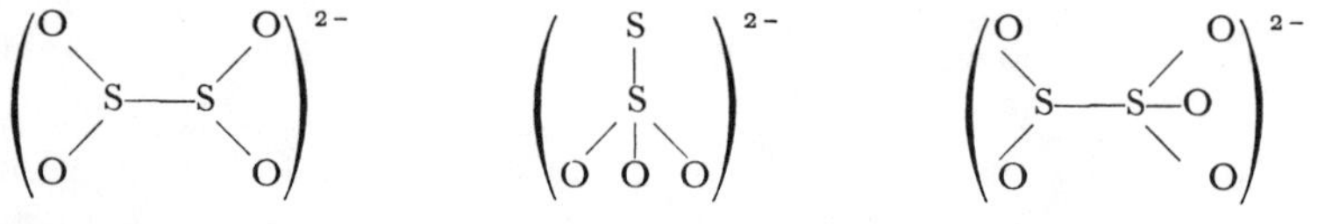

Dithionates, $S_2O_6^{2-}$, obtained by the oxidation of aqueous solutions either of sulphite or SO_2 by MnO_2, may be converted to the acid by first precipitating the barium salt and subsequently treating this with sulphuric acid. Compared with the higher thionic acids ($H_2S_nO_6$, $n = 3$ to 6), dithionic acid has considerable stability and is difficult to oxidise or reduce. As *Figure 9.4* shows, mixtures of the polythionates are to be found in Wackenroder's solution; some may be selectively prepared as indicated.

In contrast to the higher thionates, $S_2O_6^{2-}$ has no sulphur atom bonded solely to sulphur; it has the structure

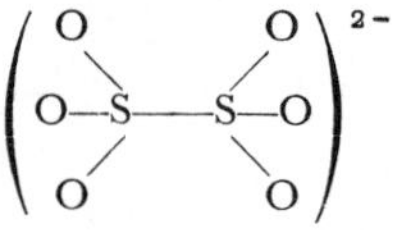

The polythionates contain unbranched sulphur — sulphur chains. Thus, in tetrathionate the arrangement is

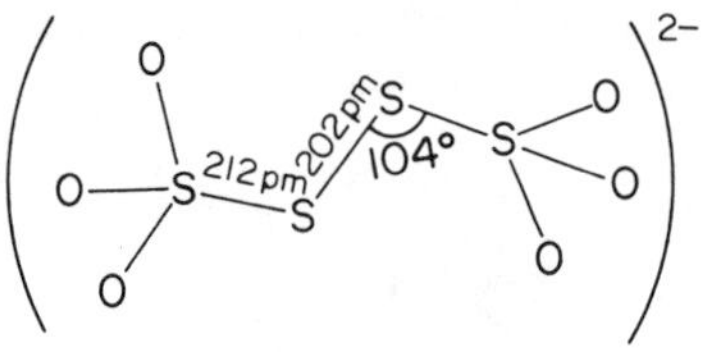

and in pentathionate it is

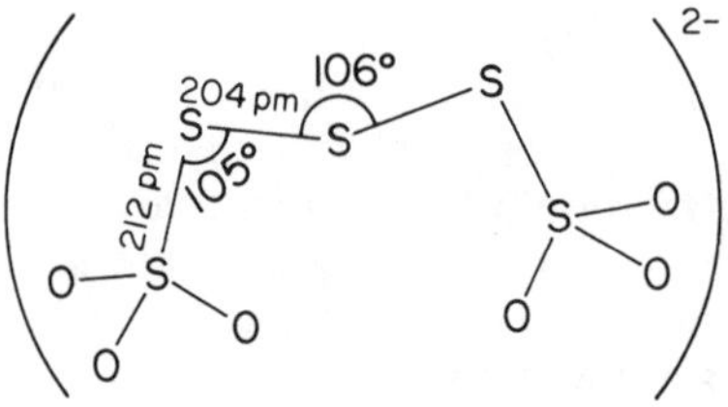

Disulphate, $S_2O_7^{2-}$, is produced by the condensation of two sulphate ions:

$$2SO_4^{2-} + 2H^+ \rightleftharpoons H_2O + S_2O_7^{2-}$$

In the ion, one oxygen is shared by two sulphur atoms:

$$\left(\begin{array}{c} O \qquad O \qquad O \\ O-S \diagup \diagdown S-O \\ O \qquad\qquad O \end{array} \right)^{2-}$$

$$S_2O_7^{2-}$$

Finally, two peroxo-acids are known. Peroxomonosulphuric acid, H_2SO_5, is formally derived from H_2SO_4 by the replacement of one oxygen by a peroxo-group, —O—O—. Peroxodisulphuric acid, $H_2S_2O_8$, is a condensed oxo-acid

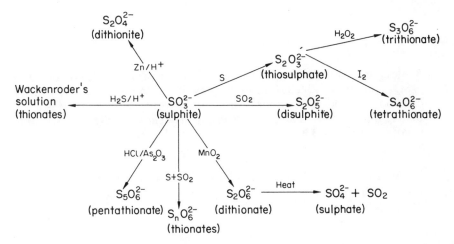

Figure 9.4. Interrelationships of the oxo-acids of sulphur

in which the sulphur atoms are joined by the peroxo-group. The peroxodisul-phate ion, $S_2O_8^{2-}$, is

$$\left(\begin{array}{c} O \qquad O \qquad S-O \\ O-S \diagup \diagdown O \diagdown O \\ O \end{array} \right)^{2-}$$

Oxo-halides

SO_2 reacts with PCl_5 to form thionyl chloride, $SOCl_2$. This is a colourless liquid, b.p. 348 K. It reacts with several fluorides to form thionyl fluoride, SOF_2, and with HBr at low temperatures to give the unstable bromide, $SOBr_2$. SOF_2 is slowly hydrolysed but $SOCl_2$ is immediately decomposed:

$$SOCl_2 + H_2O = SO_2 + 2HCi$$

Use of this reaction has been made to prepare anhydrous metal chlorides.

The $SOCl_2$ molecule is pyramidal: the S atom is sp^3 hybridised, forming 3 σ-bonds and carrying one lone pair—

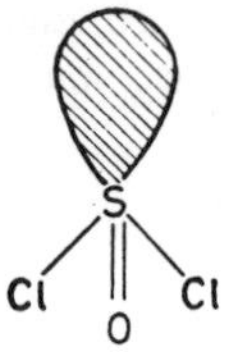

Analogous selenyl halides, $SeOX_2$ (X = F, Cl or Br) have been prepared.

SO_2 combines with chlorine in the presence of an organic catalyst such as camphor or acetic acid to form sulphuryl chloride, SO_2Cl_2. This is a low-boiling liquid which is almost completely dissociated to SO_2 and Cl_2 in the vapour phase at 373 K. It is readily hydrolysed:

$$SO_2Cl_2 + 2H_2O = 2HCl + H_2SO_4$$

The molecule is a distorted tetrahedron. The S atom (sp^3 hybridised) forms 4 σ- and 2 π-bonds:

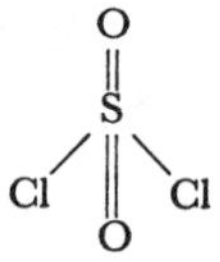

$$\underset{\underset{O}{\|}}{\overset{\overset{O}{\|}}{Cl-S-Cl}}$$

Halides

The binary halides of S, Se and Te are:

S_2F_2	S_2Cl_2	S_2Br_2	—
—	Se_2Cl_2	Se_2Br_2	—
SF_2	SCl_2	—	—
—	$SeCl_2$	$SeBr_2$	—
—	$TeCl_2$	$TeBr_2$	—
SF_4	SCl_4	—	—
SeF_4	$SeCl_4$	$SeBr_4$	—
TeF_4	$TeCl_4$	$TeBr_4$	TeI_4
SF_6	—	—	—
SeF_6	—	—	—
TeF_6	—	—	—
S_2F_{10}	—	—	—
Te_2F_{10}	—	—	—

The lower halides of the types A_2X_2, AX_2 and AX_4 are formed by direct union of the elements. They are molecular, covalent compounds which are chemically reactive and are readily hydrolysed. S_2Cl_2 has a non-planar structure similar to that of H_2O_2. The boiling points of the di- and tetra-halides of Te are relatively high and $TeCl_4$ conducts electricity in the fused state. This could be evidence of ionic character or of self-ionisation according to

$$2TeCl_4 \rightleftharpoons TeCl_3^+ + TeCl_5^-$$

The hexafluorides are all highly exothermic compounds. Although SF_6 is generally considered to be inert to chemical attack, it has recently been shown to react with sodium

$$8Na + SF_6 = Na_2S + 6NaF$$

As the electronegativity of the central atom decreases, the hexafluoride becomes more reactive. Thus, even at red heat, SF_6, is unattacked by O_2, NH_3 and many other chemical substances. SeF_6 reacts with ammonia above 473 K:

$$SeF_6 + 2NH_3 = N_2 + Se + 6HF$$

and TeF_6 is decomposed slowly by water at room temperature:

$$TeF_6 + 6H_2O = H_6TeO_6 + 6HF$$

The molecules are regular octahedra.

Nitrogen compounds

The binary compounds of oxygen with nitrogen have been discussed in Group V.

A number of interesting sulphur compounds is known in which sulphur and nitrogen are combined. The reaction between S_2Cl_2 and ammonia gives rise to the ring compound, tetrasulphur tetranitride, S_4N_4. This is an orange–yellow solid, m.p. 451 K, which is hydrolysed in alkaline solution:

$$2S_4N_4 + 6OH^- + 9H_2O = 2S_3O_6^{2-} + S_2O_3^{2-} + 8NH_3$$

Reduction with tin (II) chloride or dithionite results in the formation of tetrasulphur tetraimide, $S_4(NH)_4$. In both these sulphur compounds, 8-membered rings are present:

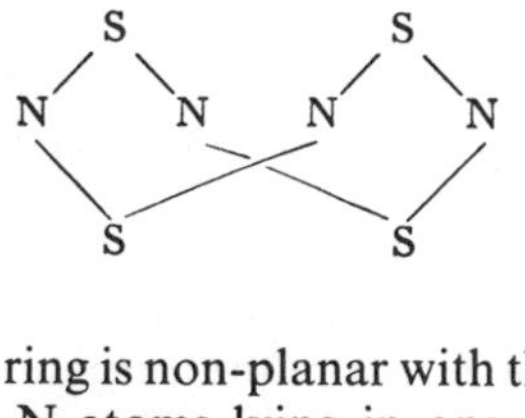

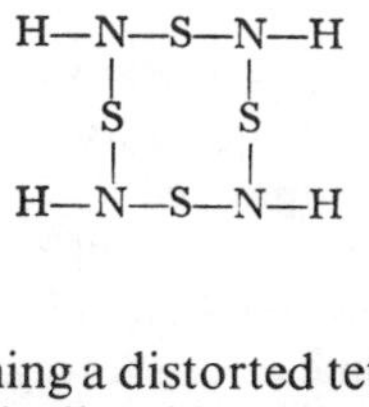

The S_4N_4 ring is non-planar with the S atoms forming a distorted tetrahedron and the 4 N atoms lying in one plane. Electronically, this compound is a resonance hybrid, all the S atoms being exactly equivalent.

The S_4N_4 ring can be easily cleaved into two molecules of disulphur dinitride. For example, on treatment with ammonia, the ammoniate, S_2N_2. NH_3 (or $H_2N.SN.SNH$), is produced. A number of metallic derivatives of this is known such as $Pb(NS)_2$, $Tl(NS)_2$ and $Cu(NS)_2$. Complex compounds of formula $M(NS)_2$ are obtained from S_4N_4 and metals of Group VIII by the interaction between the nitride and the appropriate metal carbonyl in an inert solvent.

A second sulphur imide, heptasulphur imide, S_7NH, is also found in small quantities amongst the reaction products of S_2Cl_2 and NH_3. Structurally this compound is related to the S_6 ring. The hydrogen atom can be replaced by a number of metals, by acetyl and benzoyl groups and by the sulphonic acid group, $-SO_3H$.

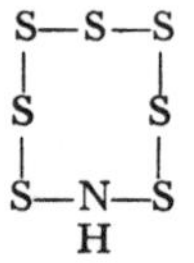

$$\begin{array}{ccc} S & - S - & S \\ | & & | \\ S & & S \\ | & & | \\ S & - N - & S \\ & H & \end{array} \qquad S_7NH$$

The other elements of Group VI do not form compounds of this type.

Polonium

This radioactive element occurs as an intermediate in the various decay series. It is made by neutron irradiation of ^{209}Bi:

$$^{209}_{83}Bi \; (n, \gamma) \; ^{210}_{83}Bi \longrightarrow \; ^{210}_{84}Po + \; _{-1}^{0}e$$

^{210}Po decays by α-emission and has a half-life of 138·4 d. Its high specific activity means that the handling of the element in quantity is hazardous.

The element itself is a soft, low-melting metal which resembles lead in appearance. It is dimorphic, and cubic and rhombohedral modifications are known. In this respect Po behaves as a true metal. The compounds of polonium which have been prepared include PoO, PoO_2, various halides ($PoCl_2$, $PoCl_4$, $PoBr_2$, $PoBr_4$ and PoI_4) and the sulphates ($PoO_2)SO_3$ and $Po(SO_4)_2$.

The general properties of polonium confirm its classification as the most electropositive element in Group VI B.

GROUP VII—REPRESENTATIVE ELEMENTS

The halogens are the non-metals fluorine (F), chlorine (Cl), bromine (Br), iodine (I) and the radioactive element, astatine (At). Physical and chemical data are summarised in *Table 9.17*.

The atomic number of each halogen is one less than that of the nearest noble gas. A completed octet of electrons is reached by a halogen atom X either by accepting one electron to give the halide ion X$^-$, as in the formation of ionic halides, or by forming a covalent bond with another atom, as in the diatomic halogen molecules X_2 and the halides of hydrogen and other non-metals.

The valency of fluorine is restricted to one because there is no possibility of expanding the second quantum shell to contain more than eight electrons. In the other halogens, however, one or more d orbitals in the outermost shell can be used in bonding. For example, a pair of electrons in a p orbital may be split up and one promoted to a vacant d orbital. This gives an excited state of

Table 9.17. THE PROPERTIES OF THE HALOGENS

	F	Cl	Br	I	At
Atomic number	9	17	35	53	85
Electron configuration	$2s^2\,2p^5$	$3s^2\,3p^5$	$4s^2\,4p^5$	$5s^2\,5p^5$	$6s^2\,6p^5$
m.p. T/K	50·2	171·2	265·9	387·2	
b.p. T/K	86·2	307·8	332·0	457·6	
Electron affinity $\Delta H_E^{\ominus}/\text{kJ mol}^{-1}$	−335·6	−356·1	−332·6	−303·8	
Ionisation energy $\Delta U_o/\text{kJ mol}^{-1}$	1 679·9	1 255·2	1 141·8	1 008·3	
Electronegativity	4·0	3·0	2·8	2·5	
Atomic radius r/pm	64	99	114	135	
Ionic radius r/pm X^-	133	181	196	220	
Dissociation energy $\Delta H^{\circ}/\text{kJ mol}^{-1}$ $(X_2(g) \rightarrow 2X(g))$	154·8	242·7	192·5	150·6	

the atom which has three unpaired electrons and is therefore capable of forming three covalent bonds. Similarly, by the involvement of the two remaining pairs of electrons in the valency shell of the halogen, valencies of five and seven can be shown. The energy necessary to split up the electron pairs and to promote electrons to higher energy levels is available from the energy released when covalent bond formation occurs.

The higher valencies of iodine, for example, are illustrated by the formation of various compounds with another halogen. The electronic configurations of some of these are given in *Table 9.18*.

Table 9.18. THE ELECTRONIC CONFIGURATION OF IODINE IN VARIOUS COMPOUNDS

Although fluorine shows the greatest tendency to form a negative ion, its electron affinity is less than that of chlorine. This is a consequence of the unusually low heat of dissociation of the F_2 molecule; this quantity and the electron affinity are interrelated by the appropriate Born–Fajans–Haber cycle.

The high values of the first ionisation potential are in accordance with the observation that cationic species are not encountered with either fluorine or chlorine. There is a considerable body of experimental evidence to support the existence of uni- and tri-positive iodine, and it has recently been reported that, in suitable non-aqueous solvents, bromine may also give rise to compounds in which it appears to be tripositive.

Fluorine is the most electronegative element. One property associated with it is the occurrence of hydrogen bonding in a number of its compounds. For example, hydrogen fluoride behaves as an associated substance due to the presence of hydrogen bonds (the m.p. and b.p. of HF are 190·2 K and 292·5 K respectively compared with the values of 158·8 K and 188·2 K for HCl).

Another result of the high electronegativity of fluorine is that metal fluorides are predominantly ionic. The chloride ion, being the larger, is more readily polarised than fluoride and, as a result, the fluoride and chloride of the same metal often have quite different structures. Similarly, the iodide ion is more polarisable than the bromide ion and this accounts for the fact that, whilst AgF, AgCl and AgBr have the sodium chloride structure, AgI has that of zinc blende, a structure typical of compounds with appreciable covalent character.

The value for the dissociation energy of F_2 is unexpectedly small. The data for Cl_2, Br_2 and I_2 indicate a steady decrease in stability of the molecule as the size of the halogen atom increases and one would, indeed, expect the covalent bond between two large atoms to be weak because of the relatively small region of overlap possible between the two big bonding orbitals. One would predict, by extrapolation of the data for the heavier halogens, the greatest stability of all for the F_2 molecule. The observed weakness of the bond in F_2 is attributable to the repulsion between non-bonding pairs of electrons which becomes important for small atoms. Supporting evidence for this hypothesis is found in the low values for the dissociation energies of hydrazine and hydrogen peroxide (87·9 and 146·4 kJ mol^{-1} respectively) where similar repulsion effects would be expected.

Fluorine is remarkable for its great chemical reactivity and this may be attributed to the ease of breakage of the F—F bond. In contrast, the bonds between fluorine and many other elements are noteworthy for their great stability. Thus the C—F bonds in fluorinated hydrocarbons and the S—F bonds in SF_6 are extremely resistant to attack.

Fluorine can also bring out the maximum valency of other elements with which it is combined, for example in IF_7 and SF_6. In an exothermic reaction between two elements, the energy released in bond formation must be more than sufficient to break up the reacting molecules and to raise the atoms to excited valency states prior to chemical combination. The reaction between sulphur and fluorine may be represented by a sequence of (hypothetical) steps in which the isolated atoms are formed from solid sulphur and molecular fluorine and then combined to form SF_6:

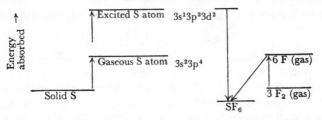

The F_2 molecule is more easily dissociated than Cl_2 and hence the formation of a chlorine atom requires more energy than the formation of a fluorine atom. As a result, SF_6 is known but the highest chloride of sulphur is SCl_4.

Reactions of the halogens

They are all chemically reactive and combine directly with many other elements to form the halides. Various reactions are possible between the halogens and water. In the case of chlorine, bromine and iodine the hypohalous acid is formed, for example,

$$Cl_2 + H_2O \rightleftharpoons H^+ + Cl^- + HClO$$

The equilibrium constants for this type of reaction are $4 \cdot 8 \times 10^{-4}$, 5×10^{-9} and 3×10^{-3} for HClO, HBrO and HIO respectively.

The reaction between fluorine and water leads to oxidation according to:

$$2F_2 + 2H_2O \rightleftharpoons 4HF + O_2$$

Recently, hypofluorous acid, HOF, has been isolated as another product of this reaction. This is a volatile compound but stable enough to be handled *in vacuo* or in a gas stream at ordinary temperature.

Some H_2O_2 and oxygen difluoride, OF_2, are produced in secondary reactions. With iodine, the reaction

$$4H^+ + 4I^- + O_2 \rightleftharpoons 2I_2 + 2H_2O$$

is important and acid solutions of an iodide are slowly oxidised by oxygen to iodine.

The oxidising properties of the halogens decrease with increase in atomic number. A lighter halogen will always displace a heavier halogen from its simple anion. An important application of this property is in the extraction of bromine from sea-water by means of the reaction

$$2\,Br^- + Cl_2 = 2\,Cl^- + Br_2$$

Compounds

Hydrogen halides

These are prepared by direct combination between the elements or by displacement from their salts by a less volatile acid:

$$CaF_2 + H_2SO_4 = CaSO_4 + 2\,HF$$

HBr and HI are often made by hydrolysis of a phosphorus trihalide:

$$PX_3 + 3 H_2O = H_3PO_3 + 3 HX$$

Their stability decreases as the size of the halogen increases as shown by the energy required to rupture the hydrogen–halogen bond. This varies as follows: HF, 565; HCl, 431; HBr, 364; HI, 297 kJ mol^{-1}. HF is very stable and cannot be chemically oxidised to fluorine in aqueous solution, hence the industrial preparation of this halogen by the electrolysis of an anhydrous fluoride melt. Chlorine can be made by the action of strong oxidising agents like MnO_2 and $KMnO_4$ on HCl. Bromine is even more readily formed from HBr; the weak oxidising action of concentrated sulphuric acid is sufficient to form bromine as well as HBr on reaction with a metallic bromide.

The hydrogen halides all act as acids in aqueous solution: approximate values for their dissociation constants are HF, 6.7×10^{-4}; HCl, 10^7; HBr, $> 10^7$; HI, $> 10^7$. In dilute solution, HF acts as a weak acid but in concentrated solution, over the range 5 to 15 mol dm^{-3} it behaves as a much stronger acid. Conductivity measurements on the more concentrated solutions have indicated that the predominant anion is F_2H^-. The striking difference between HF and the other hydrogen halides is due largely to the very high dissociation energy of the HF molecule, the unusually low electron affinity of fluorine and the extensive hydrogen bonding shown by HF in aqueous solution.

Oxides and oxo-acids

Binary compounds between oxygen and fluorine are:

	m.p. T/K	b.p. T/K	*Properties*
OF_2	49·4	127·9	Decomposition to $F_2 + O_2$ at 523–553 K
O_2F_2	109·7	216·2	Decomposition to $F_2 + O_2$ near b.p.
O_3F_2	Liquid stable only below 110·2K		Decomposition to $O_2F_2 + O_2$
O_4F_2	Liquid stable only below 90·2K		Decomposition to $O_3F_2 + O_2$

OF_2 is an oxidising agent and converts Br^- and I^- to the free halogen in aqueous solution. The OF_2 molecule is V-shaped, with a bond angle of 103·2 degrees and a F—O distance of 142 pm.

O_2F_2, O_3F_2 and O_4F_2 can be made by the action of a glow discharge on F_2—O_2 mixtures at liquid air temperatures. These compounds are not stable at normal temperatures and little is known of their reactions and structures.

Four binary compounds between oxygen and chlorine are known:

	m.p. T/K	b.p. T/K	*Properties*
Cl_2O	157·2	275·2	Highly endothermic. Exploded by shock
ClO_2	214·2	284·2	Highly endothermic. Explodes violently
Cl_2O_6	276·2		Decomposition at m.p. to ClO_2 and O_2
Cl_2O_7	182·2	353·2	

Dichlorine oxide, Cl_2O, is prepared as a brown solid by passing chlorine and dry air over mercuric oxide and condensing the products in a vessel cooled in liquid air. It reacts with water and alkali to form the hypochlorite ion, OCl^-.

Reaction with N_2O_5 gives chlorine nitrate, from which other halogen nitrates can be prepared:

$$BrCl + ClNO_3 = BrNO_3 + Cl_2$$
$$ICl + ClNO_3 = INO_3 + Cl_2$$

Cl_2O reacts with metallic chlorides to give anhydrous oxochlorides:

$$SnCl_4 + Cl_2O = SnOCl_2 + 2Cl_2$$

The Cl_2O molecule is V-shaped with a bond angle of $110 \cdot 8°$ and a Cl—O distance of 170 pm.

Chlorine dioxide, ClO_2, is prepared by the action of a reducing agent, such as oxalic acid, on a solution of chloric acid, generated from a chlorate and sulphuric acid:

$$2HClO_3 + H_2C_2O_4 = 2ClO_2 + 2CO_2 + 2H_2O$$

Reaction with water and alkali gives a mixtures of chlorite (ClO_2^-) and chlorate (ClO_3^-). Reaction with aqueous hydrogen peroxide gives a solution of chlorous acid only:

$$2ClO_2 + H_2O_2 = 2HClO_2 + O_2$$

ClO_2 is yellow in the vapour state and condenses to a red–brown liquid. This colour is associated with the presence of one unpaired electron. Although the molecule is thus paramagnetic, it apparently shows little tendency to dimerise even in the liquid state.

The ClO_2 molecule is V-shaped, with a bond angle of $118 \cdot 5°$ and a Cl—O distance of 149 pm.

Dichlorine hexaoxide, Cl_2O_6, is made by the photochemical reaction of chlorine with ozone. Like the lower oxides it is highly endothermic and reacts explosively with reducing agents and organic compounds. Reaction with water and alkali gives a mixture of chlorate and perchlorate (ClO_4^-) ions. In the liquid state, the compound is mainly dimeric, as indicated by its formula. The monomeric molecule, ClO_3, is paramagnetic with one unpaired electron but its structure is not known.

Dichlorine heptaoxide, Cl_2O_7, is the least unstable of the oxides of chlorine, although it is still liable to explode when heated or subjected to shock. It is made by the dehydration of perchloric acid with P_2O_5, followed by distillation at 133 Pa between 235 and 239 K. Reaction with water and alkali gives the perchlorate ion.

In the molecule, each chlorine is bound by a bridging oxygen atom to the other chlorine and to three other terminal oxygens:

The oxides of bromine are difficult to prepare and very unstable:

	Properties
Br_2O	Brown solid, m.p. 256·2 K. Slow decomposition to Br_2 and O_2 above 223·2 K
BrO_2	Yellow solid. Decomposition above 233·2 K
$(Br_3O_8)_n$ to $(BrO_3)_n$	White solids. Unstable above 203·2 K, except in presence of O_3

Dibromine oxide, Br_2O, is prepared by a method analogous to that used for Cl_2O, or as indicated below. It is hydrolysed to OBr^-, hypobromite, and so behaves as the true anhydride of hypobromous acid.

Bromine dioxide, BrO_2, is formed by the low temperature ozonisation of bromine. A precooled O_3—O_2 mixture is passed into a solution of bromine in a fluorine-containing solvent such as CF_2Cl_2 at 223 K:

$$Br_2 + 4O_3 = 2BrO_2 + 4O_2$$

The dioxide is then obtained by distilling off the solvent. When allowed to warm up slowly *in vacuo*, Br_2O is one of the products. Rapid heating gives bromine and oxygen.

The action of ozone on bromine at room temperature gives higher bromine oxides of variable composition. These appear to be polymeric but detailed structures are not known.

More marked evidence for polymeric structures is found in the binary compounds between iodine and oxygen:

	Properties
I_2O_4	Brown solid. Decomposition at 373·2 K to I_2 and I_2O_5
I_4O_9	Yellow solid. Decomposition at 348·2 K to I_2O_5, I_2 and O_2
I_2O_5	White solid. Decomposition at 573·2 K to I_2 and O_2
I_2O_7	Orange solid precipitated from periodic acid dissolved in 65 per cent oleum

Diiodine tetraoxide is a product of the action of water on diiodosyl sulphate, $(IO)_2SO_4$, itself prepared by the reaction between iodic acid and sulphuric acid. Prolonged treatment with water decomposes I_2O_4 according to:

$$5I_2O_4 + 4H_2O = I_2 + 8HIO_3$$

From its reactions and relationship with $(IO)_2SO_4$, I_2O_4 is formulated as $(IO)_n(IO_3)_n$. No discrete IO^+ ions are present and the structure consists of a network of IO_3 groups and polymeric . . . I—O—I—O—I—O—I—O . . . chains.

Tetraiodine enneaoxide, I_4O_9, has been made by the reaction between orthophosphoric acid and iodic acid. Hydrolysis produces HIO_3 and I_2:

$$5I_4O_9 + 9H_2O = 18HIO_3 + I_2$$

This and other reactions suggest that the compound should be formulated as iodine (III) iodate, $I(IO_3)_3$, but the structure has not yet been reported.

Diiodine pentaoxide, I_2O_5, can be prepared by heating iodic acid to 473 K:

$$2HIO_3 = H_2O + I_2O_5$$

It is a crystalline, non-volatile substance and is stable to decomposition below 573 K. These properties are completely opposite to those associated with the oxides of the lighter halogens. It is, however, a vigorous oxidising agent. One useful application is to the determination of carbon monoxide in gaseous mixtures. The reaction:

$$I_2O_5 + 5CO = I_2 + 5CO_2$$

is quantitative at 343 K and the iodine produced can be easily determined by standard volumetric procedures.

I_2O_5 reacts with water and alkali to form the iodate, IO_3^-, ion. In the solid state, i.r. spectral evidence suggests the presence of I_2O_5 units with one oxygen atom bridging two IO_2 groups.

The oxo-acids, known either in the free state or as their salts, are:

HClO	HBrO	HIO
$HClO_2$	—	—
$HClO_3$	$HBrO_3$	HIO_3
$HClO_4$	$HBrO_4$	HIO_4, H_3IO_5, H_5IO_6 and $H_4I_2O_9$

The acids, like the oxides, have oxidising properties. The three hypohalous acids, HXO, may be made in the form of their salts by the direct action of halogen on cold aqueous alkali. The acids are less stable than their salts and the stability and the acid strength both decrease as the size of the halogen increases. Many practical applications of the oxidising properties of the hypohalites are known, for example, sodium hypochlorite, NaClO, is used as a bleaching agent and as a reagent in volumetric analysis.

Chlorous acid, $HClO_2$, and the chlorites are not very stable. The free acid in solution decomposes readily:

$$8HClO_2 = 6ClO_2 + Cl_2 + 4H_2O$$

The salts, chlorites, are active oxidising agents.

Free chloric, $HClO_3$, and bromic, $HBrO_3$, acids exist only in solution and concentration causes decomposition. The alkali metal salts are conveniently prepared by direct reaction between the halogen and hot aqueous alkali. Iodic acid, HIO_3, can be prepared as a stable, white crystalline solid by the oxidation of iodine with chloric acid solution:

$$HClO_3 + I_2 = HIO_3 + ICl$$
$$HClO_3 + ICl = HIO_3 + Cl_2$$

Although perchloric acid, $HClO_4$, is stable in dilute cold aqueous solutions, it has strong oxidising properties in hot concentrated solutions. It is a remarkably strong acid and many of its salts are freely soluble in water. The K^+, NH_4^+, Rb^+ and Cs^+ salts are sparingly soluble and are of analytical importance.

Perbromates have recently been prepared for the first time. They can be made by oxidation of bromates electrolytically or with xenon difluoride or fluorine. Alkali metal and ammonium salts appear to be quite stable. Free perbromic acid is strong and is stable in aqueous solution up to at least 6 mol dm^{-3}.

Several periodic acids have been prepared. Orthoperiodic acid, H_5IO_6, can be made by the action of a strong oxidising agent such as peroxodisulphate or hypochlorite on an iodate. This contains 6-covalent iodine:

$$
\begin{array}{c}
\text{O} \\
\text{HO} \quad \| \quad \text{OH} \\
\text{\textbackslash} \| / \\
\text{I} \\
/ | \text{\textbackslash} \\
\text{HO} \quad | \quad \text{OH} \\
\text{HO}
\end{array}
$$

On heating, H_5IO_6 loses water to give firstly pyroperiodic acid, $H_4I_2O_9$, then periodic acid, HIO_4, and finally it decomposes to iodic acid. These changes may be represented as the removal of water from the pentahydrate of the hypothetical oxide I_2O_7, although there is no evidence for the formation of the anhydrous oxide at any stage.

$$2H_5IO_6 \xrightarrow[\text{reduced pressure}]{353\ K} H_4I_2O_9 \xrightarrow{373\ K} 2HIO_4 \xrightarrow{411\ K} 2HIO_3 + O_2$$
$$(I_2O_7 . 5H_2O) \qquad (I_2O_7 . 2H_2O) \qquad (I_2O_7 . H_2O)$$

The interhalogens

These are binary compounds of the type AB_n where $n = 1$, 3, 5 or 7. They are usually prepared by direct reaction between the elements. When n is 3, 5 or 7, excess of halogen B must be present in the reaction mixture. The known interhalogens are listed in *Table 9.19*. Some of these compounds are not well-characterised because of their instability, thus accurate physical data are not available for BrF and BrCl, and because of the radioactivity of astatine, little is known about AtI. While many of the compounds containing

Table 9.19. THE INTERHALOGENS

Compound	Physical state (at 293 K)	Stability
ClF	gas	Stable to at least 523 K
BrF	gas	Complete decomp. to Br_2, BrF_3 and BrF_5 at 323 K
BrCl	gas	Instability prevents accurate determination of m.p. and b.p.
ICl	solid	Decomp. ≈ 370 K
IBr	solid	Decomp. ≈ 315 K
ClF$_3$	gas	Stable
BrF$_3$	liquid	Stable
IF$_3$	solid	Decomp. above 238·2 K
ICl$_3$	solid	Complete dissociation to ICl and Cl_2 at 350·2 K
ClF$_5$	—	Stable solid below 78·2 K
BrF$_5$	liquid	Stable to at least 733·2 K
IF$_5$	liquid	Stable to at least 673·2 K
IF$_7$	gas	Stable to at least 773·2 K
AtI	—	

fluorine are colourless, those containing two of the heavier halogens are coloured and the colour becomes more pronounced the greater the molecular weight (as with the halogens themselves).

Whether or not a conceivable interhalogen can be prepared depends on its stability relative to the elements of which it is composed and to other possible interhalogens.

For example, although IF is the most stable of the interhalogens of type AB with respect to the halogens from which it is made, it cannot be isolated in the pure state because it disproportionates according to:

$$5IF = 2I_2 + IF_5; \Delta G^0 = -166\cdot5 \text{ kJ mol}^{-1} \text{ IF}_5 \text{ formed}$$

The large negative value for ΔG^0 means that IF decomposes very easily according to the above equation.

Similarly, BrF_3 tends to disproportionate thus:

$$2BrF_3 = BrF + BrF_5; \Delta G^0 = -26 \text{ kJ mol}^{-1} \text{ BrF}_5 \text{ formed}$$

but the same compound is stable towards the dissociation

$$BrF_3 = BrF + F_2 \text{ for which } \Delta G^0 = +162 \text{ kJ mol}^{-1}$$

In contrast, ICl_3 is stable only at relatively low temperatures, otherwise it dissociates:

$$ICl_3 = ICl + Cl_2; \Delta G^0 = +8.4 \text{ kJ mol}^{-1}$$

The interhalogens are noted for their great chemical reactivity and they can be used instead of the halogens in preparative chemistry. Thus BrF_3 and ClF_3 convert metals or their oxides to the fluorides. In the liquid state, some interhalogens show electrical conductivity and this is attributed to self-ionisation. For example,

$$3ICl \rightleftharpoons I_2Cl^+ + ICl_2^-$$

Use of these compounds as solvents has made possible the preparation of a whole new range of halogen compounds.

The stereochemistry of the polyatomic molecules has been interpreted in the light of the Sidgwick–Powell concept.

Polyhalides

Many complex salts are known which contain polyhalide ions consisting of a central halogen joined to other halogen atoms. Iodine is the most common central atom: it may be combined with other iodines or with the atoms of other halogens.

The well-characterised polyiodides contain the ions I_3^-, I_5^-, I_7^- and I_9^-. The hydrated triiodide of sodium, $NaI_3.H_2O$, has been isolated. The higher polyiodides are formed only with larger cations, for example $KI_7.H_2O$, $(C_2H_5)_4NI_7$ and $RbI_9.3C_6H_6$. These compounds are prepared by reaction between the iodide and the stoichiometric amount of iodine in the appropriate solvent.

Mixed polyhalides, that is, those which contain two or more halogens, are prepared by reaction between the simple halide and the appropriate halogen or interhalogen in the absence of a solvent.

$$RbI + Br_2 \quad = RbIBr_2$$
$$CsBr + 2ICl = CsICl_2 + IBr$$

ICl_2^- and IBr_2^- are linear with iodine as the central atom.

The polyhalides all decompose to some extent into a halogen or inter-halogen compound and a metal halide. Dissociation always takes place so that the heaviest halogen atoms are present in the halogen or interhalogen compound formed. Thus:

$$CsICl_2 = CsCl + ICl$$

The reason for this is that the metal halide with the highest lattice enthalpy is thereby formed.

In the solid state, stability towards dissociation is determined by the size of the cation. The stability is greater the lower the charge density on the cation and so is a maximum for large cations carrying a single charge.

X-ray analyses have shown the existence of both symmetrical and unsymmetrical trihalide ions. In $As(C_6H_5)_4I_3$ the triiodide ion is

$$I \overset{290\,pm}{\diagdown} I \overset{290\,pm}{\diagup} I$$
$$176°$$

but in NH_4I_3 it is

$$I \overset{310\,pm}{\diagdown} I \overset{282\,pm}{\diagup} I$$
$$177°$$

In $N(CH_3)_4ICl_2$, the anion is linear and symmetrical.

$$Cl \overset{234\,pm}{\text{———}} I \overset{234\,pm}{\text{———}} Cl$$

The tetrachloroiodates are among the most stable of this class. These are prepared by adding iodine to solutions of the simple chloride in HCl, followed by saturation with Cl_2. All the alkali metals form orange–red polyhalides of the type $MICl_4$ and a series of salts with dipositive cations ($Mg^{2+}, Ca^{2+}, Co^{2+}, Ni^{2+}$ and Mn^{2+}) is also known. In aqueous solution, dissociation occurs to form the metal chloride and ICl_3, the latter then hydrolyses to iodate.

The $[ICl_4]^-$ ion is square planar. This shape is derived from an octahedral arrangement in the iodine atom of the six electron pairs: two of these are lone pairs at the apices of the octahedron and the remaining four, directed towards the corners of a square, are bonding pairs.

The higher polyiodides can be regarded as arising from an I_3^- ion and one or more I_2 molecules. Thus the I_5^- ion in $N(CH_3)_4I_5$ is L-shaped:

$$\underset{I}{\overset{\overset{\displaystyle I\overset{290pm}{\rule{1.5cm}{0.4pt}}I\overset{290pm}{\rule{1.2cm}{0.4pt}}I}{}}{\underset{\displaystyle I}{\overset{\displaystyle I}{\big|}}}}$$

Positive chlorine, bromine and iodine

For some years it was believed that the cations Cl^+, Br^+ and I^+ were important as stable species or as reaction intermediates in many systems. For example, the characteristic blue colour of iodine in 65 per cent oleum was attributed to the presence of I^+. Again, the observation that iodine migrates to the cathode when solutions of ICl and IBr in acetic acid are electrolysed was interpreted in terms of the movement of I^+ ions. However, it was subsequently established by Gillespie and Milne in 1965 that the blue species in oleum is actually the I_2^+ cation, formed by the oxidation of I_2 by SO_3, and the electrolytic studies show merely that positively charged species containing iodine are involved but give no information about their complexity. There appears now to be no evidence to substantiate the existence of monatomic ions of any of these halogens.

In addition to I_2^+ some other polyatomic cations of the halogens do exist. They are all highly electrophilic and can only be made in the presence of molecules and anions of very low basicity, for example, in solvents such as fluorosulphonic acid, disulphuric acid, antimony pentafluoride and sulphur dioxide. Thus a solution of iodic acid and iodine in 100 per cent sulphuric acid is believed to contain I_3^+ and I_5^+ cations. I_2^+ dimerises to I_4^{2+} when its solutions in fluorosulphonic acid are cooled. The cations Br_3^+ and Br_2^+ have been identified and there is some evidence that unstable Cl_3^+ ions can exist at low temperatures (197 K).

Astatine

Our knowledge of the chemistry of this element is based on radiochemical studies of its co-precipitation behaviour with iodine and the heavy elements of adjacent Periodic Groups. The longest-lived isotope of astatine is ^{211}At, with a half-life of 8·3 h. This has such a high specific activity that chemical operations are not feasible even with milligram quantities because of the intense heating and radiation effects.

Astatine was first identified as a product of the bombardment of bismuth with α-particles: ^{209}Bi$(\alpha, 2n)$ ^{211}At. As many as 21 isotopes of astatine have been identified*. Preparation of these isotopes is more difficult than with most elements because they cannot be prepared by neutron irradiation and positive ion accelerators must be used.

For the preparation of astatine, bismuth is irradiated either as the metal or its oxide. The astatine is isolated after irradiation by vacuum sublimation and

* Ed. EMELEUS and SHARPE. *Advances in Inorganic Chemistry and Radiochemistry*, Academic Press, New York, 1964.

collection on a cold finger. Alternatively, irradiated bismuth oxide is dissolved in aqueous perchloric acid (containing iodine) and the bismuth precipitated with phosphate. The solution remaining contains astatine iodide, AtI, and the astatine can be precipitated from this by the precipitation of AgI or PdI_2 in the presence of a reducing agent.

The astatide ion, At^-, can be prepared by reduction of AtI with sulphur dioxide. Ion migration experiments have shown the presence of a negative charge on this ion and its properties are very similar to those of iodide.

There is evidence for the existence of At^+ and AtO_3^- and for some other oxidation state (either AtO^- or AtO_2^-). Polyhalides have also been prepared. For example, CsI_3 containing $CsAtI_2$, has been made from an aqueous solution of CsI into which I_2 and AtI were extracted from chloroform solution. On decomposition, AtI vapour is formed (cf other polyhalides). Astatine is co-precipitated with bismuth and antimony from HCl solutions by H_2S. This suggests an insoluble sulphide may be formed and hence is evidence for greater metallic character than found in iodine.

SUGGESTED REFERENCES FOR FURTHER READING

BAGNALL, K. W. *The Chemistry of Selenium, Tellurium and Polonium*, Elsevier, Amsterdam, 1966.

EVEREST, D. A. *The Chemistry of Beryllium*, Elsevier, Amsterdam, 1964.

JOLLY, W. L. *The Inorganic Chemistry of Nitrogen*, Benjamin, Inc., New York, 1964.

JOLLY, W. L. *The Synthesis and Characterisation of Inorganic Compounds*, Prentice-Hall, New Jersey, 1970.

MACKAY, K. M. and MACKAY, R. A. *Introduction to Modern Inorganic Chemistry*, Intertext, London, 1968.

MOODY, G. J. and THOMAS, J. D. R. *Noble Gases and their Compounds*, Pergamon, Oxford, 1964.

10 The comparative chemistry of the transition elements

The transition and inner transition elements consist of the four series

Sc (21) to Cu (29),
Y (39) to Ag (47),
La (57) to Au (79)
and Ac (89) to Lr (103).

In each of these series the electronic configuration in the occupied shell of highest principal quantum number remains constant whilst an inner quantum shell is progressively filled up as the atomic number increases. This is shown in *Table 2.4* (p. 38).

In the elements from Sc to Cu, Y to Ag, and Hf (72) to Au, a penultimate d sub-shell is filled up and many of the properties associated with transition metals are due to the presence of at least one d electron. The Group III A elements Sc, Y and La, by definition transitional in the atomic state, are present in many of their compounds as tripositive ions which have a noble gas configuration. They do not show many of the properties typical of the transition elements as a class and it is therefore appropriate to consider their chemistry separately.

The elements immediately following La constitute the inner transition series. These are the lanthanoids, Ce(58) to Lu (71), in which the 4f sub-shell is being filled. The lanthanoids have a strong tendency to form tripositive ions and in doing so are like the other Group III A elements, Sc, Y and La. These three elements and the lanthanoids are collectively known as the rare earths.

Chemical and physical evidence supports the existence of a second series of inner transition elements following Ac (89). These are called the actinoids, in which the 5f sub-shell is incomplete, and they include the artificially prepared transuranic elements from Np (93) onward. The actinoids, by analogy with the lanthanoids, conclude with Lr (103).

PROPERTIES OF THE d-TRANSITION ELEMENTS

The d-transition elements are all metals, generally of high density, low atomic volume and high melting and boiling points. Their structures have been described earlier.

In *Table 10.1* are listed the melting and boiling points, and the atomic and some of the ionic radii of the three series of d-transition metals. The metals copper, silver and gold are conveniently considered here as well for, although their mono positive ions possess completely filled d sub-shells and so are not strictly transitional, each of these metals can show at least one oxidation state higher than one in which the d sub-shell is incomplete. Immediately following this section, features of the chemistry of the Group

Table 10.1. THE PROPERTIES OF THE d-TRANSITION METALS

	Ti	V	Cr	Mn	Fe	Co	Ni	Cu
Atomic number	22	23	24	25	26	27	28	29
m.p. T/K	1 998	1 988	2 103	1 520	1 801	1 763	1 725	1 356
b.p. T/K	3 533	3 773	2 573	2 303	3 008	3 373	3 113	2 853
Atomic radius r/pm	132	122	117	117	116	116	115	117
Ionic radius M^{2+}								
r/pm	86	79	73	67	61	65	70	
			LS	LS	LS	LS		
			82	82	77	73		
			HS	HS	HS	HS		
Ionic radius M^{3+}								
r/pm	67	64	62	58	55	52	56	
				LS	LS	LS	LS	
				65	64	61	60	
				HS	HS	HS	HS	

	Zr	Nb	Mo	Tc	Ru	Rh	Pd	Ag
Atomic number	40	41	42	43	44	45	46	47
m.p. T/K	2 373	2 223	2 870		≈2 670	≈2 240	1 828	1 233
b.p. T/K	≈3 870	≈5 370	≈5 070		≈4 470	≈4 170	3 443	2 453
Atomic radius r/pm	145	134	129		124	125	128	134

	Hf	Ta	W	Re	Os	Ir	Pt	Au
Atomic number	72	73	74	75	76	77	78	79
m.p. T/K	2 570	3 280	3 670	3 420	≈2 970	2 727	2 047	1 336
b.p. T/K	≈5 470	≈6 270	≈6 070		≈4 870	≈4 770	≈4 070	2 970

II B metals, zinc, cadmium and mercury, are described although these are not, of course, transition elements.

The ionic radii quoted for the first series of metals refer to 6 co-ordination. As we have already seen, the radius of an ion varies with its co-ordination number and, in the case of certain transition metal ions, crystal field effects and the spin state of the ion also markedly affect the value assigned to this parameter. For these reasons, the radii given for M^{2+} in *Table 10.1* are usually slightly different from the values calculated from internuclear distances in the binary oxides, MO (p. 208). In general, ions of the same charge in a given transition series show a progressive decrease in radius with increase in atomic number. The reason for this is that the extra electron enters a d orbital each time the nuclear charge increases by unity. The screening power of a d electron is small, the net electrostatic field between the nuclear charge and the outermost electrons is increased by increase in nuclear charge and so the ionic radius falls. The atomic radius varies in the same way. The variation within a series is quite small and one consequence is that ionic compounds of the same formula type are often isostructural. There are other resemblances within a series which can be attributed to similarities in ionic size. Similar relationships exist in the second and third series elements but are less well-defined because relatively few compounds are known in which simple monatomic cations of these elements exist.

The electronic configurations of the atoms are generally $d^x s^2$. The energies of the $(n-1)d$ and ns orbitals are, however, close to one another and there are some instances where an alternative ground state configuration is found. For example, chromium and copper have the configurations $3d^5 4s^1$ and $3d^{10} 4s^1$ respectively. These are an indication that there is an extra degree of stability associated with half-filled and completely filled d sub-shells.

Oxidation states

A very characteristic property of a d-transition metal is its ability to show several different oxidation states. The particular oxidation state is dependent on the nature of the other elements with which it is combined. The highest states are found in compounds with fluorine and/or oxygen. These are the two most electronegative elements and a simple argument based on the Electroneutrality Principle indicates why they act in this way. Bond formation between a metal carrying a high formal positive charge and negative oxide or fluoride ions serves to reduce the charge on the metal. If there were a complete transfer of one or more electrons to the metal ion this would be reduced and the non-metal would be oxidised. Only those elements which retain a very strong hold on their electrons, that is, are highly electronegative, can therefore combine with a metal and preserve its high oxidation state. It is important to recognise that covalent bonding plays a significant and sometimes predominant part in these compounds. Oxygen is particularly effective at stabilising high oxidation states because, as well as forming a σ-bond, another pair of oxygen electrons can be involved in the formation of a donor π-bond. When double bonding like this is possible, only a small number of oxygen atoms suffices to reduce nearly to zero the net charge on a metal in a high oxidation state.

Much of the familiar chemistry of any element concerns its compounds which are stable in water and in ordinary atmospheres. Where more than one oxidation state is possible, standard electrode potential data provide a guide to relative stabilities of the different states with respect to oxidation, reduction or disproportionation. Many factors contribute towards the stability of a particular oxidation state in aqueous solution. These include the size and electronic structure of the metal atom and its ions, the magnitude of hydration energies and of other relevant enthalpy changes such as the ionisation potentials of the metal and the lattice enthalpies of its compounds.

The range of oxidation states in water is restricted by the nature of the solvent which is itself susceptible to oxidation or reduction under appropriate conditions. Similarly, reaction with aerial oxygen will inhibit the attainment of certain oxidation states. Each transition metal has a much more extensive chemistry than we might expect from those compounds which are stable in the presence of water and air. This has been amply demonstrated, particularly in recent years, by experimental studies in non-aqueous media, in the molten state, in the vapour phase at high temperatures and in the absence of air.

Low oxidation states such as $+1, 0$ or below can generally be reached only in the absence of air. Notable exceptions are Cu^+, Ag^+ and Au^+. Even when air is excluded, some form of stabilisation, involving complex-formation, is usually necessary. The bonding within compounds of a transition metal in a low oxidation state is largely covalent. The formation of several bonds by one metal atom would normally result in a build-up of negative charge on the metal. This would contravene the Electroneutrality Principle and we would not expect the compounds formed to be stable. If, however, the ligand is of such a kind that it has one or more vacant acceptor orbitals and the metal has electrons in d orbitals of appropriate symmetry so that bonding can occur, the metal may in this way be able to donate some of its electronic charge to the ligands. The build-up of negative charge would be avoided and the compound stabilised. In this category are ligands like CO, CN^-, NO^+ and certain compounds of phosphorus, arsenic, sulphur and selenium. In addition, organic ligands like 1,10-phenanthroline and 2,2'-bipyridyl can receive electrons because they have empty π^* orbitals. They are also chelating agents and this adds to their effectiveness at stabilising low oxidation states.

For the metals titanium to manganese inclusive the highest oxidation state shown corresponds with the total number of 3d and 4s electrons in the atom. The oxidising ability of the highest state increases from Ti(IV) to Mn(VII). For the metals of the first series beyond manganese, the highest oxidation state stable in aqueous solution is $+3$. There is also a marked change in the properties of the dipositive ion from manganese onward. Thus the ions Ti^{2+} to Cr^{2+} inclusive are strongly reducing, but Mn^{2+} and the corresponding ions beyond form stable compounds. Mn^{2+} has the configuration $3d^5$ and this represents the half-way stage in the filling-up of the 3d sub-shell. In most manganese (II) compounds, the five d electrons are distributed singly with parallel spins in the available orbitals. This, according to Hund's rule of maximum multiplicity, constitutes a particularly stable arrangement. The few exceptions are found in complexes like $[Mn(CN)_6]^{4-}$, where the ligand exerts a strong field and the electrons are forced to pair up in the t_{2g} orbitals. Beyond manganese, the $+2$ state becomes increasingly important in aqueous solutions and it is evidently much more difficult for the 3d electrons to become involved in compound formation.

Whenever the highest oxidation state shown is the same as the number of the Periodic Group, there are resemblances between the chemistry of the transition metal and the non-transition elements of the same Group. Thus titanium (IV) compounds resemble those of Sn(IV) and Si(IV), vanadium (V) is similar to phosphorus (V) and so on as far as Group VII. Such similarities become less marked as the non-metallic character of the non-transition element increases and, by the time Group VII is reached they are mainly formal.

In the second and third series of d-transition metals, the highest oxidation state attained by the first few metals again corresponds with the number of d and s electrons in the atom. This generalisation is true up to the Group VIII metals, ruthenium and osmium, both of which form tetraoxides, MO_4. Beyond these, there is a decrease in the highest possible oxidation state, even though the platinum metals generally show higher oxidation states than are found in the iron, cobalt and nickel sub-group.

The oxides and oxo-anions of a second or third series metal in a high oxidation state are not strongly oxidising like the analogous compounds of the first row metal in the same Group. For example, chromates (VI) are much stronger oxidising agents than either molybdates or tungstates (VI). This reflects the general trend to more stable oxidation states as we move from the first to the second and third transition series.

On the other hand, a first series metal may form a stable, monatomic ion carrying a charge of $+2$ or $+3$ which has no counterpart in the chemistry of the heavier metals of that Group. For example, there is an extensive chemistry of Mn^{2+} but Tc and Re in the $+2$ state are found only in a few complexes. Again, Cr^{3+} and its complexes are well-known and stable entities but Mo and W in the $+3$ state are much less stable. The higher oxidation states of molybdenum and tungsten up to and including $+6$ are much more important.

As a consequence of the lanthanide contraction, there is little or no difference between the atomic radii or between the ionic radii of the second and third series metals in a Group. This is particularly significant in the early transition metal groups such as IV and V and accounts for close resemblances between pairs of metals in these series and for differences between them and the corresponding metal in the first series.

Compounds

Oxides

The binary oxides formed by the d-transition metals are listed in *Table 10.2*. Compounds with the metal in a low oxidation state have predominantly basic properties; those with the metal in a high oxidation state are amphoteric or acidic. As we have already noted, combination with oxygen stabilises the very high oxidation states of these metals. When the metal in this state has the same valency as the representative elements of the same Periodic Group there is often some similarity between the compounds with oxygen.

The table shows that the highest oxidation state towards oxygen is reached with manganese in the first series but is found in Group VIII in the second and third series.

Binary oxides and oxo-acids of the first series metals are considered individually in the succeeding pages but the oxygen compounds of the heavier transition metals are not considered here.

Fluorides

The binary fluorides, together with some mixed oxide fluorides and complex fluorides, of the transition metals are listed in *Table 10.3*. This illustrates how the highest oxidation states are brought out in combination with fluorine. As we move along a transition metal series, the highest oxidation state shown in a binary fluoride reaches a maximum and then falls off quite sharply. Thus, in the first series, the highest oxidation state of $+6$ is reached with chromium (albeit in a very unstable compound); thereafter manganese

Table 10.2. THE OXIDES OF THE TRANSITION METALS

IV A	V A	VI A	VII A	VIII			I B
							Cu_2O (b, w.a.)
TiO (b)	VO (b)	CrO (b)	MnO (b)	FeO (b, w.a.)	CoO (b, w.a.)	NiO (b)	CuO (am.)
Ti_2O_3 (b)	V_2O_3 (b)	Cr_2O_3 (am.)	Mn_2O_3 (b)	Fe_2O_3 (am.)	(Co_2O_3)	(Ni_2O_3)	
			Mn_3O_4	Fe_3O_4	Co_3O_4		
TiO_2 (am.)	VO_2 (am.)	CrO_2 (a)	MnO_2 (am.)		(CoO_2)	(NiO_2)	
	V_2O_5 (am.)						
		CrO_3 (a)					
			Mn_2O_7 (a)				
							Ag_2O (b, w.a.)
	NbO				RhO	PdO (b)	AgO
		(Mo_2O_3)			Rh_2O_3	Pd_2O_3	
ZrO_2 (am.)	NbO_2 (a)	MoO_2		RuO_2	(RhO_2)	(PdO_2)	
	Nb_2O_5 (a)	Mo_2O_5 (am.)					
		MoO_3 (a)	Tc_2O_7 (a)				
				RuO_4			
			(Re_2O_3)		Ir_2O_3	Pt_2O_3	Au_2O Au_2O_3 (a, w.b.)
HfO_2 (am.)	TaO_2 (a)	WO_2	ReO_2 (am.)	OsO_2	IrO_2 (b)	PtO_2 (am.)	
	Ta_2O_5 (a)						
		WO_3 (a)	ReO_3 Re_2O_7 (a)	OsO_3		PtO_3	
				OsO_4 (a)			

b = basic;　w.b. = weak basic;　a = acidic;　w.a. = weak acidic　am. = amphoteric
Brackets around a formula indicate the compound has not been prepared in a pure state.

forms unstable MnF_4 and iron and cobalt their trifluorides as the highest oxidation states in combination with fluorine.

Chemical reactivity and thermal instability show a general correlation. Thus those fluorides which are unstable are good fluorinating agents. The fluorides of the first series metals are described in more detail later on.

Chlorides

The binary chlorides and some mixed oxide chlorides and complex chlorides of the transition metals are listed in *Table 10.4.*

Table 10.3. TRANSITION METAL FLUORIDES

Group	IVA	VA	VIA	VIIA	VIII			IB
		VF_2	CrF_2	MnF_2	FeF_2	CoF_2	NiF_2	CuF_2
	TiF_3	VF_3	CrF_3	MnF_3	FeF_3	CoF_3	NiF_6^{3-}	CuF_6^{3-}
	TiF_4	VF_4	CrF_4	MnF_4^*		CoF_6^{2-}	NiF_6^{2-}	
		VF_5	CrF_5					
			CrF_6^*					
				MnO_3F^*				
								Ag_2F
								AgF
	ZrF_2					PdF_2		AgF_2
		$NbF_{2.5}$						
	ZrF_3	NbF_3	MoF_3		RuF_3	RhF_3		AgF_6^{3-}
	ZrF_4		MoF_4	TcF_6^{2-}	RuF_4	RhF_4	PdF_4	
		NbF_5	MoF_5	TcF_5	RuF_5	RhF_5		
			MoF_6	TcF_6	RuF_6^*	RhF_6^*		
				TcO_3F				
						IrF_3		AuF_3
	HfF_4		WF_4	ReF_4	OsF_4	IrF_6^{2-}	PtF_4	
		TaF_5	WF_5	ReF_5	OsF_5	IrF_5	PtF_5	
			WF_6	ReF_6	OsF_6	ItF_6	PtF_6^*	
				ReF_7	OsF_7^*			
					OsO_3F_2			

* Denotes a compound which is thermally unstable at or below 293 K.

Table 10.4. TRANSITION METAL CHLORIDES

Group	IVA	VA	VIA	VIIA	VIII			IB
								$CuCl$
	$TiCl_2$	VCl_2	$CrCl_2$	$MnCl_2$	$FeCl_2$	$CoCl_2$	$NiCl_2$	$CuCl_2$
	$TiCl_3$	VCl_3	$CrCl_3$	$(MnCl_3)$	$FeCl_3$			
	$TiCl_4$	VCl_4	$(CrCl_4)$	$(MnCl_6^{2-})$				
		$VOCl_3$	$CrOCl_3$					
			CrO_2Cl_2					
				MnO_3Cl				
	$ZrCl$							$AgCl$
	$ZrCl_2$		$MoCl_2$				$PdCl_2$	
		Nb_6Cl_{14}						
	$ZrCl_3$	$NbCl_3$	$MoCl_3$		$RuCl_3$	$RhCl_3$		
	$ZrCl_4$	$NbCl_4$	$MoCl_4$	$TcCl_4$	$RuCl_4$			
		$NbCl_5$	$MoCl_5$	$TcOCl_3$				
			$MoCl_6$,	$TcCl_6$				
			MoO_2Cl_2					
				TcO_3Cl				
	$HfCl$							$AuCl$
	$HfCl_2$	Ta_6Cl_{14}	WCl_2				$PtCl_2$	
	$HfCl_3$	$TaCl_3$	WCl_3	$ReCl_3$	$OsCl_3$	$IrCl_3$		$AuCl_3$
	$HfCl_4$	$TaCl_4$	WCl_4	$ReCl_4$	$OsCl_4$		$PtCl_4$	
		$TaCl_5$	WCl_5	$ReCl_5$	$OsOCl_4$			
			WCl_6	$ReCl_6$				
				ReO_3Cl				

For some low oxidation states, the binary chloride is known but the fluoride is not. Conversely, for some high oxidation states, the fluoride, but not the chloride, can be made.

The highest oxidation state in a binary chloride is $+4$ in the first series (with vanadium and chromium). In the second and third series it is $+6$ (in the case of tungsten and rhenium). The same general trend is evident as for fluorides, namely that in any sub-group, the highest oxidation state tends to increase as the atomic number increases.

The compounds of the first series metals with chlorine are discussed later.

Bromides and iodides

These are listed in *Tables 10.5* and *10.6* so that a comparison of the range of oxidation states shown towards each halogen may be made. Much less is known about the bromides and iodides and they are not considered further here.

GROUP IV A—TITANIUM

Oxidation states

Titanium has four electrons in its valency shell and its highest oxidation state is $+4$. This is the most stable state and all naturally-occurring titanium compounds have the metal in this state. Titanium compounds are also known wherein the metal is in oxidation states -1, 0, $+2$ and $+3$.

Formal oxidation states of -1 and 0 are found in the complexes of titanium with 2,2'-bipyridyl. The reduction of 1 mol of $TiCl_4$ with 2 or 2·5 mol of (2,2'-bipyridyl)dilithium (Li_2bipy) in tetrahydrofuran in the presence of excess bipyridyl gives respectively $Ti(bipy)_3$ or $[Ti(bipy)_3]^-$.

Titanium(II) is known as the dichloride, dibromide and diiodide and in the oxide, TiO. The $+2$ state is unstable in aqueous solution, the Ti^{2+} ion being so strongly reducing that hydrogen ion is immediately reduced to hydrogen:

$$Ti^{3+} + e = Ti^{2+} : E° \approx -0.37\ V$$

Titanium(III) compounds include the oxide, the sulphate, double sulphates like $CsTi(SO_4)_2 . 12H_2O$ and the trihalides. The tripositive titanium ion has a similar radius to that of V^{3+}, Cr^{3+}, Mn^{3+}, Fe^{3+} and Al^{3+}. This accounts for similarities between salts containing these ions. For example, all the caesium alums are isomorphous.

Titanium(III) has moderately strong reducing properties (for $Ti^4 + e = Ti^{3+}$, $E^0 = +0.04\ V$) and is easily oxidised by atmospheric oxygen. Its salts are used in volumetric analysis but solutions must be stored in sealed containers to avoid oxidation.

Many compounds of titanium(IV) are known. The dioxide, TiO_2, is very important and so are the tetrahalides. The latter, from the tetrachloride

Table 10.5. TRANSITION METAL BROMIDES

Group	IVA	VA	VIA	VIIA	VIII			IB
								$CuBr$
	$TiBr_2$	VBr_2	$CrBr_2$	$MnBr_2$	$FeBr_2$	$CoBr_2$	$NiBr_2$	$CuBr_2$
	$TiBr_3$	VBr_3	$CrBr_3$		$FeBr_3$			
	$TiBr_4$	VBr_4						
		$VOBr_3$						
			CrO_2Br_2					
								$AgBr$
	$ZrBr_2$		$MoBr_2$				$PdBr_2$	
		$NbBr_{2\cdot67}$						
	$ZrBr_3$	$NbBr_3$	$MoBr_3$		$RuBr_3$	$RhBr_6^{3-}$		
	$ZrBr_4$	$NbBr_4$	$MoBr_4$					
		$NbBr_5$	$MoOBr_3$					
			MoO_2Br_2					
								$AuBr$
	$HfBr_2$		WBr_2				$PtBr_2$	
		Ta_6Br_{14}						
	$HfBr_3$	$TaBr_3$	WBr_3	$ReBr_3$	$OsBr_3$	$IrBr_3$		$AuBr_3$
	$HfBr_4$	$TaBr_4$	WBr_4	$ReBr_4$	$OsBr_4$	$IrBr_6^{2-}$	$PtBr_4$	
		$TaBr_5$	WBr_5	$ReBr_5$				
			WBr_6					

Table 10.6. TRANSITION METAL IODIDES

Group	IVA	VA	VIA	VIIA	VIII			IB
								CuI
	TiI_2	VI_2	CrI_2	MnI_2	FeI_2	CoI_2	NiI_2	
	TiI_3	VI_3	CrI_3					
	TiI_4							
								AgI
	ZrI_2		MoI_2				PdI_2	
	ZrI_3	NbI_3	MoI_3		RuI_3			
	ZrI_4	NbI_4						
		NbI_5						
				ReI	OsI			AuI
				ReI_2	OsI_2		PtI_2	
		Ta_6I_{14}						
	HfI_3		WI_3	ReI_3	OsI_3		PtI_3	
	HfI_4	TaI_4		ReI_4			PtI_4	
		TaI_5						

onward show covalent properties consistent with compounds of a metal in a high oxidation state.

Oxides and oxo-compounds

Titanium (II) oxide is made by the reduction of TiO_2 with titanium metal:

$$Ti + TiO_2 = 2TiO$$

This compound shows marked non-stoichiometry in its composition. Above 1170 K, TiO has a defect sodium chloride lattice over the composition range $TiO_{0.64}$ to $TiO_{1.27}$. Whatever the composition in this range, there are always lattice sites unfilled and the variation possible in composition is due to changes in the ratio of titanium to oxygen site vacancies. For example, in $TiO_{0.86}$, 10·5 per cent of the titanium and 22·8 per cent of the oxygen sites are vacant. In TiO, 15 per cent of both sites are vacant, and in $TiO_{1.2}$ there are 22·0 per cent titanium vacancies and 6·3 per cent oxygen vacancies. Electrical neutrality is preserved in the crystal by changes in the charges on the titanium ions. Below 1170 K, titanium (II) oxide changes to a mixture of much more ordered structures.

Titanium (III) oxide is prepared by heating TiO_2 with carbon or by direct reaction between TiO_2 and titanium metal in the correct proportions:

$$Ti + 3TiO_2 = 2Ti_2O_3 \text{ (at 1870 K)}$$

The oxide is a violet powder, m.p. 2102 K, with a corundum (Al_2O_3) structure. In this, the oxide ions form a hexagonally close-packed array and the titanium ions are distributed symmetrically among the octahedral sites. It behaves as a basic oxide; reaction with sulphuric acid produces the sulphate, $Ti_2(SO_4)_3$.

Titanium (IV) oxide is found in nature in three forms: rutile, anatase and brookite. Rutile is the most common and the other two are comparatively rare. Rutile and anatase are tetragonal and brookite is rhombic.

Titanium (IV) oxide has important industrial applications as a white pigment for paints. It is also the chief source of titanium metal, produced via the tetrachloride.

TiO_2 is insoluble in water and dilute acids but slowly dissolves in concentrated sulphuric acid or fused alkali hydrogen sulphates to give titanium (IV) sulphate. This salt is hydrolysed by water to an unreactive hydrous titanium (IV) oxide. The addition of ammonia to an acidic solution containing Ti(IV) precipitates a much more reactive form which readily dissolves, for example, in caustic alkali to give alkali metal titanates.

Oxo-cations

In acidic solutions of titanium (IV), the Ti^{4+} ion does not exist and, for many years, it was believed that the stable species was the titanyl ion, TiO^{2+}. This ion has no independent existence in the solid state for compounds such as 'titanyl' sulphate, $TiOSO_4.H_2O$, actually contain polymeric chains of alternate titanium and oxygen atoms, —Ti—O—Ti—O—Ti—. Each titanium atom is also bound to an oxygen from each of three sulphato groups and to a water molecule, building up a distorted octahedron of oxygens around it. In acidic aqueous solution it is likely that octahedral co-ordination is preserved in the form of cationic species like $[Ti(OH)_2(H_2O)_4]^{2+}$. If the pH of the solution is raised, loss of protons and polymerisation occur and hydrous titanium dioxide is precipitated.

Titanates

The simple hydroxide, $Ti(OH)_4$, does not appear to exist nor, with one exception, do discrete titanate ions, $[TiO_4]^{4-}$.

Metal titanates, $M_2^ITiO_3$ (M = Li, Na or K) can be made by dissolving hydrous titanium dioxide in concentrated aqueous alkali hydroxide. Their structures are not known.

Two series of titanates of divalent metals have been made. Their general formulae are $M^{II}TiO_3$ (M = Ca, Mg, Mn, Co, Fe, Ni) and $M_2^{II}TiO_4$ (M = Mg, Zn, Mn, Co). They are prepared by heating the stoichiometric amounts of metal oxide and titanium dioxide in a sealed tube above 1270 K for several hours. Structurally they are all mixed oxides and not salts of hypothetical 'titanic' acids.

For the compounds $M^{II}TiO_3$ two structural types are found. If the metal M is of similar ionic size to Ti^{4+}, the titanate has the *ilmenite* ($FeTiO_3$) structure. In this the oxide ions are hexagonally close-packed with one-third of the octahedral holes occupied by M^{2+} ions and another third by titanium ions. If M is much larger than Ti^{4+}, the *perovskite* ($CaTiO_3$) structure is found. This has a cubic close-packed array of oxide ions and the larger cations and the Ti^{4+} ions are distributed in the octahedral holes in an ordered fashion. In general, mixed metal oxides are found in one or other of these two structures.

Titanates of formula $M_2^{II}TiO_4$ have the *spinel* structure. The oxide ions are close-packed and the cations occupy both tetrahedral and octahedral holes.

Barium titanate, Ba_2TiO_4, is the only compound which appears to contain discrete TiO_4 groups. These are in the shape of a somewhat distorted tetrahedron.

Peroxo-compounds

The addition of hydrogen peroxide to solutions of Ti(IV) produces intense colourations. In strongly acidic solutions the colour is red; in dilute acid or in neutral solution, it is orange–yellow. The reactions in aqueous solutions form the basis of a sensitive test for either reagent. Below pH = 1, the coloured complex is believed to be a mononuclear ion, $Ti\,O_2(OH)^+$. This condenses between pH 1 and 3 to a dinuclear species containing a Ti_2O_5 unit. The structure of this is probably in which the titanium is 6 co-ordinated by water molecules or other ligands present in solution. The addition of alkali deprotonates the co-ordinated water and these deprotonated products condense to form an insoluble peroxo-titanium hydrate, $TiO_3(H_2O)_x$ (x lies between 1 and 2). The colour of the complexes arises from electron transfer processes from the O_2^{2-} ligand to the metal ion.

Fluorides

Titanium (III) fluoride is formed in high yield in the reaction between titanium and titanium tetrafluoride when these are heated together at 1170 K in a nickel boat in an argon atmosphere.

$$Ti + 3TiF_4 = 4TiF_3$$

The trifluoride is a blue solid, stable in air at room temperature but decomposing to TiO_2 when heated in air to 373 K. It sublimes at 1203 K *in vacuo*. It is insoluble in water, alcohol and dilute acids and bases. These properties are in keeping with a polymeric structure: each titanium atom is located at the centre of a slightly distorted octahedron of fluorines, all of which are shared with other titanium atoms.

Titanium (IV) fluoride is made by the fluorination of titanium sponge at 473 K or by reaction between $TiCl_4$ and anhydrous liquid HF. The compound is a white, hygroscopic solid, which volatilises at 557 K. Its structure is likely to be polymeric with the titanium atoms linked by fluorine bridges. Its properties are markedly different from those of the tetrafluorides of C, Si and Ge, all of which are monomeric and molecular compounds. TiF_4 is reduced to the metal by sodium, magnesium or calcium. The metal has a tendency to expand its co-ordination shell to 6 as in the 1:1 and 1:2 adducts formed with nitriles, ketones, alcohols, amines and other organic donor molecules. The 1:1 adducts are probably polymers: the 1:2 adducts are monomeric. Both appear to contain 6 co-ordinate titanium.

Dissolution of TiO_2 in aqueous HF produces hexafluoro-titanate (IV) ions, TiF_6^{2-}. Well-defined salts containing TiF_6^{2-} are known, titanium being very similar in this respect to other elements of Group IV, silicon, germanium and tin. The parent acid, H_2TiF_6, is strong and resistant to hydrolysis.

Chlorides

Titanium (II) chloride is most conveniently obtained by the disproportionation reaction:

$$2TiCl_3 = TiCl_4 + TiCl_2$$

which occurs at 723 K. It is a red–brown solid, m.p. 1308 K, which is believed to have the cadmium iodide structure. It reacts violently with water to reduce it to hydrogen.

Titanium (III) chloride can be made by reducing $TiCl_4$ with hydrogen at temperatures above 770 K. Alternatively, titanium itself may be used as reducing agent:

$$3TiCl_4 + Ti = 4TiCl_3$$

although some dichloride is produced at the same time.

These reactions give one form of $TiCl_3$, the violet (α) form. This has a layer-lattice type of structure composed of two-dimensional sheets in which all the metal atoms are co-ordinated octahedrally. Three other modifications of $TiCl_3$ (β, γ and δ) have been isolated. These all also contain octahedral titanium and differ from the α-form in the mode of packing and linking the $TiCl_6$ octahedra.

The solid sublimes *in vacuo* at 700–715 K and disproportionates as indicated earlier. It is oxidised by moist air to form hydrated titanium dioxide. Electrolytic reduction of Ti(IV) solutions in hydrochloric acid leads to the formation of the hexahydrate, which is formulated as *trans*-$[TiCl_2(H_2O)_4]$Cl.$2H_2O$.

$TiCl_3$ is one of a number of transition metal trihalides which are useful catalysts for the polymerisation of olefins to isotactic polymers.

Titanium (IV) chloride can be prepared by the action of chlorine on TiO_2 at 1170–1270 K in the presence of carbon as reducing agent:

$$TiO_2 + 2Cl_2 + 2C = TiCl_4 + 2CO$$

This reaction is of great importance in the extraction metallurgy of titanium because the tetrachloride can be conveniently reduced to the metal by magnesium or sodium. Reactions of the dioxide with chlorinating agents like carbonyl chloride or thionyl chloride may also be used to make $TiCl_4$.

The pure compound is a liquid at ordinary temperature, b.p. 409·5 K. The vapour consists of tetrahedral molecules. $TiCl_4$ is highly reactive towards water and is rapidly hydrolysed to oxo-chlorides and hydrochloric acid.

Like TiF_4, the tetrachloride reacts with many donor molecules to form 1:1 and 1:2 adducts. For example, the compounds $TiCl_4 . POCl_3$ and $TiCl_4 . 2POCl_3$ are produced by reaction with $POCl_3$. The oxygen atom acts as donor for each ligand molecule. In both adducts the titanium atom has expanded its co-ordination shell to six atoms. The 1:1 adduct is composed of dimeric molecules. The two titanium atoms are bridged by two chlorine atoms and each titanium is also bound to three non-bridging chlorines and one $POCl_3$ molecule, thus completing the co-ordination number of 6. The 1:2 adduct is monomeric and also contains octahedral titanium.

Chlorines in $TiCl_4$ are easily replaced successively by oxygen in reactions with alcohols and phenols. These give rise respectively to alkoxides and ·phenoxides of general formula $TiCl_mX_n$, where X is the alkoxide or phenoxide group and m is 3, 2, 1 and 0 when n is 1, 2, 3 and 4 respectively. For example:

$$TiCl_4 + 3C_2H_5OH = TiCl_2(OC_2H_5)_2 . C_2H_5OH + 2HCl$$

When the reaction is carried out with an ethanolic solution of sodium ethoxide, complete replacement of chlorine leads to the formation of titanium tetraethoxide.

ZIRCONIUM AND HAFNIUM

These two elements are difficult to separate chemically because of their very close similarity in atomic and ionic radii. Indeed hafnium was not discovered until 1922 when it was identified spectroscopically in zirconium minerals. Ion-exchange and solvent extraction methods have proved to be most successful in effecting separation.

Zirconium and hafnium are tetravalent in almost all their compounds. Some lower-valent compounds have been prepared, such as $ZrCl_3$ and $ZrCl_2$, but these are only stable in the solid state and no reaction in aqueous solution comparable with the reduction of Ti^{4+} to Ti^{3+} takes place with either Zr or Hf.

The dioxides ZrO_2 and HfO_2 are amphoteric, with acidic properties less pronounced than those of TiO_2. The chemistry of both the elements in aqueous solution is dominated by the marked tendency to form complex ions, e.g. $[ZrF_6]^{2-}$, $[ZrF_8]^{4-}$, $[ZrO(SO_4)_2]^{2-}$, $[Zr(C_2O_4)_4]^{4-}$, etc.

GROUP V A—VANADIUM

Oxidation states

Vanadium has five electrons in its valency shell and its highest oxidation state is $+5$. All oxidation states between -1 and $+4$ inclusive are also known. The $+5$ state is more oxidising than the $+4$ state of titanium and some of the lower oxidation states of vanadium, notably $+4$, are important.

The formal oxidation state of -1 is represented in the complex anions $[V(CO)_6]^-$, $[V(bipy)_3]^-$ and $[V(CN)_5(NO)]^{5-}$. The ion $[V(bipy)_3]^-$ is obtained by reduction of the vanadium (II) complex, $[V(bipy)_3]^{2+}$, with lithium tetrahydroaluminate in tetrahydrofuran. The zero oxidation state is found in $V(CO)_6$ and $V(bipy)_3$. The latter is also prepared from $[V(bipy)_3]^{2+}$ as starting material: this time the reducing agent is magnesium in 50 per cent methanol. $V(bipy)_3$ is oxidised to $[V(bipy)_3]^+$ by treatment with 1 mol of iodine.

In aqueous solution, the lowest attainable oxidation state is $+2$. This is strongly reducing;

$$V^{3+} + e = V^{2+} \qquad E^0 = -0.255 \text{ V}$$

and liberates hydrogen from dilute acids:

$$V^{2+} + H_3O^+ = V^{3+} + \tfrac{1}{2}H_2 + H_2O$$

Vanadium (II) is obtained in aqueous solution as $[V(H_2O)_6]^{2+}$ by the reduction of vanadium (V) with zinc amalgam or electrolytically. It rapidly oxidises in air. Vanadium (II) salts are often isomorphous with the corresponding iron (II) salts, for example, $VSO_4 \cdot 7H_2O$ with $FeSO_4 \cdot 7H_2O$ and

$K_4[V(CN)_6]3H_2O$ with $K_4[Fe(CN)_6] . 3H_2O$. Other vanadium (II) compounds have been prepared and studied in non-aqueous or molten media.

Vanadium (III) is much less reducing than vanadium (II). For the half-reaction:

$$VO^{2+} + e + 2H^+ = V^{3+} + 2H_2O \qquad E° = +0·36 \text{ V}.$$

This oxidation state is typified by all four halides, the alums $M^I V(SO_4)_2$. $12H_2O$, and many complexes, including $[V(C_2O_4)_3]^{3-}$, $[V(CN)_6]^{3-}$, $[V(SCN)_6]^{3-}$ and $[VF_6]^{3-}$.

For the half-reaction

$$VO_2^+ + 2H + e = VO^{2+} + H_2O \qquad E° = +1·00 \text{ V}.$$

This means that acidic solutions of vanadium (V) are moderately strong oxidising agents. Mild reducing agents like sulphur dioxide and oxalic acid can effect reduction to vanadium (IV). In aqueous media, the most important vanadium (IV) species is the oxo-vanadium (IV) ion, VO^{2+}. Many stable compounds containing this have been isolated, including all four oxo-halides, VOX_2. The simple binary halides VF_4 and VCl_4 also exist but are unstable in the presence of water because they undergo rapid hydrolysis.

With the exception of VF_5, all vanadium (V) compounds contain combined oxygen. Examples of these include VOF_3, $VOCl_3$, $VOBr_3$, VO_2F and VO_2Cl. The most familiar compounds are the oxide, V_2O_5, and the vanadates, salts of hypothetical vanadic acids like H_3VO_4 and HVO_3.

Compounds

Oxides and oxo-compounds

Vanadium (II) oxide is obtained by the reduction of higher oxides with various reducing agents. It is a grey powder which shows non-stoichiometry similar to that of TiO. Thus it has a defect sodium chloride structure over the range of composition $VO_{1·20}$ to $VO_{0·75}$. The oxide is basic and dissolves in mineral acids to give violet solutions containing vanadium (II).

Vanadium (III) oxide is prepared by the reduction of V_2O_5 with hydrogen. It is a greyish-black solid, m.p. 2243 K. It crystallises with the corundum structure but has a marked tendency to show oxygen deficiency and its composition is generally below that represented by V_2O_3.

Vanadium (III) oxide is basic and reacts with acids to give vanadium (III) salts. In aqueous solution, the green hexa-aquo ion, $[V(H_2O)_6]^{3+}$, is formed.

Vanadium (IV) oxide is made by the reduction of V_2O_5 with sulphur dioxide, oxalic acid or V_2O_3. It is a deep-blue solid, m.p. 1913 K. At temperatures above 341 K, the solid has the rutile structure. Below 341 K, there is a less regular arrangement with alternate long and short metal–metal distances.

The amphoteric nature of VO_2 is shown by its reaction with acids to form blue oxovanadium (IV) ('vanadyl') salts and with caustic alkalis to form vanadates (IV).

The oxovanadium (IV) ion, VO^{2+}, is one of the most stable diatomic ions known. It forms many compounds and anionic, cationic and neutral complexes. The ion has a discrete existence in the solid, liquid and vapour states and in solution. It is characterised by a particularly short V—O bond length, within the range 157–168 pm. One of the best examples of a 5 co-ordinate complex, *bis*(acetylacetonato)oxovanadium (IV) contains the VO^{2+} ion.

Vanadium (V) oxide is the final product of oxidising vanadium or one of its lower oxides. It is also formed when ammonium vanadate, NH_4VO_3, is heated in air. The compound contains octahedrally co-ordinated vanadium atoms. The arrangement of the six oxygens is a highly distorted one and the V—O distances range from 158 to 278 pm.

Vanadium (V) oxide is an important catalyst for the large-scale oxidation of SO_2 to SO_3. Its usefulness is probably associated with the dissociation which it undergoes above 970 K:

$$2V_2O_5 \rightleftharpoons 4VO_2 + O_2$$

It is predominantly an acidic oxide which, whilst only slightly soluble in water, dissolves readily in alkali to form vanadates (V). The simple ortho-vanadate ion, VO_4^{3-}, exists only in strongly alkaline solution. The addition of H^+ produces condensed oxo-anions:

$$2VO_4^{3-} + 2H^+ \rightleftharpoons 2HVO_4^{2-} \rightleftharpoons V_2O_7^{2-} + H_2O \quad (pH \geqslant 13)$$
$$3V_2O_7^{4-} + 6H^+ \rightleftharpoons 2V_3O_9^{3-} + 3H_2O \quad (pH \geqslant 8\cdot4)$$
$$10V_3O_9^{3-} + 12H^+ \rightleftharpoons 3V_{10}O_{28}^{6-} + 6H_2O \quad (8 > pH > 3)$$

$$[V_{10}O_{28}]^{6-} + H^+ \rightleftharpoons [HV_{10}O_{28}]^{5-}$$
$$[HV_{10}O_{28}]^{5-} + H^+ \rightleftharpoons [H_2V_{10}O_{28}]^{4-}$$
$$[H_2V_{10}O_{28}]^{4-} + 14H^+ \rightleftharpoons 10VO_2^+ + 8H_2O \quad (pH < 3)$$

The greater the degree of polymerisation the more intense the colour; there is a progressive deepening of colour from pale yellow to deep red as the pH is decreased.

Below pH = 2·0 a reddish-brown precipitate of hydrous vanadium (V) oxide is formed. This dissolves if further acid is added to give a yellow solution which contains the VO_2^+ ion.

The above set of equilibria are those which are believed to exist in solutions of vanadium (V) of concentrations above 10^{-4} mol dm^{-3}. Below this concentration, monomeric vanadate and acid vanadate ions exist. In these dilute solutions, the equilibria predominantly found are:

$$VO_4^{3-} + H^+ \rightleftharpoons HVO_4^{2-} \qquad (pH \approx 13)$$
$$HVO_4^{2-} + H^+ \rightleftharpoons VO_3^- + H_2O \qquad (pH \approx 8\cdot4)$$
$$VO_3^- + H^+ \rightleftharpoons HVO_3 \qquad (pH = 4\cdot5)$$
$$HVO_3 + H^+ \rightleftharpoons VO_2^+ + H_2O \qquad (pH = 3\cdot7)$$

The vanadates (V) resemble the phosphates in their marked tendency to condense (although the method of formation of condensed oxo-anions is different).

Peroxo-compounds

The addition of H_2O_2 to a dilute solution of a trivanadate $[V_3O_9]^{3-}$ produces a yellow colour. This is attributed to the formation of a diperoxo-anion, $[VO(O_2)_2]^-$. The reaction between a vanadium (V) salt in acid solution or a decavanadate ($[V_{10}O_{28}]^{6-}$) and hydrogen peroxide produces a red colour believed to be due to the presence of a monoperoxo-(oxovanadium) cation $[V(O)(O_2)]^+$. Other peroxo complexes are formed, depending on the experimental conditions. The solution chemistry is further complicated by the existence of polymeric ions.

Fluorides

Vanadium (II) fluoride is formed by the reduction of vanadium (III) fluoride using a mixture of hydrogen and anhydrous hydrogen fluoride at 1423 K. It is a blue solid which has the rutile structure and is isomorphous with the fluorides of manganese, iron, cobalt and nickel.

Vanadium (III) fluoride is conveniently prepared by the thermal decomposition of ammonium hexafluorovanadate (III), $(NH_4)_3VF_6$, at 770–870 K in an inert atmosphere. The ammonium salt itself is made by heating V_2O_3 and ammonium hydrogen fluoride at 520 K.

The trifluoride is a yellowish-green solid, sparingly soluble in water. It melts around 1670 K without decomposition. In the solid there are almost regular VF_6 octahedra and the fluorines are in a distorted form of hexagonal close-packing. All fluorines are shared between vanadium atoms.

Vanadium (IV) fluoride can be made by the reaction between anhydrous hydrogen fluoride and vanadium tetrachloride at 195 K in trichlorofluoromethane solution. Removal of the solvent and excess HF at room temperature leaves VF_4.

It is a lime-green solid readily hydrolysed to vanadium oxide difluoride, VOF_2. The solid sublimes with disproportionation at temperatures above 373 K. The structure is believed to be polymeric and to contain VF_6 octahedra which each share four fluorines. VF_4 acts as a fluorinating agent although less reactive than VF_5.

Vanadium (V) fluoride is obtained when vanadium metal is fluorinated at 570 K. It also results when VF_4 undergoes disproportionation at 420 K:

$$2VF_4 = VF_3 + VF_5$$

It is a white solid which melts at 292·6 K to a pale-yellow viscous liquid, b.p. 321 K. In the vapour state the compound is monomeric and the molecule has the shape of a trigonal bipyramid. The liquid shows some electrical conductance, probably due to partial ionisation to VF_4^+ and VF_6^-. The solid has a chain-like structure composed of octahedrally co-ordinated vanadium atoms, the octahedra being joined by sharing two fluorines. The increase in volatility from VF_3 to VF_5 is directly related to the decrease in polymeric nature as the extent of fluorine bridging decreases.

VF_5 is a violent oxidising and fluorinating agent and is readily reduced to VF_4.

Chlorides

Vanadium (II) chloride results from the reduction of vanadium (III) chloride with hydrogen at 770 K. It is a green solid, m.p. 1623 K, which has the cadmium iodide structure. A hydrated form, $VCl_2 . 4H_2O$, can be made by the electrolytic reduction of vanadium (V) solutions followed by the addition of hydrochloric acid and evaporation to dryness.

Vanadium (III) chloride is produced by the reduction of vanadium (IV) chloride with H_2S, CS_2 or S_2Cl_2 or by the thermal decomposition of VCl_4 alone *in vacuo* at 420–440 K. It is a violet solid which dissolves readily in water to give an acidic solution. When hydrochloric acid is added, the green hexahydrate can be crystallised out. This contains $[V(H_2O)_4Cl_2]^+$ ions. Anhydrous VCl_3 disproportionates above 700 K to a mixture of VCl_4 and VCl_2. In the solid state, VCl_3 has a hexagonal layer lattice resembling that of $FeCl_3$.

VCl_3 reacts with a number of ligands forming 1 : 2 adducts like $VCl_3 . 2N(CH_3)_3$ and $VCl_3 . 2S(CH_3)_2$. These contain 5 co-ordinate vanadium.

Vanadium (IV) chloride is most easily prepared by the passage of chlorine over vanadium heated above 470 K. It is a red liquid, b.p. 425 K, which readily hydrolyses in contact with water. When heated *in vacuo* it decomposes into a mixture of VCl_3 and chlorine. The compound has a molecular structure composed of tetrahedral molecules. The magneton number is 1·61, corresponding with the d^1 configuration of vanadium (IV). There is some experimental evidence to suggest that some Jahn–Teller distortion is present in the VCl_4 molecule, as we would expect from its electronic configuration.

Vanadium (V) oxide trichloride, $VOCl_3$, is formed when V_2O_5 is chlorinated in the presence of carbon. It is a yellow liquid, b.p. 400 K. In moist air it is rapidly hydrolysed to V_2O_5. The physical properties are consistent with a covalent structure. In the vapour state $VOCl_3$ exists as tetrahedral molecules.

No binary chloride of vanadium (V) is known. It appears that oxygen must be present as well when vanadium combines in this oxidation state with chlorine.

NIOBIUM AND TANTALUM

Like zirconium and hafnium, these two metals are difficult to separate from one another. Isolation is best achieved using modern selective techniques.

The +5 state is the most stable and the lower-valent states are obtained with increasing difficulty on passing from vanadium to tantalum. Blue Nb^{III} solutions are obtained by the reduction of Nb^V solutions with zinc and HCl. Ta^V, however, is unaffected in these conditions.

The pentoxides, like V_2O_5, are acidic. Pentahalides such as NbF_5, TaF_5,

$NbCl_5$, $TaCl_5$ and $NbBr_5$ are known. These are chiefly covalent and are easily hydrolysed to give hydrated pentoxides. A variety of halo-complexes are also known, including $[NbOF_5]^{2-}$, $[NbOF_6]^{3-}$, $[NbF_7]^{2-}$, $[TaF_7]^{2-}$ and $[TaF_8]^{3-}$.

GROUP VI A—CHROMIUM

Oxidation states

Chromium shows oxidation states from -2 to $+6$ inclusive in its compounds. The highest has marked oxidising properties and the $+3$ state is more important in the range and stability of its compounds.

Formal oxidation states of -2 and -1 are found respectively in the complex anions $[Cr(CO)_5]^{2-}$ and $[Cr_2(CO)_{10}]^{2-}$. The 0 state is found in $Cr(CO)_6$ and the $+1$ in $[Cr(bipy)_3]^+$. The 2,2'-bipyridyl complex is obtained from $[Cr(bipy)_3]^{2+}$ by reducing this in aqueous solution.

Chromium (II) can be made by the reduction of chromium (III) with zinc in acidic solution but an inert atmosphere is necessary to prevent oxidation back to the $+3$ state. Cr^{2+} is strongly reducing:

$$Cr^{3+} + e = Cr^{2+} \quad E° = -0.41 \text{ V}$$

E^0 is sufficiently negative for Cr^{2+} to be able to reduce acidic aqueous solutions to hydrogen. If, however, chromium (II) is obtained in the form of one of its water-insoluble complexes, then this oxidation state is quite stable towards aerial oxidation, for example in $Cr(CH_3COO)_2 . 2H_2O$ and $CrSO_4$ $(N_2H_4)_2 . H_2SO_4$. Other compounds of chromium (II) include the oxide and the four binary halides.

Many compounds of chromium (III) are known and have been well characterised because they are kinetically inert; in other words they are extremely slow to react and can be isolated and studied at leisure. From a structural point of view chromium (III) compounds are often isomorphous with the corresponding iron (III) and aluminium (III) compounds.

Chromium (IV) and (V) are generally found in combination with oxygen or fluorine but not many compounds of either oxidation state are known.

In the $+6$ state, chromium is always combined with oxygen (a possible exception is CrF_6 which may exist at low temperatures). The best-known compounds are the oxide and salts of chromic, H_2CrO_4, and dichromic, $H_2Cr_2O_7$, acids. Other chromium (VI) compounds include CrO_2Cl_2 and CrO_2F_2.

Compounds

Oxides and oxo-compounds

Chromium (II) oxide has been made by treating chromium amalgam with dilute nitric acid to remove the mercury. It is the parent oxide of chromium (II) salts. When these are prepared in aqueous media they contain the blue $[Cr(H_2O)_6]^{2+}$ ion.

Chromium (III) oxide is formed when ammonium dichromate, $(NH_4)_2$ Cr_2O_7, is heated in air. It is a green solid and has the corundum structure. A hydrated form, $Cr_2O_3 . nH_2O$, is precipitated by the addition of alkali to a solution of a chromium (III) salt. This hydrate is soluble in acids to form chromium (III) salts and in caustic alkalis to give chromates (III). Chromium (III) salts may be either green or violet depending on the nature of the ligands co-ordinated to the metal. The species present in chromate solutions are most likely to be $[Cr(H_2O)(OH)_5]^{2-}$ and $[Cr(OH)_6]^{3-}$.

Cr_2O_3 is an intermediate in the large-scale production of chromium metal. Sodium dichromate is reduced by carbon to Cr_2O_3:

$$Na_2Cr_2O_7 + 2C = Cr_2O_3 + Na_2CO_3 + CO$$

This is further reduced to the metal by aluminium or silicon:

$$Cr_2O_3 + 2Al = Al_2O_3 + 2Cr$$

Chromium (IV) oxide is produced when CrO_3 is heated at 620 K under a high pressure of oxygen. It is a black, ferromagnetic solid with the rutile structure. Its main use is in the preparation of magnetic tapes.

Although formulated as an oxide of chromium (IV), some at least of its properties are consistent with a formulation as a mixed-valency compound. namely $Cr(III)Cr(V)O_2$.

A number of mixed oxides are formed with alkaline-earth metal oxides. These are of general formulae M_2CrO_4, M_3CrO_5 and M_4CrO_6.

Other chromium oxides are formed by the decomposition of CrO_3. One well-defined compound has the stoichiometry Cr_5O_{13} and is believed to contain the metal in two valency states. It may be formulated as $Cr_3^{VI}Cr_2^{IV}O_{13}$.

Chromium (VI) oxide is prepared by the acidification of a saturated solution of sodium or potassium dichromate using concentrated sulphuric acid. It is an orange–red solid and is an acidic oxide, reacting with water to form chromic acids. In solutions of low pH, dichromic acid predominates. When the pH is increased, dichromate ion is formed and then this reacts with water to form chromate in accordance with the equilibrium:

$$2[CrO_4]^{2-} + 2H^+ = [Cr_2O_7]^{2-} + H_2O \text{ for which } K = 4 \cdot 2 \times 10^{14}.$$

Acidified dichromate solutions are strongly oxidising:

$$[Cr_2O_7]^{2-} + 6e + 14H^+ = 2Cr^{3+} + 7H_2O \quad E° = +1 \cdot 33 \text{ V}$$

Under alkaline conditions, chromium (VI) is much more stable:

$$[CrO_4]^{2-} + 4H_2O + 3e = Cr(OH)_3(s) + 5OH^- \quad E° = -0 \cdot 13 \text{ V}$$

This means that the most effective way of preparing chromium (VI) from chromium (III) in aqueous media is by oxidation under alkaline conditions.

The chromate ion is tetrahedral. The dichromate ion is composed of two CrO_4 tetrahedra joined by sharing one oxygen:

$$\left(\begin{array}{c} O \\ \diagdown \\ O-\overset{\displaystyle |}{Cr}-O-\overset{\displaystyle |}{Cr}-O \\ \diagup \quad\quad \diagdown \\ O \quad\quad\quad O \end{array} \right)^{2-}$$

In concentrated acid solution, trichromate $[Cr_3O_{10}]^{2-}$ and tetrachromate $[Cr_4O_{13}]^{2-}$ are formed. The extent of polymerisation is therefore quite limited compared with the extensive polymerisation of oxo-anions which is so characteristic of vanadium (V) and molybdenum and tungsten (VI).

Peroxo-compounds

Various peroxides of chromium have been characterised. For example, when an alkaline solution of a chromate is treated with hydrogen peroxide, the colour changes from yellow to red–brown. At 273 K dark-brown crystals separate out slowly. The alkali metal salts isolated in this way have the general formula $M_3^ICrO_8$ and volumetric reaction with permanganate indicates the presence of four peroxo groups for each chromium atom. It should therefore be formulated as $M_3^ICr(O_2)_4$ and hence contains Cr(V). This is confirmed by the paramagnetism of the potassium salt (magneton number of 1·80 at 293 K).

X-ray analysis shows the metal atom is surrounded by four equivalent peroxo groups. The oxygen atoms can be regarded as arranged in a distorted dodecahedron around the chromium.

When H_2O_2 is added to an acid solution of a chromate, an unstable blue complex is obtained. This can be extracted with ether or other oxygen-containing organic solvents and is stabilised by co-ordination with solvent molecules. If an electron-donor molecule such as pyridine is present, then solid addition compounds can be isolated by evaporation of the solvent. The pyridine compound is formulated as a diperoxo derivative of Cr(IV), $pyCr(O)(O_2)_2$. The blue pentaoxide itself, CrO_5, has not been isolated.

Fluorides

Chromium (II) fluoride can be made by the reduction of chromium (III) fluoride with the stoichiometric quantity of metallic chromium in a nickel bomb at 1270 K. It is a bluish-green solid which melts at 1167 K. The structure is a distorted version of rutile and there are pairs of fluorine atoms around each chromium at three different distances, 198, 201 and 243 pm. This distortion from a regular octahedral arrangement to four short and two long bonds provides one of the best examples of the Jahn–Teller effect.

Chromium (III) fluoride is formed when hydrogen fluoride is passed over $CrCl_3$ heated between 720 and 870 K. Above 870 K some disproportionation to CrF_2 and CrF_5 occurs. CrF_3 is a green solid, m.p. 1677 K (in a closed system). The solid is composed of CrF_6 octahedra joined by bridging fluorine atoms.

Chromium (IV) fluorine results from the fluorination of chromium at 623 K. It is a green solid which sublimes above 373 K *in vacuo*. Its structure is similar to that of TiF_4 and VF_4. The compound undergoes rapid hydrolysis but is generally less reactive as a fluorinating agent than CrF_5.

Chromium (V) fluoride is the main product of fluorinating chromium or CrF_3 between 623 and 773 K. It is a red solid, m.p. 303 K, which has powerful fluorinating properties.

Chromium (VI) fluoride is an unstable compound which can only be made under extreme conditions, the reaction between a mixture of chromium and fluorine sealed into a bomb which is then raised to a high temperature. It decomposes above 173 K to CrF_5 and F_2.

Chromium dioxide difluoride, CrO_2F_2, is produced by the reaction at room temperature between CrO_3 and anhydrous HF. It is a dark-red solid, m.p. 304·6 K. The compound is highly reactive and readily attacks glass and silica.

Chlorides

Chromium (II) chloride is formed when $CrCl_3$ is reduced by hydrogen at 773 K or when chromium metal reacts with hydrogen chloride at 1173 K. It is a white, hygroscopic solid, m.p. 1097 K. $CrCl_2$ is readily soluble in water to give a blue solution which has marked reducing properties. The solid has an ionic lattice in which each chromium is surrounded by a distorted octahedron of chloride ions. Cr^{2+} is a spin-free d^4 ion and the tetragonal distortion (four chlorines at 239·5 and two more at 291·5 pm) is accounted for by crystal field theory.

The hydrate, $CrCl_2 \cdot 4H_2O$, is obtained by evaporation of the aqueous solution prepared by dissolving metallic chromium in dilute hydrochloric acid or by the reduction of acidified chromium (III) solutions with zinc amalgam.

Chromium (III) chloride is obtained in the anhydrous state by the direct chlorination of chromium or by reaction between chlorine and a mixture of Cr_2O_3 and carbon. It is a violet solid, m.p. 1423 K. It has a layer lattice, each layer being composed of chromium atoms octahedrally co-ordinated by chlorines.

Three isomeric hydrates of $CrCl_3$ are known. The common form is a dark-green compound, $[CrCl_2(H_2O)_4]Cl \cdot 2H_2O$, which contains octahedral *trans*-$[CrCl_2(H_2O)_4]^+$ ions, chloride ions and water molecules. Each co-ordinated water molecule is hydrogen-bonded to free chloride ions and unco-ordinated water. Other hydrates which can be made are a pale-green isomer, $[CrCl(H_2O)_5]Cl_2 \cdot H_2O$, and a violet form, $[Cr(H_2O)_6]Cl_3$. When any of the three hydrates is heated, loss of water is accompanied by partial hydrolysis. Dehydration to give the anhydrous chloride can be accomplished by refluxing a hydrate with thionyl chloride:

$$Cr(H_2O)_6Cl_3 + 6SOCl_2 = CrCl_3 + 12HCl + 6SO_2$$

Chromium (IV) chloride appears to be produced when $CrCl_3$ is chlorinated at 973 K but has never been isolated as a pure compound.

Chromium (V) and (VI) are only found in combination with chlorine if oxygen is present as well. For example, chromium (V) is found in *chromium oxide trichloride*, $CrOCl_3$, made by the interaction of CrO_3 with SO_2Cl_2 or $SOCl_2$. This is an unstable compound at ordinary temperature.

Chromium dioxide dichloride, CrO_2Cl_2, also known as chromyl chloride, can be distilled from a reaction mixture consisting of an alkali metal chloride, concentrated sulphuric acid and chromium (VI) oxide or an alkali metal dichromate. The compound is a dark-red liquid, b.p. 390 K. It hydrolyses slowly in cold water:

$$2CrO_2Cl_2 + 3H_2O = H_2Cr_2O_7 + 4HCl$$

On heating, CrO_2Cl_2 decomposes to chlorine, oxygen and intermediate oxo-chlorides before finally losing all halogen and leaving an oxide, Cr_5O_9. CrO_2Cl_2 is a covalent molecule of tetrahedral shape.

MOLYBDENUM AND TUNGSTEN

These two elements show marked resemblances to each other because of the lanthanide contraction. The most stable oxidation state is $+6$, although the lower states 0 to $+5$ inclusive are also known.

The trioxides, MO_3, are acidic. In aqueous solution they give rise to oxo-acids which show a great tendency to form polyacids of high molecular weight. Some of these are formed from one acid only and are then known as isopolyacids. Examples of their salts are $(NH_4)_6(Mo_7O_{24}) . 4H_2O$, ammonium heptamolybdate, and $K_6(H_2W_{12}O_{40}) . 18H_2O$, potassium dihydrogen-dodecatungstate. The heteropolyacids are another group which contain a second acid besides either molybdic or tungstic acid. Derivatives of these are important in quantitative analysis, for example, ammonium dodecamolybdo-phosphate, $(NH_4)_3PMo_{12}O_{40}$, which is used for the gravimetric analysis of phosphorus.

Both metals form many complexes and this property is made use of to stabilise the lower-valent states, for instance, Mo^{IV} and W^{IV} are appreciably stabilised in their octacyano-complexes, $[Mo(CN)_8]^{4-}$ and $[W(CN)_8]^{4-}$ respectively.

In general, there is a decrease in stability of the lower oxidation states as we progress down a sub-group of transition metals and a corresponding increase in the stability of the higher oxidation states. Some notable exceptions to the first trend are the lower halides of some second- and third-series metals. These are unexpectedly stable and recently much attention has been paid to their structures. One of the best-known examples is molybdenum (II) chloride. This actually contains polymeric ions or 'clusters' of the type $[Mo_6Cl_8]^{4+}$. Each of the clusters is composed of an octahedron of molybdenum atoms with chlorines situated above each face of the octahedron. Each chlorine acts as a bridge between three molybdenum atoms and there is also direct bonding between pairs of metal atoms. Four more chloride ions are bound by electrostatic forces to complete the Mo_6Cl_{12} structural unit. The $[Mo_6Cl_8]^{4+}$ cluster occurs in other compounds of molybdenum (II) and the same type of ion is found in tungsten (II) chloride.

In these and other cluster compounds (for example, Nb_6Cl_{14}, Ta_6Cl_{14}, and the ions $[Re_2Cl_8]^{2-}$, $[Re_3Cl_{12}]^{3-}$, $[W_2Cl_9]^{3-}$) the metal is in a low oxidation state but any comparison of stability with the ionic compounds of first series metals in the same state would not be really justifiable in view of the disparate nature of the units which make up the compounds.

GROUP VIIA—MANGANESE

Oxidation states

Compounds containing manganese in the oxidation states -1 to $+7$ inclusive are known. The most familiar are those with manganese in $+2$, $+4$ and $+7$ states.

The element is in a formal oxidation state of -1 in the carbonyl anion, $[Mn(CO)_5]^-$, and 0 in the carbonyl, $Mn_2(CO)_{10}$. Manganese (I) is present in the carbonyl halides, $Mn(CO)_5X$ and in the hexacyano-complex, $[Mn(CN)_6]^{5-}$. The latter is formed when an aqueous solution of $[Mn(CN)_6]^{4-}$ is reduced electrolytically or with sodium amalgam.

Manganese (II) is a very stable oxidation state in acidic solutions. Most salts containing manganese (II) are water-soluble: two noteworthy exceptions are the sulphide and the hydroxide. Many complexes of manganese (II) have been made but these are significantly less stable than the corresponding ones of iron and subsequent transition metals in this oxidation state (Irving–Williams order).

Manganese (III) is found in combination with oxygen and with fluorine and in various complexes. Manganese (IV) is best-known in manganese dioxide, MnO_2. Manganese (V) and (VI) are found in their respective oxo-anions, $[MnO_4]^{3-}$ and $[MnO_4]^{2-}$. Manganese (VII) occurs in manganates (VII), commonly known as permanganates.

The complexity of manganese chemistry is illustrated by the range of oxidation states ($+2$ to $+7$) which can exist in aqueous media. Not all of these can co-exist and a number of disproportionation reactions is possible depending on the relative magnitudes of the electrode potentials, the pH of the solution, the nature of complexing species present and the relative solubilities of the manganese compounds concerned.

We have already shown (*Figure 5.5*) the relationship between the various oxidation states of manganese in acid solution. Standard electrode potential data presented in this Figure show that Mn^{3+} is unstable with respect to disproportionation to Mn^{2+} and MnO_2. This means that the equilibrium

$$2Mn^{3+} + 2H_2O \rightleftharpoons Mn^{2+} + MnO_2 + 4H^+$$

lies very much on the right-hand side. The tendency for Mn^{3+} to decompose in this way is enhanced by the sparing solubility of MnO_2 in aqueous media. In practice, high concentrations of hydrogen ion are required to prevent complete disproportionation of Mn^{3+}

For similar reasons, $[MnO_4]^{2-}$ is unstable with respect to disproportionation to a mixture of MnO_2 and $[MnO_4]^-$:

$$3[MnO_4]^{2-} + 2H_2O \rightleftharpoons 2[MnO_4]^- + MnO_2 + 4OH^-$$

This reaction proceeds virtually to completion to the right-hand side unless sufficient hydroxyl ion (at least 1 mol dm^{-3}) is present to stabilise the $[MnO_4]^{2-}$.

The electrode potentials for various oxidation states of manganese are enumerated on p. 140. These relate to acidic solutions and account, for example, for the reduction of $[MnO_4]^-$ directly to Mn^{2+}. All intermediate species that could be formed are either strong oxidising agents themselves or tend to disproportionate under these conditions.

In alkaline conditions, the electrode potentials and hence the relative stabilities of various oxidation states are very different. For example, $E° = -0.2$ V for $Mn(OH)_3/Mn(OH)_2$ and $E° = +0.1$ V for $MnO_2/Mn(OH)_3$. These figures mean that $Mn(OH)_3$ is stable and does not disproportionate whereas $Mn(OH)_2$ is strongly reducing and very readily converted to $Mn(OH)_3$.

Again, in alkaline conditions, $E° = +0.588$ V for

$$MnO_4^- + 4H^+ + 3e \rightleftharpoons MnO_2 + 2H_2O$$

Manganate (VII) is evidently not such a strong oxidising agent as it is in acidic media and it is only reduced as far as MnO_2. One practical consequence of these data is that alkaline media are much more suited to the preparation of the high oxidation states of manganese.

Manganate (V) requires very high concentrations of alkali for its preparation. Unless the hydroxyl ion concentration is of the order of 14–15 mol dm^{-3}, manganate (V) disproportionates thus:

$$2[MnO_4]^{3-} + 2H_2O \rightleftharpoons [MnO_4]^{2-} + MnO_2 + 4OH^-$$

Compounds

Oxides and oxo-ions

Manganese (II) oxide is obtained when the higher oxides of manganese are reduced by hydrogen or carbon monoxide. It is a basic oxide and reacts with acids to form manganese (II) salts. These, like many other manganese (II) compounds, are stable in the anhydrous state and in acidic solution. The hydrated salts are coloured pale pink due to the presence of $[Mn(H_2O)_6]^{2+}$ ions.

Trimanganese tetraoxide, Mn_3O_4, is the product of heating any other oxide of manganese to temperatures above 1220 K. This compound is a mixed-valency compound and is correctly formulated as $Mn^{II}Mn_2^{III}O_4$. It has a normal spinel structure with Mn^{2+} ions in the tetrahedral and Mn^{3+} ions in the octahedral sites. There is some distortion of the MnO_6 octahedra which may be related to the d^4 configuration of Mn^{3+}.

Manganese (III) oxide occurs as the mineral, *braunite*. It is formed when MnO_2 is heated in air between 820 and 1170 K. Mn_2O_3 has basic properties and dissolves in sulphuric acid to form manganese (III) sulphate. With hydrofluoric acid, manganese (III) forms complex ions like $[MnF_4]^-$ and $[MnF_5]^{2-}$ and similar complexes are present in solutions of manganese (III) in concentrated hydrochloric acid.

Manganese (IV) oxide occurs as the mineral, *pyrolusite*. It can be prepared in the laboratory by the reduction of potassium manganate (VII) in alkaline solution or by heating manganese (II) nitrate in air to 800 K. It is a brownish-black solid with the rutile structure. It is usually non-stoichiometric, the oxygen content rarely exceeding that represented by $MnO_{1.95}$.

Although MnO_2 is a moderately strong oxidising agent, its low solubility in aqueous media confers an apparent chemical inertness upon it.

The compound is amphoteric. Thus reaction with concentrated hydrochloric acid gives solutions containing complex ions like $[MnCl_6]^{2-}$. Reaction with alkalis or fusion with other metal oxides gives manganates (IV) containing $[MnO_3]^{2-}$ ions.

Manganate (V) is present as a component of the solution obtained by dissolving MnO_2 in concentrated KOH. Manganese (IV) disproportionates to give a mixture of Mn(III) and Mn(V). The colour of the solution is blue due to the manganate (V) ion, $[MnO_4]^{3-}$. This is also obtained by the reduction, with the stoichiometric amount of sodium sulphite, of a solution of potassium manganate (VI) in strong sodium hydroxide.

Manganate (VI) is formed when manganese dioxide is oxidised in the presence of molten alkali. The colour of the manganate (VI) ion, MnO_4^{2-}, is deep green. As mentioned earlier it is only stable in alkaline media. Treatment of such a solution with acid causes disproportionation to give an alkali metal manganate (VII) as one of the products and this is, in fact, one of the most convenient procedures for preparing the latter type of compound.

Manganese (VII) oxide and manganates (VII). *Manganese (VII) oxide* is formed as an unstable oil when potassium manganate (VII) is added to concentrated sulphuric acid and then a little water is added to the mixture. Mn_2O_7 is red by transmitted light and green by reflection. It decomposes explosively to MnO_2 and oxygen. The compound acts as an acidic oxide and reacts with water to give a solution of permanganic acid, $HMnO_4$. This acid cannot be isolated in the anhydrous condition but, like perchloric acid, its counterpart in chlorine chemistry, it is very strong and acts as a powerful oxidant.

The most familiar compound of manganese (VII) is potassium permanganate, $KMnO_4$, used extensively in volumetric analysis. In the solid state and in solution this compound has a very striking deep-purple colour. The colour is due to an electronic transition of the 'charge-transfer' type. This transition takes place between two molecular orbitals, one of which is closely associated with the oxygens and the other largely localised on the manganese atom. The absorption of light-energy causes an electronic transition from the first to the second of these orbitals.

We may note that $[MnO_4]^-$ is the last of a series of isoelectronic tetrahedral ions beginning with $[TiO_4]^{4-}$. Of these, $[TiO_4]^{4-}$ shows the greatest tendency to polymerise (hydrous titanium dioxide can be regarded as a giant polymer), then $[VO_4]^{3-}$ with its complex reactions in aqueous solution, then

$[CrO_4]^{2-}$ where the tetramer is the most complex species formed, and finally, $[MnO_4]^-$, which shows no tendency to polymerise at all.

Fluorides

The only fluorine-containing compound in which manganese shows its highest oxidation state contains oxygen as well. Otherwise the highest state reached in combination with fluorine alone is $+4$.

Manganese (II) fluoride can be made by the reaction between manganese (II) carbonate and hydrofluoric acid. It is a solid, m.p. 1193 K, which is sparingly soluble in water. The compound has a very slightly distorted rutile structure.

Manganese (III) fluoride is made by the action of fluorine on manganese (II) halides or an oxide of manganese. It is a red–purple solid which is thermally stable but is immediately hydrolysed by water. It has a number of industrial applications as a fluorinating agent for organic compounds.

MnF_3 has a magneton number of 4·9. This means that the Mn^{3+} ion is high-spin d^4 and that the fourth electron is in either the $d_{x^2-y^2}$ or the d_{z^2} orbital. The solid has a structure like that of VF_3 but there is considerable distortion from a regular octahedral arrangement of the six fluorines around the manganese. There are three different Mn—F bond lengths, 179, 191 and 209 pm respectively, and the structure represents an extreme case of Jahn–Teller distortion.

Manganese (IV) fluoride is the product of fluorinating manganese powder between 870 and 970 K. The compound is an unstable blue solid which decomposes into MnF_3 and fluorine even at room temperature.

Manganese trioxide fluoride, MnO_3F, is made by the interaction between potassium permanganate and hydrogen fluoride. The compound is a dark-green solid at low temperatures, m.p. 195 K. At room temperature it decomposes explosively to MnF_2, MnO_2 and O_2.

Chlorides

Manganese (II) chloride is prepared as a tetrahydrate by crystallisation from a solution of manganese in concentrated hydrochloric acid. In contrast to the earlier transition metals of the first series, the only stable oxidation state shown by manganese in binary combination with chlorine is $+2$.

Anhydrous $MnCl_2$ is made by the dehydration of the tetrahydrate in a stream of hydrogen chloride above 670 K. It has a hexagonal crystal lattice of the cadmium chloride type. Its magneton number is 5·73, around the value expected for a high-spin d^5 ion.

$MnCl_2$ reacts with many nitrogen-containing organic molecules to form adducts. Thus reaction with pyridine in ethanolic solution gives rise to $Mnpy_2Cl_2$ or $Mnpy_4Cl_2$, depending on the proportion of pyridine to manganese chloride used. Both these adducts contain octahedral manganese. This is achieved in the bis complex of the bridging action of the chlorines which results in a polymeric structure. The pyridine ligands are in the *trans* arrangement around each manganese.

Manganese (III) chloride is made by the action of hydrogen chloride on MnO_2 in anhydrous ethanol at 210 K, followed by precipitation with carbon tetrachloride or petroleum ether. $MnCl_3$ itself is thermally unstable above 233 K but is stabilised by complex-formation with 2,2′-bipyridyl or 1 : 10-phenanthroline.

The complex *potassium hexachloromanganate (IV)*, K_2MnCl_6, has been prepared by adding potassium permanganate to a hydrochloric acid solution of potassium chloride saturated with hydrogen chloride at 273 K. Its magneton number is 3·90, confirming the presence of manganese (IV), a d^3 ion.

Manganese (VII) trioxide chloride, MnO_3Cl, is formed when hydrogen is passed through a solution of $KMnO_4$ in concentrated sulphuric acid. At ordinary temperatures it is a greenish-violet gas which tends to decompose violently.

TECHNETIUM AND RHENIUM

Traces of technetium occur naturally. It was first prepared, however, in 1947 by the following nuclear reaction:

$$^{98}_{42}Mo \, (n, \, \gamma) \, ^{99}_{42}Mo \xrightarrow{-\beta} \, ^{99}_{43}Tc \xrightarrow[2\cdot12 \times 10^5 y]{-\beta} \, ^{99}_{44}Ru$$

It is mainly obtained from the fission products of uranium and in its chemistry resembles rhenium rather than manganese.

Rhenium, an extremely rare element, was first identified in 1925. The +4 and +7 oxidation states are important. The enhanced stability of the higher valency states of rhenium compared with those of manganese is shown by the properties of one of its most stable compounds, potassium perrhenate, $KReO_4$. This is quite stable; it does not have the marked oxidising ability of $KMnO_4$ and is not decomposed by alkali.

GROUP VIII—IRON

Oxidation states

Iron has eight electrons in its valency shell but it does not show a high oxidation state of this number. In fact, once we pass the half-way stage of filling up the d sub-shell at manganese, the succeeding elements show stable low oxidation states but any oxidation state above three is comparatively quite unstable. The most well-known valencies of iron are +2 and +3. The higher ones of +4, +5 and +6 are known but are of little importance.

Iron has a formal oxidation state of −2 in the anion, $[Fe(CO)_4]^{2-}$, of 0 in $Fe(CO)_5$ and of +1 in $[Fe(H_2O)_5NO]^{2+}$.

Iron (II) is known in many stable salts and complexes. In acidic aqueous solution, iron (II) salts are quite stable, even in the presence of air. For the half-reaction

$$Fe^{3+} + e = Fe^{2+}, E° = +0\cdot771 \text{ V}$$

This means that moderately strong oxidising agents are required to convert iron (II) to iron (III).

In alkaline media, $E° = -0.6$ V for the system $Fe(OH)_3/Fe(OH)_2$. In contrast to the situation in an acidic medium, iron (II) is very easily oxidised when in the form of $Fe(OH)_2$.

The standard electrode potential between the $+2$ and $+3$ oxidation states depends very largely on the presence and nature of any complexing ligand. Thus the half-reaction given above refers to aquo complexes of Fe^{2+} and Fe^{3+}. When both states are complexed with cyanide, we may write the half-reaction as

$$[Fe(CN)_6]^{3-} + e = [Fe(CN)_6]^{4-}$$

For this, $E° = +0.36$ V, and so $[Fe(CN)_6]^{4-}$ is a better reducing agent than Fe^{2+} (aquo). Alternatively, we may say that the change in standard electrode potential on complexing with cyanide has resulted in a stabilisation of iron (III).

When 1 : 10-phenanthroline is the ligand, the appropriate half-reaction is:

$$[Fe(phen)_3]^{3+} + e = [Fe(phen)_3]^{2+} \text{ and } E° = +1.12 \text{ V}$$

In this case, complexing with the phenanthroline has stabilised the $+2$ state relative to the aquo complex.

Iron (IV) is found in the complex $(FeCl_2 . 2C_6H_4(As(CH_3)_2)_2][FeCl_4]_2$ in which the cation is

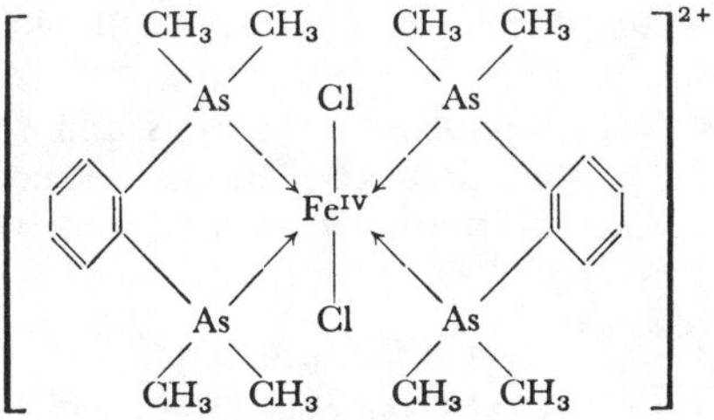

Iron (VI) is found in ferrates (VI) which contain $[FeO_4]^{2-}$ ions. In acid solution,

$$[FeO_4]^{2-} + 8H^+ + 3e = Fe^{3+} + 4H_2O \quad E° = +1.9 \text{ V}$$

Ferrate (VI) is therefore a powerful oxidising agent. Indeed a comparison of $E°$ value shows it is even stronger than manganate (VII).

In alkaline conditions, $E° = +0.9$ V for the system $[FeO_4]^{2-}/Fe(OH)_3$. This means that iron(III) is much more readily oxidised to iron (VI) in alkaline than in acid media.

Compounds

Oxides and oxo-ions

Iron (II) oxide may be prepared by heating iron (II) oxalate in the absence of air. It reacts with acids to give iron (II) salts. When alkali is added to a solution of an iron (II) compound, the hydroxide $Fe(OH)_2$ is precipitated. This is amphoteric because it re-dissolves in acids and also reacts with alkali to form ferrates (II), containing $[Fe(OH)_6]^{4-}$ ions.

FeO has the sodium chloride structure although the compound invariably contains less oxygen than required by the formula. The structure may be described as a cubic close-packed array of oxide ions with the octahedral holes occupied by iron. The composition of the oxide normally approximates to $Fe_{0.95}O$ and hence there is a stoichiometric deficit of iron. Some of the octahedral sites are empty and a proportion of the ions are Fe^{3+} so that electrical neutrality is maintained.

Triiron tetraoxide, Fe_3O_4, occurs in nature as *magnetite*. It may be prepared in the laboratory by heating Fe_2O_3 above 1670 K. This is another mixed-valency compound like Mn_3O_4 and should be formulated as $Fe^{II}Fe_2^{III}O_4$. The compound has the inverse spinel structure in which half of the Fe^{3+} ions occupy tetrahedral sites and the Fe^{2+} and the other half of the Fe^{3+} ions are in octahedral sites. Although there is no crystal field stabilisation energy for Fe^{3+} in either type of site because it is a d^5 ion, $Fe^{2+}(d^6)$ shows a preference for octahedral co-ordination because the stabilisation energy is greater here than for tetrahedral. In addition, the inverse spinel structure is favoured because Fe^{3+} tends to form bonds with marked covalent character and so to occupy sites of low co-ordination number. In magnetite, there is an essentially random arrangement of Fe^{2+} and Fe^{3+} ions in the octahedral sites. This confers high electrical conductivity on the solid because electron-transfer between iron in the two oxidation states can occur rapidly.

Iron (III) oxide occurs in nature as *haematite*. It can be obtained in the hydrous form by precipitation by the addition of alkali to an iron (III) solution. The oxide has amphoteric properties and reacts with acids to form iron (III) salts and with alkalis to give ferrates (III), containing $Fe(OH)_6^{3-}$.

Iron (III) salts in aqueous solution undergo extensive hydrolysis due to loss of protons from $Fe(H_2O)_6^{3+}$.

$$[Fe(H_2O)_6]^{3+} \rightleftharpoons [Fe(H_2O)_5(OH)]^{2+} + H^+ \quad pK = 3.05$$
$$[Fe(H_2O)_5(OH)]^{2+} \rightleftharpoons [Fe(H_2O)_4(OH)_2]^+ + H^+ \quad pK = 3.26$$

Even below pH 3 there is considerable hydrolysis. As the pH is raised beyond 3, condensed species are formed in which iron atoms are joined by hydroxo-bridges. Continued increase in pH results in further deprotonation and the precipitation of hydrous Fe_2O_3 as a gelatinous red–brown solid.

Haematite (α-Fe_2O_3) has the corundum structure. There is a slightly distorted hexagonal arrangement of oxide ions with Fe^{3+} ions in octahedral co-ordination. Oxidation of Fe_3O_4 or the thermal dehydration of hydrous Fe_2O_3 at low temperatures gives another form of iron (III) oxide, γ-Fe_2O_3. This has a spinel-type structure. The oxide ions form a face-centred cubic lattice and Fe^{3+} ions occupy all the tetrahedral sites and some of the octahedral sites.

Ferrate (VI) is obtained when a suspension of hydrous Fe_2O_3 in concentrated alkali is oxidised by chlorine. Alkali metal salts are very soluble and it is convenient to isolate the product of the reaction by precipitation as sparingly-soluble barium ferrate, $BaFeO_4$. In alkaline solution, the ferrate ion is stable but in neutral or acidic media it decomposes to iron (III) compounds and oxygen.

The oxidation state of iron is shown to be $+6$ by the paramagnetism of K_2FeO_4. This has a magneton number of $3\cdot06$ and confirms the presence of two unpaired electrons.

The ferrate ion is red–purple in colour and its potassium salt is isomorphous with potassium sulphate and chromate.

Fluorides

Only two binary fluorides are known, FeF_2 and FeF_3.

Iron (II) fluoride is produced by reaction between iron and anhydrous hydrogen fluoride. It is a white solid, sparingly soluble in water, m.p. 1293 K. The solid has a distorted rutile structure with a compressed octahedron of fluorines around the metal atom, two at 199 and four at 212 pm.

Iron (III) fluoride can be prepared by reaction between $FeCl_3$ and fluorine or anhydrous hydrogen fluoride. It is a white solid which decomposes on heating to 1270 K to FeF_2. The structure of the solid is similar to that of VF_3.

Salts containing the hexafluoroferrate (III) ion, $[FeF_6]^{3-}$, may be prepared in the absence of water but in aqueous solution appear to dissociate largely into lower complexes like $[FeF]^{2+}$ and $[FeF_2]^{+}$.

Chlorides

Only two binary chlorides are known, $FeCl_2$ and $FeCl_3$. Unlike manganese and the earlier transition metals, iron does not even show any tendency to form compounds with oxygen and chlorine together in which its oxidation state is greater than $+3$.

Iron (II) chloride is prepared by the reaction between hydrogen chloride and red-hot iron or by the reduction of $FeCl_3$ with hydrogen. Anhydrous $FeCl_2$ is a white, hygroscopic solid which dissolves readily in water. Under normal conditions it has the cadmium chloride structure but can be converted to the cadmium iodide structure by the application of high pressure.

Reaction with pyridine and substituted pyridines gives adducts containing 2, 4 or 6 ligand molecules per metal atom. In $Fepy_2Cl_2$, iron is octahedrally co-ordinated and the structure is the same as that of $Mnpy_2Cl_2$.

Iron (III) chloride is produced when a stream of chlorine is passed over iron heated between 570 and 620 K and also when Fe_2O_3 reacts with hydrogen chloride above 575 K. A hexahydrate is obtained when a solution of the chloride in dilute hydrochloric acid is evaporated to dryness. Conversion of the hydrate to the anhydrous salt requires a reflux with thionyl chloride or some similar dehydrating agent. If the hexahydrate is heated alone, some hydrolysis accompanies the loss of water and some Fe_2O_3 is formed.

Above 475 K, anhydrous $FeCl_3$ loses some chlorine and $FeCl_2$ is formed.

In the solid state, iron (III) chloride has a hexagonal lattice composed of $FeCl_6$ octahedra sharing all chlorines. In the vapour state, dimeric molecules of Fe_2Cl_6, structurally similar to Al_2Cl_6, exist.

The hexahydrate contains octahedral $[Fe(H_2O)_4Cl_2]^{+}$ ions. The two chlorines are in the *trans* arrangement with respect to each other. The structure is completed by chloride ions and water molecules and the bonding involves ionic interactions between cations and anions as well as hydrogen-bonding.

When excess chloride ions are added to an aqueous solution containing Fe(III), a series of chloro-complexes is formed, of general formula, $FeCl_x$, where x varies from 1 to 4 inclusive. The formation of $FeCl_4^-$ accounts for the extractability of iron from acidic aqueous solution into immiscible oxygen-containing organic solvents.

Iron (III) chloride forms 1:1 and 1:2 adducts with many ligands, including $POCl_3$, dimethylsulphoxide, alcohols and various substituted phosphines.

COBALT

Oxidation states

Cobalt has nine electrons in its valency shell. The highest oxidation state shown is $+4$ and the commonest ones are $+2$ and $+3$.

As with the other transition metals, the very low oxidation states are found in combination with ligands like CO and CN^-. For example, the formal oxidation state in $[Co(CO)_4]^-$ is -1, in $[Co(CN)_4]^{4-}$ is 0 and in $[Co(CN)_3(CO)]^{2-}$ and $[Cobipy_3]^+$ is $+1$.

Cobalt (II) oxide and its simple salts are well known. In aqueous solution this oxidation state is extremely stable with respect to oxidation to cobalt (III):

$$[Co(H_2O)_6]^{3+} + e = [Co(H_2O)_6]^{2+} \quad E° = +1.84\,V$$

We would therefore expect aqueous solutions of cobalt (III) to be powerful oxidising agents. In practice, solutions containing appreciable concentrations of cobalt (III) cannot be prepared because the water itself is oxidised.

When complexing agents are present the standard potential for Co^{3+}/Co^{2+} may be altered considerably. For example, with ammonia:

$$[Co(NH_3)_6]^{3+} + e = [Co(NH_3)_6]^{2+} \quad E° = +0.1\,V$$

The low value for E^0 means that cobalt (II) complexed with ammonia is very easily oxidised to cobalt (III).

Cobalt (II) forms a number of 6 co-ordinate, octahedral complexes but these, in general, are labile and oxidised readily to cobalt (III). For example, $[Co(NH_3)_6]^{2+}$ and $[Co(en)_3]^{2+}$ are both converted to the corresponding cobalt (III) complexes by molecular oxygen. This reaction forms the basis of a general preparative method.

A number of tetrahedral complexes of cobalt (II) is known. For example, with halide and thiocyanate ions, complexes of general formula $[CoX_4]^{2-}$ are formed. Mixed complexes in which a halogen and an organic donor molecule like pyridine are bound to the same metal atom are also well characterised.

Cobalt (III) complexes are kinetically rather inert. For this reason they proved to be ideally suited for studies relating to the isomerism and reaction mechanisms of 6 co-ordinate complexes. Classical work on the isomers of cobalt (III) complexes clearly established for the first time the octahedral stereochemistry of these and many other complexes.

Cobalt (IV) exists in Cs_2CoF_6, a compound obtained by the fluorination of Cs_2CoCl_4. An ill-defined oxide of cobalt (IV), CoO_2, can be made in the hydrous state by the action of chlorine on strongly alkaline solutions of cobalt (II). It appears that, in general, cobalt (IV) compounds are not particularly stable and are of little importance.

Compounds

Oxides

Cobalt (II) oxide is prepared by heating cobalt (II) hydroxide or cobalt (II) carbonate in the absence of air. It is an olive-green powder and the solid has the sodium chloride structure and is anti-ferromagnetic at ordinary temperatures.

Cobalt (II) hydroxide is precipitated from aqueous solutions of cobalt (II) salts by the addition of strong alkali. In the presence of oxygen the freshly precipitated compound is oxidised to brown hydrous cobalt (III) oxide. $Co(OH)_2$ is sparingly soluble in water but dissolves in acids to give cobalt (II) salts and in concentrated alkali to give deep-blue solutions containing $[Co(OH)_4]^{2-}$ ions. Co^{2+} ions in aqueous media have a characteristic pink colour but striking colour changes can be brought about by the addition of complexing agents.

Tricobalt tetraoxide, Co_3O_4, is prepared by heating cobalt (II) oxide to 670–770 K in an oxygen atmosphere. Like the analogous compounds of manganese and iron, this is a mixed valency oxide, $Co^{II}Co_2^{III}O_4$. It has a normal spinel structure with Co^{2+} ions in tetrahedral and Co^{3+} ions in octahedral sites. This, rather than the inverse spinel, is found because Co^{3+} (low-spin d^6) gains more stabilisation energy in going into an octahedral site than Co^{2+} loses by going into a tetrahedral site.

Cobalt (III) oxide is known as a monohydrate, $Co_2O_3 \cdot H_2O$, but the anhydrous material has not been obtained as a pure compound. The hydrate decomposes at about 570 K to Co_3O_4 with loss of water and oxygen.

Fluorides

Cobalt (II) fluoride is prepared by passing anhydrous hydrogen fluoride over $CoCl_2$ at 575 K. The compound is a pink, crystalline solid, m.p. 1400 K. It has the rutile structure.

Cobalt (III) fluoride can be made by fluorination of CoF_2 or $CoCl_2$ at 470–570 K. The compound is a brown solid, its structure being similar to that of VF_3. Cobalt (III) fluoride is widely used in the synthesis of organic fluoro-compounds.

Chloride

Cobalt (II) chloride is obtained by the chlorination of cobalt or may be prepared from the hexahydrate by dehydration of this by heating or by refluxing with thionyl chloride.

The anhydrous compound is pale blue and readily dissolves in water and organic solvents like acetone and ethanol. In the presence of water, it is rapidly converted to the pink hexahydrate. In the solid state the hydrate contains trans-$[Co(H_2O)_4Cl_2]^+$ ions.

The colour of $[Co(H_2O)_6]^{2+}$ is pink due to a broad absorption band centred at 540 nm. The addition of excess chloride ion to a solution of cobalt (II) causes an immediate change to deep blue. The change in colour is associated directly with the conversion of octahedral cobalt, in $[Co(H_2O)_6]^{2+}$, to tetrahedral, in $[CoCl_4]^{2-}$. There is a decrease in the number of ligands and this causes the shift in absorption to longer wavelengths (lower frequency) because the crystal field splitting, Δ, decreases. At the same time, the intensity of absorption increases more than 50-fold. These changes in wavelength and intensity of absorption constitute a useful criterion of the stereochemistry of cobalt (II) complexes.

Cobalt (II) chloride forms 1 : 2 adducts with pyridine, substituted pyridines and other heterocyclic ligands. Two isomeric forms of some adducts are known. For example, with pyridine itself, two forms of the complex $Copy_2Cl_2$ can be prepared. One is coloured violet and this is the product of the direct reaction between $CoCl_2$ and pyridine. Its colour is typical of octahedral cobalt and the metal achieves this stereochemistry because the chlorines can function as bridges to build up a chain-like polymeric structure. Violet-$Copy_2Cl_2$ is converted to a second isomer, blue in colour, by heating in a sealed tube to 390 K or by solution in polar solvents followed by re-precipitation. The blue isomer, because of its colour, contains tetrahedral cobalt. The processes by which it is formed are evidently sufficient to break down the polymeric structure and to give discrete tetrahedral molecules of $Copy_2Cl_2$.

Blue—$Copy_2Cl_2$ Violet—$Copy_2Cl_2$

When $CoCl_2$ is fused with RbCl or CsCl in the molar ratio of 1 : 3, a blue, crystalline solid of formula M_3CoCl_5 is formed. This stoichiometry could suggest the presence of a complex anion containing 5 co-ordinate cobalt. X-ray analysis and the characteristic colour show that tetrahedral $[CoCl_4]^{2-}$

ions are, in fact, present. The remainder of the halogen is present as chloride ion.

NICKEL

Oxidation States

Nickel shows a limited range of oxidation states, from 0 to +4 inclusive. In keeping with the general trend in the first transition series, the oxidation states above +2 are of relatively little importance.

Examples of very low oxidation states include nickel (0) in $Ni(CO)_4$, $Ni(PF_3)_4$ and $[Ni(CN)_4]^{4-}$ and nickel (I) in $[Ni_2(CN)_6]^{4-}$.

By far the most important state is nickel (II). The only well-defined oxide is NiO and nickel (II) is the only state of real significance in aqueous solution.

The stereochemistry of nickel (II) compounds has points of special interest for Ni^{2+} can form 4 co-ordinate complexes which are either square-planar or tetrahedral as well as 6 co-ordinate octahedral complexes. Examples are: square-planar $-$ $[Ni(CN)_4]^{2-}$ and $NiBr_2[P(C_2H_5)_3]_2$; tetrahedral $-$ $[NiCl_4]^{2-}$; octahedral $-[Ni(NH_3)_6]^{2+}$, $[Ni(en)_3]^{2+}$ and $[Ni(NSC)_6]^{4-}$. The complex, bis(dimethylglyoximato)nickel (II), has an unusual stereochemistry. Four nitrogen atoms from the two ligands are in a square-planar configuration around the nickel. Each molecule of complex is attached to two adjacent ones by nickel–nickel bonds perpendicular to the planes of the chelate rings. The nickel–nickel bonds are longer than the nickel–nitrogen bonds and so the co-ordination around each metal atom is a distorted octahedron.

Nickel (III) is found in the hydrous oxide, Ni_2O_3, and in a few complexes. The latter include tribromobis(triethylphosphine)nickel (III) which is of interest as an example of a square-pyramidal complex. Also known is the complex with o-phenylenebis(dimethylarsine) and chloride which contains octahedral nickel (III): $[NiCl_2 . 2C_6H_4(As(CH_3)_2)_2]^+$.

Nickel (IV) is found in the hydrous oxide, NiO_2, which has not, however, been made in the pure state, and in a few complexes. This oxidation state appears to exist in combination with a highly electronegative element like fluorine, as in hexafluoroniccolate (IV), $[NiF_6]^{2-}$, and in certain chelate complexes, including $[NiCl_2 . 2C_6H_4(As(CH_3)_2)_2]^{2+}$.

Compounds

Oxides

Nickel (II) oxide is formed by the thermal decomposition of nickel nitrate, carbonate or oxalate. It is a basic oxide, dissolving in acids to form nickel (II) salts. The aquated ion, $[Ni(H_2O)_6]^{2+}$, is deep green, and once again the colour shows striking changes when water is replaced by other ligands.

NiO is a green solid which has the sodium chloride structure.

Nickel (III) oxide has not been isolated as an anhydrous compound although a hydrated form can be prepared by oxidising nickel (II) in potassium hydroxide solution with bromine. The hydrate is a black solid and it approximates in composition to $Ni_2O_3 . 2H_2O$. On heating, this loses oxygen and water and forms NiO.

Nickel (IV) oxide results from the action of chlorine on a suspension of $Ni(OH)_2$ in aqueous alkali. The product is a dark-coloured precipitate which contains up to 1·9 atoms of oxygen per atom of nickel. The metal is therefore substantially in the +4 oxidation state. The compound has powerful oxidising properties and will convert, for example, Mn^{2+} to MnO_4^- in acid solution. Barium niccolate (IV), $BaNiO_3$, has been made by the reaction at 970 K between oxygen and an equimolar mixture of $Ba(OH)_2$ and NiO.

Fluorides

Although only one binary fluoride is known, NiF_2, the +3 and +4 oxidation states are represented in the complex hexafluoroniccolate ions, NiF_6^{3-} and NiF_6^{2-}. Alkali metal salts containing these ions are prepared by fluorination of mixtures of the alkali metal chloride and nickel (II) chloride in appropriate proportions.

Nickel (II) fluoride is prepared by the reaction between nickel and anhydrous hydrogen fluoride or by the fluorination of $NiCl_2$ or NiO at elevated temperatures. The compound is a yellow solid, m.p. 1723 K, and it has the rutile structure.

Chloride

Like cobalt, nickel forms only one binary compound with chlorine.

Nickel (II) chloride results from the chlorination of nickel at elevated temperatures or the reaction between nickel oxide, carbon and chlorine above 570 K. The hexahydrate can be prepared from aqueous solution and may be dehydrated by the same methods used for the lower chlorides of the other transition metals.

Anhydrous $NiCl_2$ has the cadmium chloride structure. It resembles $CoCl_2$ in the range of adducts which it forms with ligands containing donor nitrogen, phosphorus, arsenic or oxygen atoms. Adducts of general formula $NiCl_2L_2$ (where L = monodentate ligand) are numerous. In the case of pyridine ligands, only one form of each complex has been made. With other ligands, two isomers can be prepared and these correspond to different stereochemistry around the metal. For example, $Ni(quinoline)_2Cl_2$, is known as yellow (octahedral) and as blue (tetrahedral) isomers.

THE PLATINUM METALS

These are ruthenium (Ru), rhodium (Rh), palladium (Pd), osmium (Os), iridium (Ir) and platinum (Pt). They almost invariably occur with one another and are often found native. They are rare because of their siderophile nature.

Important valency states are as follows (those of iron, cobalt and nickel are also included for comparison):

Fe (II), (III)	Co (II), (III)	Ni (II)
Ru (IV), (VIII)	Rh (III)	Pd (II)
Os (VI), (VIII)	Ir (IV)	Pt (II), (IV)

A number of other states is also known.

The affinity for oxygen decreases from left to right as we approach I B, the sub-group containing the noble metals silver and gold. All six elements have a relatively much higher affinity for sulphur. As well as the simple sulphides, many complexes are formed containing sulphur as the ligand atom.

GROUP I B—COPPER

Oxidation states

Copper has the electronic configuration $3d^{10}4s^1$. Ionisation of the single s electron gives copper (I) and a number of very stable compounds of this oxidation state are known. In these, copper shows a formal resemblance to the alkali metals. However, the first ionisation energy of copper is much higher than that of any of the alkali metals and, as a result, copper (I) compounds show a considerable amount of covalent bonding and so their properties are very different from those of the corresponding compounds of the alkali metals.

The ionisation energy for the process $Cu \rightarrow Cu^+$ is $745 \cdot 2 \, kJ \, mol^{-1}$ and that for $Cu \rightarrow Cu^{2+}$ is $2702 \cdot 9 \, kJ \, mol^{-1}$. Hence $1957 \cdot 7 \, kJ \, mol^{-1}$ more energy is required to produce Cu^{2+} than Cu^+ from atomic copper. In many circumstances, copper (I) compounds are more stable than copper (II) compounds, as we would expect from these figures. Thus both CuO and CuS decompose when heated to give Cu_2O and Cu_2S respectively. Again the greater stability of copper (I) is indicated by the widespread occurrence in the Earth's crust of Cu_2S and Cu_2O compared with the scarcity of CuS and CuO.

In aqueous solution, copper (II) is the stable species. The reason for this may be appreciated from a consideration of hydration energies. The hydration energies for Cu^+ and Cu^{2+} are respectively $482 \cdot 4$ and $2248 \cdot 1 \, kJ \, mol^{-1}$. Using the value of $339 \, kJ \, mol^{-1}$ for the enthalpy of atomisation of copper, the enthalpy change for the process

$$2Cu^+(aq) \longrightarrow 2Cu(s) \text{ is } \Delta H = -1203 \cdot 6 \, kJ$$

and for

$$Cu(s) \longrightarrow Cu^{2+}(aq), \Delta H = +793 \cdot 2 \, kJ$$

Therefore for

$$2Cu^+(aq) \longrightarrow Cu^{2+}(aq) + Cu(s), \Delta H = -410 \cdot 4 \, kJ$$

In other words, copper (I) disproportionates to copper (II) and metallic copper in aqueous solution. We see that this is primarily because the hydration energy for Cu^{2+} is very much greater than that for Cu^+ so that the expected order of stabilities for these two oxidation states is reversed. Copper (I) compounds are, however, quite stable when they are insoluble.

Copper (II) is the most important state of the element. Many salts and complexes are known.

The hydrated ion has a characteristic blue colour and the complexes are variously blue, yellow or green depending on the strength of the ligand field. Copper (II) has a d^9 configuration and, in regular octahedral co-ordination, the d orbitals should be split into two sets; only one electronic transition should be possible and hence only one absorption band should be observed. In fact, copper (II) complexes show marked Jahn–Teller distortions, being tetragonally distorted or, in the extreme case, when two ligands are removed completely, square-planar. In the less symmetric field further splitting of the d orbitals occurs to give four separate energy levels. Up to three absorption bands should be observed but, in practice, these overlap considerably and only a single, very broad absorption band is seen.

A few copper (III) compounds are known. They generally involve combination of the metal with fluorine or oxygen.

Compounds

Oxides

Copper (I) oxide, Cu_2O, occurs naturally as *cuprite*. It may be prepared as a yellowish-red powder by reduction of an alkaline solution of a copper (II) compound with hydrazine. It has a structural resemblance to silica in that each copper atom is co-ordinated by two oxygens and each oxygen is surrounded by four copper atoms. The low co-ordination number of the metal is indicative of the extent of covalent character in the bonding.

Copper (II) oxide is obtained by the direct oxidation of the metal or by heating $Cu(OH)_2$, $Cu(NO_3)_2$ or $CuCO_3$. This is reduced back to the metal by hydrogen or carbon monoxide at 520 K. When heated to 1170 K, copper (II) oxide loses some of its oxygen and Cu_2O is formed. CuO has an ionic lattice with the copper ions effectively in square-planar co-ordination due to Jahn–Teller distortion.

The hydroxide, $Cu(OH)_2$, is obtained as a blue precipitate when an aqueous solution of copper (II) is made alkaline. This compound is amphoteric, dissolving in acids and also in concentrated alkali hydroxide to give deep-blue hydroxo complexes. It dissolves in ammoniacal solutions to give complexes of general formula $[Cu(NH_3)_x]^{2+}$ (x varies from 1 to 5) which are characteristically very intense blue.

Fluorides

As with the preceding elements cobalt and nickel, copper shows a higher oxidation state in a complex fluoride ($+3$ in $[CuF_6]^{3-}$) than it does in the simple binary fluoride ($+2$ in CuF_2).

Copper (II) fluoride is made by the reaction of copper metal or its oxides with hydrogen fluoride or fluorine. It is a white solid, m.p. 1058 K. The solid has a distorted rutile structure, four short Cu—F bonds of length 193 pm and two long ones of 227 pm. In the case of the complex fluoride, K_2CuF_4, the situation is reversed for there appear to be two short and four long bonds to each copper.

Potassium hexafluorocuprate (III), K_3CuF_6, is made by treating a mixture of potassium and copper chlorides with fluorine. It is paramagnetic and its magnetic moment corresponds with the presence of two unpaired electrons.

Chlorides

Two binary chlorides are known. We see for the first time a transition metal which shows an oxidation state of $+1$ in combination with chlorine.

Copper (I) chloride is obtained when a solution of a copper (II) salt in hydrochloric acid is reduced with copper metal, sulphur dioxide or ascorbic acid. The compound is a white solid, m.p. 695 K. It has the zinc-blende structure in which each copper atom and each chlorine are tetrahedrally co-ordinated.

In the vapour state, a trimeric molecule exists. This is a six-membered ring composed of alternate copper and chlorine atoms.

Copper (II) chloride is readily made by the dehydration of $CuCl_2 . 2H_2O$ by methods already mentioned for other transition-metal dichlorides. The dihydrate is obtained by crystallisation from aqueous solution.

Anhydrous $CuCl_2$ is a yellow–brown solid, m.p. 895 K, which is very hygroscopic and dissolves readily in water. The solid is composed of infinite chains of planar $CuCl_4$ groups. The chlorine atoms bridge adjacent metal atoms as shown.

The Cu—Cl distance within the chain is 230 pm. Chains of this type are parallel to one another in the crystal and each metal atom has two chlorines as next nearest neighbours, one above and the other below the plane, with a Cu—Cl distance of 295 pm. Stereochemically, this arrangement is in the form of tetragonally elongated octahedron.

In the dihydrate, there are also infinite chains of square-planar $CuCl_4$ groups linked as before. Each copper atom is further co-ordinated by two water molecules in the *trans* position.

SILVER AND GOLD

Silver shows oxidation states of $+1$, $+2$ and $+3$. Silver (I) is the stable state and many of its compounds are well known. These include the nitrate, perchlorate and sulphate and the four halides. Complexes of silver (I) are noteworthy for their low co-ordination number. This often does not exceed two.

Examples are $[Ag(CN)_2]^-$ and $[Ag(NH_3)_2]^+$. Although higher oxidation states than 1 can be made these have strong oxidising properties. For example, the couple Ag^{2+}/Ag^+ has a standard electrode potential of $+1.98$ V. Silver (II) is therefore too strongly oxidising to exist in aqueous solution. It is known in the fluoride and in various complexes. The compound of formula AgO is actually a mixed-valency compound and is correctly represented as $Ag^IAg^{III}O_2$.

Gold shows oxidation states of $+1$ and $+3$. Gold (I) is similar to Cu(I) in undergoing disproportionation to Au(III) and metallic gold Even gold (I) chloride, AuCl, decomposes in this way in the presence of water. The chemistry of gold (I) is largely that of its anionic complexes such as $[AuCl_2]^{2-}$ and $[Au(CN)_2]^-$. Again, no simple monatomic ion exists for gold (III) and this state is always found in complexes.

GROUP II B

This sub-group comprises the three metals zinc (Zn), cadmium (Cd) and mercury (Hg). The atoms have the electron configuration $(n-1)d^{10}ns^2$ and the removal of one or two electrons might be expected to give rise to the formation of monopositive and dipositive oxidation states respectively. The existence of the $+1$ oxidation state in the case of mercury has been well established but there is no reliable evidence for this state with either zinc or cadmium.

Compared with the elements of the A sub-group, compounds of zinc, cadmium and mercury in the dipositive state are considerably more covalent in character. The stability of the inner d^{10} core is such that oxidation states higher than two are not formed. Complexes are, however, very common; these are tetrahedral where 4 co-ordinate and octahedral where 6 co-ordinate.

Table 10.7 summarises the general physical properties of the group.

Table 10.7. THE PROPERTIES OF THE GROUP II B METALS

		Zn	Cd	Hg
Atomic number		30	48	80
Electron configuration		$3d^{10}4s^2$	$4d^{10}5s^2$	$5d^{10}6s^2$
m.p. T/K		692·6	594·1	234·3
b.p. T/K		1 179	1 038	629·8
Ionisation energy $\Delta U_o/kJ\ mol^{-1}$	1st	906·3	867·3	1 006·3
	2nd	1 733·0	1 630·5	1 809·2
Atomic radius r/pm		125	141	144
Ionic radius $r/pm\ M^{2+}$		74	95	102

Mercury shows a few peculiarities:
  (i) it is a liquid at normal temperatures,
 (ii) it is capable of forming amalgams,
(iii) it has an ionisation energy comparable with that of radon, showing the increased inert character of the outermost s electrons.

Compounds

Mercury (I) compounds

Compounds of this oxidation state are numerous and most oxidising agents are capable of converting mercury to either Hg^I (the mercurous) or Hg^{II} (the mercuric) state. The electrode potentials for these two couples are

$$Hg^I \rightarrow Hg \qquad E^\circ = +0.789 \text{ V}$$
$$\text{(at 298.15 K)}$$
$$Hg^{II} \rightarrow Hg \qquad E^0 = +0.854 \text{ V}$$

The corresponding electrode potential for the couple Hg^{II}/Hg^I is $+0.92$ V. In the presence of excess mercury the product of oxidation is Hg^I since the equilibrium constant of the reaction

$$Hg + Hg^{II} \rightleftharpoons 2 Hg^I$$

is approximately 160. In the absence of excess mercury, or in reactions where the mercury (II) compound is less soluble than the mercury (I) compound, the formation of Hg^{II} is favoured.

The oxide, hydroxide and cyanide of Hg^I have not been isolated, being unstable with respect to disproportionation to Hg and the corresponding Hg^{II} compound.

Most Hg^I compounds are only sparingly soluble in water, the main exceptions to this being the nitrate, chlorate and perchlorate.

The constitution of the mercurous ion is of considerable interest. In contrast to the cuprous ion, evidence is conclusive in formulating the mercurous ion as a diatomic species:

$$(Hg-Hg)^{2+}$$

There are several pieces of experimental evidence to support this contention:

(i) X-ray crystal analysis of mercurous chloride indicates the presence of distinct Cl—Hg—Hg—Cl units.

(ii) Electrochemical data from the concentration cell

Hg	Hg^I nitrate $0.1 \text{ mol dm}^{-3} (c_2)$ in dil. HNO_3		Hg^I nitrate $0.01 \text{ mol dm}^{-3} (c_1)$ in dil. HNO_3	Hg

The potential of the cell at 298.15 K is 0.028 V. The theoretical equation relating potential (E) to the concentrations (c_1 and c_2) is

$$E = \frac{RT}{zF} \ln \frac{c_2}{c_1}$$

where z is the electron change. Using this expression and substituting for the constant R gives $z = 2$. Hence in the cell the reaction is $2Hg = (Hg)_2^{2+} + 2e^-$.

(iii) Conductivity measurements on solutions of mercurous nitrate indicate the presence of a bi–univalent salt and not a uni–univalent salt.

(iv) Determinations of the concentration of Hg^I and Hg^{II} in solutions obtained by shaking excess mercury with mercuric nitrate until equilibrium is attained. The two possible reactions are

$$Hg + Hg^{2+} \rightleftharpoons 2\,Hg^+$$

and

$$Hg + Hg^{2+} \rightleftharpoons Hg_2^{2+}$$

Evaluation of the equilibrium constants for each reaction using the experimental data gives consistent values for the second reaction only.

Compounds of the dipositive oxidation states

Oxides

Both zinc and cadmium on exposure to air yield surface coatings of the oxide, in contrast to mercury. Mercury forms an oxide only on heating in air. The thermal stability of the oxides is noteworthy; on heating, zinc and cadmium oxides sublime without decomposition but mercury oxide decomposes to mercury and oxygen.

The basic character varies in the same manner as the monoxides of sub-group I B; thus cadmium oxide is most basic. Compared with the corresponding oxides of sub-group II A elements, they are more resistant to attack by dilute acids. Zinc oxide shows amphoteric character and dissolves in alkali with the formation of the zincate ion $[ZnO_2]^{2-}$.

Halides

The halides show less ionic character than those of sub-group II A. The increase in covalent character as the group is descended is illustrated by the small conductivity of zinc chloride in the fused state, compared with mercuric chloride which is non-conducting. This increase in covalent character is also shown by the properties of the halides in aqueous solution; zinc and cadmium halides are soluble to form ions but mercuric chloride, for example, although soluble in water, is not ionised and is present as undissociated $HgCl_2$.

Cadmium halides show a great tendency to autocomplex as indicated by conductivity measurements on the concentrated aqueous solutions. Ions such as $[CdX_3]^-$ and $[CdX_4]^{2-}$ are formed. Zinc shows less tendency to complex in this way and shows a great inclination to form hydrated ions in aqueous solution. Mercury forms 4 co-ordinate complexes of the type $[HgX_4]^{2-}$ in the presence of excess of the anion.

Complexes

Zinc (II) and cadmium (II) form a wide variety of tetrahedral complexes particularly where nitrogen is present to act as donor atom. Thus strong complexes are formed with ammonia, amines such as 1,10-phenanthroline and also molecules where both nitrogen and oxygen are present to act as donor atoms, e.g. 8-hydroxyquinoline and ethylenediaminetetra-acetic acid.

Mercury (II) also forms stable ammine complexes. The general products of interaction between mercury (II) and ammonia gas under anhydrous conditions are ammonobasic compounds. Mercuric chloride reacts with aqueous ammonia to form 'infusible white precipitate' or mercury (II) amidochloride, $HgNH_2Cl$:

$$HgCl_2 + 2NH_3 = HgNH_2Cl + NH_4Cl$$

If the reaction is carried out in the presence of excess ammonium chloride the product is diamminemercury (II) dichloride, $Hg(NH_3)_2Cl_2$.

Mercuric oxide reacts with ammonia to form Millon's base:

$$2HgO + NH_3 + H_2O = (HOHg)_2NH_2OH$$

the anhydride of which is the parent compound of the precipitate, OHg_2NH_2I, formed by Nessler's reagent in the detection of small amounts of ammonia.

GROUP III A

The elements of this sub-group are called the rare earths and comprise scandium (Sc), yttrium (Y), lanthanum (La) and the elements from cerium (Ce) to lutetium (Lu). The latter group, cerium, etc., constitute an inner transition series corresponding to the filling up of the 4f quantum level and are termed the *lanthanoid elements* (*Table 10.8*).

Table 10.8. PHYSICAL PROPERTIES OF THE LANTHANOID ELEMENTS

Element	Atomic number	Symbol	Electron configuration	Ionic radius r/pm (M^{3+})
Cerium	58	Ce	$4f^2 6s^2$	103·4
Praseodymium	59	Pr	$4f^3 6s^2$	101·3
Neodymium	60	Nd	$4f^4 6s^2$	99·5
Promethium	61	Pm	$4f^5 6s^2$	97·9
Samarium	62	Sm	$4f^6 6s^2$	96·4
Europium	63	Eu	$4f^7 6s^2$	95·0
Gadolinium	64	Gd	$4f^7 5d^1 6s^2$	93·8
Terbium	65	Tb	$4f^9 6s^2$	92·3
Dysprosium	66	Dy	$4f^{10} 6s^2$	90·3
Holmium	67	Ho	$4f^{11} 6s^2$	89·4
Erbium	68	Er	$4f^{12} 6s^2$	88·1
Thulium	69	Tm	$4f^{13} 6s^2$	86·9
Ytterbium	70	Yb	$4f^{14} 6s^2$	85·8
Lutetium	71	Lu	$4f^{14} 5d^1 6s^2$	84·8

In their chemical behaviour, scandium and yttrium show strong resemblances to the lanthanoids. The elements are electropositive metals and the characteristic oxidation state found for all of them is $+3$; certain lanthanoids also show either $+2$ or $+4$. These additional states are usually associated with the elements where electron loss occurs to form ions which have a particularly stable electron configuration. Three arrangements of especial stability should be noted. These are $4f^0$ (in La^{3+}, Ce^{4+}), $4f^7$ (in Eu^{2+}, Gd^{3+}, Tb^{4+}) and $4f^{14}$ (in Yb^{2+}, Lu^{3+}).

In the lanthanoid elements, the addition of the electrons to the 4f shell has little effect upon the size of the ions which show a gradual decrease with increase in atomic number. This lanthanide contraction is responsible for the great chemical similarity within the group and has an important influence on succeeding groups. Because of the close similarity in both chemical and

Table 10.9. THE PROPERTIES OF SCANDIUM, YTTRIUM AND LANTHANUM

		Sc	Y	La
Atomic number		21	39	57
Electron configuration		$4s^2\,3d^1$	$5s^2\,4d^1$	$6s^2\,5d^1$
m.p. T/K		1 693	1 773	1 193
b.p. T/K		2 753	3 503	3 643
Ionisation energy $\Delta U_o/kJ\ mol^{-1}$	1st	633·0	615·0	541·4
	2nd	1 235·1	1 179·9	1 102·9
	3rd	2 387·8	1 986·2	1 968·2
Atomic radius r/pm		144	162	169
Ionic radius r/pm M^{3+}		73	89	106

physical properties, the separation of the lanthanoids had proved a difficult task for many years. In recent years the development of ion-exchange and solvent extraction methods has resulted in the efficient separation of the lanthanoids.

Some physical properties of the first three elements of the rare earth group are given in *Table 10.9*.

Oxidation states

In the tripositive oxidation state, which in many cases is the only state found, there is a close similarity in chemical properties between the rare earths. The oxides in this state are strong bases, the strength being related to ionic size. Thus scandium oxide is the least basic and the basic power falls from La_2O_3 to Lu_2O_3.

Oxidation states other than +3 are found; cerium, praseodymium and terbium form the +4 oxidation state; samarium, europium and ytterbium form the +2 oxidation state.

In the +4 state, the rare earths tend to resemble the metals of Group IV A. In the +2 state, their chemical properties are like those of the alkaline-earth metals.

Magnetic susceptibility and colour

The magnetic susceptibilities of the trivalent ions do not obey the spin-only formula because of the contribution made to the magnetic moment by the orbital moments of the 4f electrons. *Figure 10.1* shows the magnetic moments of the ions M^{3+} from lanthanum to lutetium.

Solutions of many of the trivalent ions are distinctly coloured and correlation between the number of f electrons present and colour has been found. The ions with n more electrons than La^{3+} are observed to have similar colours to those with an extra $(14-n)$ electrons; thus, praseodymium and

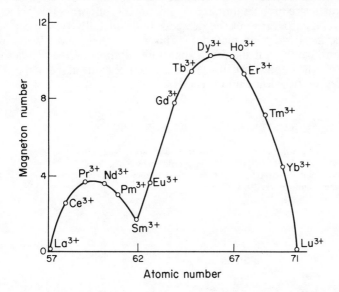

Figure 10.1. Magneton numbers of the lanthanoid ions (M^{3+})

thulium are both green in their $+3$ state and neodymium and erbium are red-coloured in the $+3$ state.

The actinoid series

These are actinium (Ac), thorium (Th), protoactinium (Pa), uranium (U) and the elements beyond uranium up to atomic number 103. The preparation of the transuranic elements has been dealt with earlier. There is little doubt that in this series of elements the 5f level is being filled. In contrast to the lanthanoids, several oxidation states are found for each of the earlier members of this series and chemically some of the elements show a strong resemblance to certain families of transition elements. Thus thorium shows a stable oxidation state of $+4$ and in many ways behaves as a sub-group IV A metal, and is more electropositive than hafnium. Similarly, uranium, as it forms many compounds in which its oxidation state is $+6$, shows a strong resemblance to sub-group VI A.

The $+3$ oxidation state becomes increasingly important as the atomic number increases. For example, Np^{III} and Pu^{III} are immediately oxidised in aqueous solution when access of air is permitted. However, Cm^{III} is very stable and is indeed the only oxidation state observed for this element in aqueous solution. It is this fact, together with the strong paramagnetism of Cm^{3+} (comparable with Gd^{3+}), which suggests that the 5f shell is half filled at curium (96) and hence that a series of inner transition elements begins at actinium (89).

The separation of uranium from plutonium and fission products by solvent extraction and the use of ion-exchange resins for the production of uranium concentrates are important large-scale separation processes which have been developed because of the demand, both for military and for peaceful purposes, for fissionable materials. This requirement has also stimulated intensive research with the result that, for a number of the heavier metals, hitherto of little interest because few uses could be made of them, the chemical properties are now known in great detail.

SUGGESTED REFERENCES FOR FURTHER READING

MOELLER, T. *The Chemistry of the Lanthanides*, Reinhold, New York, 1963.
SANDERSON, R. T. *Chemical Periodicity*, Reinhold, New York, 1960.
SNEED, M. C. and BRASTED, R. C. *Comprehensive Inorganic Chemistry*, vols. I–VIII, Van Nostrand, New York, 1953–61.

Index

DATE DUE			
GAYLORD			PRINTED IN U.S.A